A Guide to
UNDERGRADUATE PROJECTIVE GEOMETRY

A Guide to

UNDERGRADUATE PROJECTIVE GEOMETRY

A. F. HORADAM

Department of Mathematics,
University of New England,
Armidale, N.S.W., Australia

'And gladly wolde he lerne and gladly teche.'
(Chaucer)

PERGAMON PRESS AUSTRALIA

Pergamon Press (Australia) Pty Limited, 19a Boundary Street, Rushcutters Bay, NSW 2011
Pergamon Press Ltd, Headington Hill Hall, Oxford OX3 OBW
Pergamon Press Inc., Fairview Park, Elmsford, N.Y. 10523
Pergamon of Canada Ltd, 207 Queen's Quay West, Toronto 1

First published 1970

Typeset in Australia by Typemasters Pty Ltd in Sydney
Printed in Australia by The Specialty Press in Melbourne
Registered in Australia for transmission by post as a book
Library of Congress Catalog Card No. 71-110243
SBN 08 017479 5

To my wife and children

PREFACE

Value of Geometry. With the advent of the modern technological revolution and the accompanying rapid expansion of knowledge, the study of Mathematics today has assumed a position of central significance in our culture. In keeping with this forward movement of scientific ideas, many new branches of Mathematics have been developed in the past two or three decades, while Geometry has to some extent fallen into neglect. Against this background, the geometer has the challenging task of reassessing his subject.

Geometry, the most ancient and satisfying of disciplines, must, in a competitive world, justify its being. On the one hand, it has a utilitarian relevance (as evidenced by its applications in Science generally) quite apart from its value as a fundamental ingredient of Mathematics. On the other hand, it possesses a unique beauty which demands for it a place of aesthetic value beside the other creative arts, along with music and poetry. In this important respect, one feels that an appreciation of the cultural worth of Geometry will last while ever man is civilised. Yet another aspect stirs the author's sensibilities. It is simply that Geometry is one of the major influences which inspires in the student that feeling of power which is one of the chief attractions of Mathematics.

Recently (1960), Herbert Busemann made the following statement regarding the inclusion of Projective Geometry in the undergraduate Mathematics curriculum:

'The subject is beautiful, it required major efforts of some of the best mathematicians during and after the last century to understand its structure thoroughly, it still gives a deep insight into geometry. The projective duality principle, besides being constantly used by geometers, is an exciting experience for any interested student. But most of all, do we have the right to completely disrupt historical continuity whenever a subject moves out of the focus of contemporary interests? . . . A mathematics major should not obtain his degree without knowing projective geometry.'

Aim of the Book. My aims in this book are (i) to present a broad picture of the variety of types of Projective Geometry, (ii) to communicate an understanding of some of its implications for the philosophers and the scientists (as well as for the mathematicians), (iii) to illustrate some of its techniques, and (iv) to try to impart to the reader some of my own feeling for the subject.

Accordingly, no section of the theory is pursued with full rigour, but the several lines of development are indicated. Appeal is directed more to the stimulation of the reader's imagination than to the accumulation of facts (the memory of which is probably annihilated in the fullness of time, in any case).

A major concern is the presentation of both Euclidean Geometry and the non-Euclidean geometries as special cases of Projective Geometry. Historically, though, the process was just the reverse, that is to say, (non-metrical) Projective Geometry developed as a generalisation of (metrical) Euclidean Geometry.

Another major concern is to exhibit the vital connection between geometries and groups of transformations, a discovery originating in the brilliant mind of the 19th century German mathematician, Felix Klein.

A third major concern is with Finite Geometry as a special case of Projective Geometry and in particular, with its relation to the many-faceted study of Combinatorial Analysis, one of the fastest developing areas of mathematical research. Indeed, the reader is taken to the frontiers of several current developments in Combinatorial Mathematics.

Background Knowledge Required. Little background knowledge is necessary for an understanding of the geometrical aspect of this work. All that is required here is some fairly elementary Euclidean Geometry and some elementary Analytical Geometry. When it comes to the algebraic side of things, however, more is demanded of the reader, namely some knowledge of matrices, complex numbers and the structures of such algebraic systems as groups, rings, fields and vector spaces.

To a large extent the material of this book is intended for upper level undergraduate classes. Most of it has been class-tested over a number of years.

Structure of the Book. In case it may guide the reader in discerning an inter-relationship among the chapters, the following scheme is provided. Certainly, it is at best only a rough guide. Interested readers could perhaps compose variations of it.

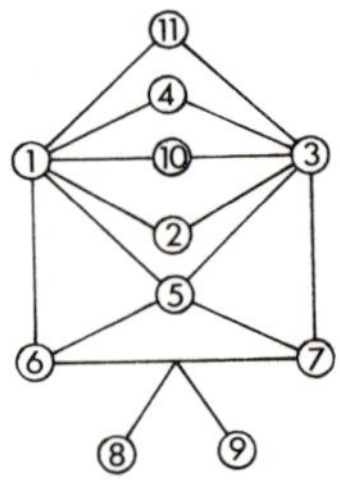

Inter-relationship of the chapters in this book

It is strongly suggested that the introductory chapter, Part 1, should be carefully read before the formal development in Part 2 is attempted. Informal in its treatment, Part 1 is intended to provide a meaning and a motive for the subsequent theory. Admittedly the introduction is long, but it seems essential to offer a detailed historical and metrical background for some of the challenging and intriguing projective concepts which must be mastered.

References and Suggested Topics. Literature available on classical Geometry and Algebraic Geometry is indeed abundant. Together with the creative surge of recent fundamental research into the nature of projective planes, this confronts us with such a vast amount of material that we can only hope that our selection is reasonably balanced and representative.

Suggested topics for further study are meant to provide for a wide range of interests, but there are clearly many others which the reader will seek for himself.

Journals, it is hoped, will be readily consulted, along with the other sources of original research which lie at the heart of Projective Geometry. Those journals chosen for mention are intended to indicate a far-reaching geographical distribution of talent, so that the reader will be left in no doubt of the vigour and international nature of Mathematics.

Symbolism. No attempt has been made to prepare a separate list of symbols used in this book, as most of these are in common usage and will be known to the reader. Unfamiliar symbolism is carefully explained as it is introduced. Since there are relatively few new symbols, it is confidently expected that the reader will have no difficulty at all in absorbing them into his mathematical bloodstream.

Exercises in Part 2 are designated by two numbers separated by a dot, the first number referring to the chapter and the second to the exercise itself. Thus, Exercise 6.13 means the thirteenth exercise in Chapter 6.

Acknowledgements. Reading the many splendid books which deal with Projective Geometry, one is consciously or unconsciously influenced by the thoughts and techniques of presentation of their authors. It is hoped that due recognition of these influences has always been acknowledged by me. They will certainly be evident in the list of references.

It is a pleasure to record the typing assistance given by my daughter, Kathryn, Miss Deirdre Robertson and Mrs Marion Brown. It is likewise a pleasure to acknowledge my appreciation to the friendly efficiency of the Pergamon staff. My thanks go also to my post-graduate students who willingly assisted in checking chapters of the manuscript, and particularly to Mr N. Leeson who read carefully through the galleys and offered helpful comments.

Finally, it is a great satisfaction to me to acknowledge my indebtedness over a number of years to those undergraduates who, by their interest and their perceptive questions, have largely been responsible for any merits which this book may possess. Certainly their positive attitude has relieved the original version of some of its grosser imperfections.

A. F. Horadam,
Armidale 1970

CONTENTS

PART 1

INTRODUCTION. PHYSICAL AND METRICAL PRELIMINARIES

§ 1. Euclidean and non-Euclidean Geometry 1
§ 2. Projective Geometry 14
§ 3. Analytical and algebraic treatment 27
§ 4. Axiomatic approach 36
Exercises on the Introduction 41

PART 2

CHAPTER 1. AXIOMS OF INCIDENCE

§ 5. Axioms of incidence 47
§ 6. Projective space of n dimensions ($n \geqslant 3$) 61
Exercises 1 69

CHAPTER 2. THEOREMS OF DESARGUES AND PAPPUS

§ 7. Desargues' Theorem 72
§ 8. Pappus' Theorem and Desargues' Theorem 79
§ 9. Non-Desarguesian Geometry. Moulton planes 85
Exercises 2 88

CHAPTER 3. COORDINATES IN THE PLANE

§ 10. General homogeneous co-ordinates 91
§ 11. Applications of analytical methods to basic theorems 99
§ 12. Coordinates in n dimensions 105
§ 13. Geometric description of addition and multiplication 107
Exercises 3 117

CHAPTER 4. COLLINEATIONS

§ 14. Transformations in the plane 119
§ 15. Collineations. Groups 123
§ 16. Central collineations. Homologies and elations 133
§ 17. Transitivity. Translation planes 137
Exercises 4 138

CHAPTER 5. PROJECTIVITIES, CROSS-RATIO AND INVOLUTION

§ 18. Projectivities 140
§ 19. Cross-ratio 150
§ 20. Involution 158
Exercises 5 164

CHAPTER 6. THE CONIC: GEOMETRIC THEORY

§ 21. Projective generation and elementary properties of the conic 167
§ 22. Involution on a conic 173

§ 23. Desargues' Involution Theorem 175
§ 24. Pascal's Theorem 179
§ 25. Pole and polar 183
§ 26. Von Staudt and Braikenridge-Maclaurin definitions of a conic 185
Exercises 6 187

CHAPTER 7. ALGEBRAIC TREATMENT OF THE CONIC
§ 27. Elementary theory 189
§ 28. Special forms of the conic 193
§ 29. Invariants of a pair of conics 196
§ 30. Apolarity 200
§ 31. Harmonic conics 201
Exercises 7 204

CHAPTER 8. AFFINE AND EUCLIDEAN GEOMETRY
§ 32. Parallelism. Perpendicularity and bisection 207
§ 33. The circle 211
§ 34. The conic 213
§ 35. Projective metric 218
§ 36. Euclidean Geometry 222
§ 37. Euclidean interpretation of projective theorems 234
§ 38. Real transformations in the Euclidean plane 238
Exercises 8 250

CHAPTER 9. NON-EUCLIDEAN GEOMETRY
§ 39. Saccheri's Geometry 254
§ 40. Hyperbolic Geometry 257
§ 41. Elliptic Geometry 266
§ 42. Models of non-Euclidean Geometry 268
§ 43. Metric non-Euclidean Geometry 272
§ 44. Applications of non-Euclidean Geometry to the Physical Universe 275
§ 45. Philosophical considerations 278
Exercises 9 281

CHAPTER 10. FINITE GEOMETRY
§ 46. Fano planes 284
§ 47. Incidence matrices 291
§ 48. Symmetric block designs. The (v, k, λ) configuration 291
§ 49. Hadamard matrices 293
§ 50. Difference sets 294
§ 51. Orthogonal Latin squares and Graeco-Latin squares 295
§ 52. Existence of finite planes 298
§ 53. Other current research in Finite Geometry 303
Exercise 10 308

CHAPTER 11. TERNARY RINGS AND PROJECTIVE PLANES
§ 54. Ternary rings 310
§ 55. Linearity and the minor Desargues' Theorem 315
§ 56. Veblen-Wedderburn planes 317
§ 57. Moufang planes 318
§ 58. Desarguesian planes 321
§ 59. Pappian planes 322
§ 60. Comparison of ternary ring co-ordinates and homogeneous co-ordinates 324

Exercises 11 327
Miscellaneous Research Articles 328
CHAPTER 12. THROUGH THE LOOKING GLASS 330
REFERENCES 332

PART 1

INTRODUCTION

PHYSICAL AND METRICAL PRELIMINARIES

'Aristippus, a Socratic philosopher, having been cast by shipwreck on the Rhodean shore, and having noticed the drawing of a geometrical design, is said to have exclaimed to his companions: Be of good hope, for I see the footprints of men.'

(Vitruvius)

'. . . there are many geometries, each describing another world: wonderlands and Utopias, refreshingly different from the world we live in.'

(H. S. M. Coxeter)

§ 1 EUCLIDEAN AND NON-EUCLIDEAN GEOMETRY

§ 1.1 Neolithic and Pre-Euclidean Geometry. Primitive man no doubt possessed some intuitive feeling for geometric form and pattern, the sort of unanalysed feeling expressed by Christopher Fry when, in *The Lady's Not for Burning*, he makes Thomas say to Richard (with reference to the well-known constellation):

'The Great Bear is looking so geometrical
One would think that something or other could be proved.'

Just when and how this rudimentary sense arose in prehistoric man is a matter of conjecture; its origins are obscured by the darkness of antiquity. Certainly, by Neolithic times, man had developed a keen appreciation for decorative designs, particularly those which emphasised parallelism, symmetry, congruence and similarity. Examples of this ornamental use of lines, as revealed by archaeological research, may be seen on Neolithic pottery, such as that of the pre-dynastic period in Egypt, in which magic and religion were no doubt interfused (Fig. I(a)). Besides this aesthetic use of patterns by Neolithic craftsmen, early man had quite utilitarian applications of geometrical ideas, as the Pyramids and other engineering achievements of ancient Babylon and Egypt bear silent but eloquent witness. At a more elementary level, one may think of Stonehenge (or Callanish, in the Outer

Hebrides) and the lonely cromlechs and stone-circle ruins, isolated but impressive in their moorland settings (Fig. I(b)). Such practical geometry, derived intuitively from man's observations and experiences of the physical world about him, may be called *Empirical Geometry.*

FIG. I. (a) Prehistoric Egyptian pottery pattern (4,000–5,000 B.C.)

FIG. I. (b) Lanyon Quoit, Cornwall, England

Originally, in ancient times, Geometry, as its derivation from two Greek words suggests, (*ge* ≡ 'earth', *metrein* ≡ 'to measure'), was concerned with the measurement of the earth and hence found practical use in the measurement of fields for taxation purposes (though Plato, in the *Phaedrus* , attributed the art of Geometry to the ibis-headed scribe-god Thoth). It was thus an essential part of the economic organisation of ancient communities. Herodotus, in Book 2 of his *History* , recorded the practice of the Pharaoh Sesostris (14th century B.C.) of employing taxation officials to survey land after the annual flooding of the Nile had destroyed farm boundaries. 'From this practice', observed Herodotus, 'I think geometry first came to Egypt, whence it passed to Greece.'

§ **1.2 Nature of Euclidean Geometry.** Here, on account of the mental complexion of the Greeks, it blossomed into a body of deductive thought, studied for its own sake, and in time played an essential role in the education of the young. You will recall, for instance, Plato's decree over the entrance to his Academy: 'Let no one ignorant of geometry enter here'. For Geometry as a practical art, which the craftsmen of the older civilisations had needed in their basketry, pottery and metal, stone and textile work, the Greeks had little patience (though the accuracy of the 6th century tunnel on the island of Samos (Kulczŷcki [61] p. 10) reveals that its architect, Eupalinus, did not despise Applied Geometry). The geometrical theorising of the early Greek scholars, such as Thales, the Pythagorean Brotherhood and the Platonic School, was codified by the organising genius of Euclid in his *Elements* (13 books) into a logical system of deductive reasoning, the prototype of a *mathematical science.*

Written in Alexandria, Egypt, about 300 B.C., the *Elements* displays an august procession of definitions, axioms, postulates and theorems deduced from them. For over 2,000 years it was accepted by the majority of mankind as an impeccably beautiful system of logic which, moreover, described exactly the properties of space suggested by experience (though Euclid himself nowhere made any claims about physical space).

Though originating with notions received from the world of experience, Euclidean Geometry is an abstract deductive system of thought dealing with *mathematical space* which, by its very nature, is a construction of the mind. For example, a triangle is an idealisation of all the material triangles of the Egyptian surveyors and the triangular clay tablets of the Babylonian builders. Applying such a conceptual geometry to the space we inhabit, we find it gives only a rough approximation to that space. Briefly, then, Euclidean Geometry is an ideal mathematical model of the structure of the physical world—a model of human reasoning wherein the realities of the world are refined—and its mathematical validity must be tested by its own inner consistency.

By the 18th century the Euclidean conceptions of mathematical space had crystallised into a dogma for the disciples of the German philosopher, Immanuel Kant, for whom these conceptions were an inherent feature of the human mind. Unquestionably, the *Elements* has proved to be one of the most influential books ever written. Since its first printing in 1482 in Venice (a Latin version of the Arabic), over 1,000 editions of it have testified to the impact of its logical structure on scientific thinking and on the general culture of the civilised world.

This conception of Geometry as a deductive science concerned with the logical relationships existing among given assumptions was probably the finest contribution of the Greeks and the *Elements* remains as a superb monument to the Greek intellect.

The raw materials of Euclidean Geometry are the definitions, the axioms or 'common notions' (sometimes styled 'self-evident truths') and the postulates, i.e. statements which a reasonable man would accept without proof. Axioms and postulates are thus simply assumptions. Until fairly recent times, few dared to question the perfection of the noble and lofty edifice Euclid had erected on these foundations. Yet there are blemishes. Wherein do they lie?

§ 1.21 Some Weaknesses of Euclidean Geometry. Weaknesses in the Euclidean system arise largely in two ways, either as an attempt to define all terms, or as tacit assumptions derived from experience. Also, several Euclidean theorems require the kinematical notion of superimposition of triangles in which the triangle is regarded as a rigid material body. Let us examine some of the difficulties raised in Euclid's logic.

(a) *Concept of Point and Line.* Euclid defines a point as 'that which has no part'. (*Σημεῖόν ἐστιν, οὗ μέρος οὐθέν.*)

Logically, this invites the query: What is the 'that' which has no part?

In any case, it is a negative approach, as it tells us what a point is not, rather than what it is. Furthermore, he defines a line as 'length without breadth' (which presupposes distance has been defined) and a straight line as 'that which lies equally between its points'. What, if anything, can be deduced from this latter assertion? It seems clear that these definitions (which actually can be traced to pre-Euclidean scholars such as Plato and Aristotle) are clumsy attempts to idealise intuitively their physical counterparts in the world of sense perception.

Endeavouring to explain the nature of point and line in simple language to practical men, Robert Recorde (1510?–1558, English), quoted by Joy Easton ('A Tudor Euclid', *Scripta Mathematica*, 27(4), (1964) 339–355), wrote in the charming style of his time:

'A *Poynt* or a *Prycke*, is named of Geometricians that small and unsensible shape, whiche hath in it no partes, that is to say: nother length, breadth nor depth. But as this exactnes of definition is more meeter for onlye Theorike speculacion, then for practise and outwarde worke (consideringe that myne intente is to applye all these whole principles to woorke) I thynke meeter for this purpose, to call a poynt or prycke, that small printe of penne, pencyle, or other instrumente, which is not moved, nor drawen from his fyrst touche, and therefore hath no notable length nor bredthe. . .

'Now of a great numbre of these prickes, is made a *Lyne*, as you may perceive by this forme ensuying .... where as I have set a numbre of prickes, so if you with your pen will set in more other prickes betweene everye two of these, then wil it be a lyne, as here you may see——— and this lyne, is called of Geometricians, Lengthe without breadth.

'But as they in theyr theorikes (which ar only mind workes) do precisely understand these definitions . . .' (Recorde goes on to offer practical advice to his readers.)

Nowadays, a cross-hair might be thought of as a representative of a point, or a ray of light as representing a straight line. (But Einstein's Relativity teaches us that this attribute of straightness depends on the absence of matter from the space in which the light travels, e.g. a ray of light is deflected when passing through the heavy gravitational field of the Sun.)

(b) *Unstated assumptions.* Again, using intuitive processes which are not disallowed by Euclid's axioms, we may prove quite ridiculous results, such as: 'Every triangle is isosceles' (Exercise 8). Basically, the difficulty here is that Euclid's axioms alone are insufficient to establish that a line cannot cut internally all three sides of a triangle. To eliminate this deficiency of Euclidean logic, Moritz Pasch (1843–1930, German) in 1926 devised the following hypothesis, used by David Hilbert (1862–1943, German) [50]:

(*Pasch's Axiom.*) Let A, B, C be three points not lying in the same straight line and let l be a straight line lying in the plane ABC and not passing through any of the points A, B, C. Then, if the straight line l passes through a point of the segment AB, it will also pass through either a point of the segment BC or a point of the segment AC (Fig. II).

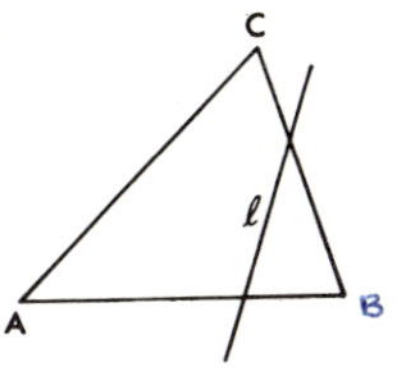

FIG. II. Pasch's Axiom

Notice that the idea of the order of points on a line is involved here.

(c) *Parallelism. Euclid's Postulate* 5. A crucial feature of Euclid's approach to Geometry is his theory of parallels and, in particular, his Postulate 5, which runs thus:

If a straight line falling on two straight lines makes the interior angles on the same side less than two right angles, the two straight lines, if produced indefinitely, meet on that side on which are the angles less than the two right angles. (See Fig. III.)

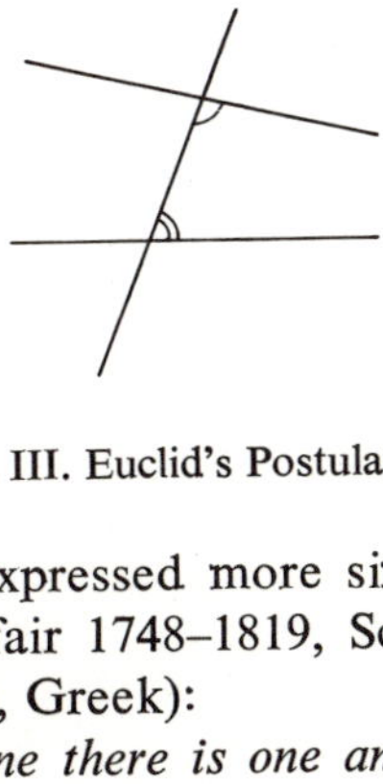

FIG. III. Euclid's Postulate 5

Alternatively, it may be expressed more simply in the form known as *Playfair's Axiom* (John Playfair 1748–1819, Scottish), though this form is traceable to Proclus (410–485, Greek):

Through any point in the plane there is one and only one line parallel to a given line.

Significantly, Postulate 5 is an assertion about a non-finite line, a fact which distinguishes it from Euclid's other statements. (But note that Euclid nowhere deals with non-finite points.)

Even in antiquity, before the birth of Christ, some doubts were entertained about the justification of Postulate 5 on logical grounds, e.g. by Posidonius (1st century B.C., Greek)—see Tuller [148] p. 4. Later, Proclus criticised it as a 'merely plausible and unreasoned statement'. (Remember that our lines may be separated by astronomical distances and not just by a few inches on a piece of paper.) Nevertheless, its essential validity seemed beyond question—it seemed in agreement with commonsense, and no finite measuring instruments could experimentally verify or dispute it. Intent on vindicating Euclid, philosophers and mathematicians endeavoured to show that Postulate 5 was a theorem which could be deduced from Euclid's other assumptions. 'The amount of trouble that has been taken in that chimerical hope is truly beyond imagination', remarked Henri Poincaré (1854–1912, French). (Some faulty 'solutions' of the problem are offered for criticism by Eves [36] 1, p. 332 *et seq.*)

When the Graeco-Roman civilisation declined and the light of mathematical scholarship shone among the Arabs, Hindus and Persians, further fruitless attempts were made by some of the finest intellects of their times to justify this postulate. One important attempt was made by Nasîraddîn (1201–1274), the Persian astronomer and mathematician who flourished during the reign of the Mongol conqueror, Hulagu (who freed Nasîraddîn from enforced service with the sect of the Assassins). Wiser than all these believing scholars, Euclid apparently realised that it was an assumption and hence unprovable—one small step for man, one giant step for Geometry!

§ 1.3 Non-Euclidean Geometry. Of the countless failures to demonstrate that Postulate 5 was a consequence of Euclid's preceding assumptions, the most spectacular was that of the Jesuit priest, Girolamo Saccheri (1667–1733, Italian), who, in 1733, published a book the purpose of which was to vindicate Euclid from all flaws. To accomplish this objective, Saccheri constructed a quadrilateral by erecting perpendiculars $AD = BC$ at the extremities of the segment AB and joining CD (Fig. IV). According to Euclid, $\angle ADC = \angle BCD = \frac{\pi}{2}$ (to prove which Nasîraddîn constructed the quadrilateral in Fig. IV(a)).

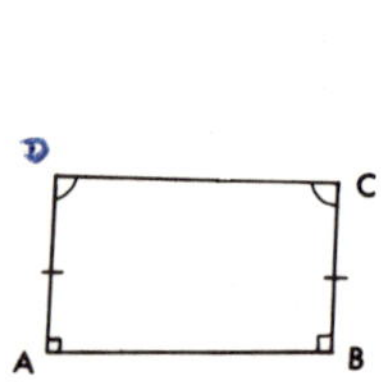

FIG. IV. Saccheri's quadrilateral

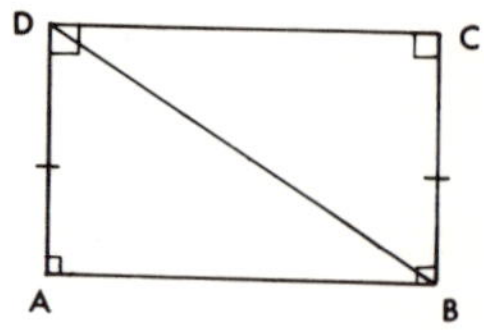

FIG. IV. (a) Nasîraddîn's quadrilateral

Now Saccheri, who was familiar with Nasîraddîn's work, assumed two other possibilities (acute or obtuse) for these equal angles, namely

(a) the Hypothesis of the Acute Angle (implying that the angle-sum of a triangle is less than two right angles), and

(b) the Hypothesis of the Obtuse Angle (implying that the angle-sum of a triangle is greater than two right angles).

Rigidly believing in the truth of Euclid's arguments, Saccheri desired to show that both these Hypotheses led to contradictions (and hence the validity of Euclid's reasoning would be vindicated). Try as he might, Saccheri could find no logical error in the consequences which flowed from his assumptions and his attempted *reductio ad absurdum* argument was thus negated. Alarmed that he had created a Frankenstein monster, Saccheri decided that he must have made a mistake somewhere, as it was inconceivable to him, a respecter of traditional habits of thought and of authority, that an alternative to Euclidean Geometry could exist. Orthodox opinion held that to go against Euclid was to 'go against Nature', and Saccheri dared not cross the threshold. Unwittingly he became the first man to glimpse the revolutionary lines of development which were to occur a century later.

Full advantage of the breakthrough made by Saccheri was taken quite independently and at much the same time in the 1820's by Nicolai Lobatschewsky* (1793–1856, Russian), Professor of Mathematics at Kazan, Russia, and by Johann Bolyai, (1802–1860, Hungarian), an officer of the Austro-Hungarian Army. Concerning the weaknesses in Euclid, Lobatschewsky observed: 'In geometry, I find certain imperfections which I hold to be the reason why this science, apart from transition into analytics, can as yet make no advance from that state into which it has come to us from Euclid. As belonging to these imperfections I consider . . . the momentous gap in the theory of parallels, to fill which all efforts of mathematicians have been so far in vain.' Lobatschewsky's subsequent discoveries represent one of the noblest achievements of the human mind.

Bolyai's father, Wolfgang, who had unsuccessfully pitted himself against Postulate 5, warned his son to 'detest it just as much as the lewd intercourse' as it would deprive him of his leisure, health and happiness. 'This abysmal darkness might perhaps devour a thousand towering Newtons,' he lamented, 'it will never be light on earth.' And again: 'It is inconceivable that this inevitable darkness, this eternal eclipse, this blot on geometry was allowed —this eternal cloud on virgin truth.' Young Bolyai, however, was soon able to declare triumphantly: 'I have created a new Universe from nothing.'

Karl Gauss (1777–1855), the brilliant German mathematician, had earlier come to the conclusions similar to those reached by Lobatschewsky and Bolyai but, fearing public ridicule for the expression of unpopular opinions, had kept silent. Nevertheless, his study of the theory of parallels gave Gauss

---

* There are many transliterations of the name of this great geometer; a widely-used transliteration is Lobachevsky.

intense personal satisfaction (*'le plus vif plaisir'*).

Though Lobatschewsky and Bolyai were the first men to publish a theory of non-Euclidean Geometry, they were not alone in voyaging across these strange seas of thought. In 1854, Georg Riemann (1826–1866, German) delivered his famous dissertation on the hypotheses underlying the foundation of Geometry where the principles of another, different non-Euclidean Geometry were enunciated. He asserted that space is finite but unbounded (as on the surface of a sphere), a concept which Einstein was to seize upon to describe his relativistic theory of the Universe.

The *non-Euclidean Geometries* of Riemann and of Lobatschewsky and Bolyai (corresponding to Saccheri's Hypothesis (b), (a), respectively) are logically sound systems which assume the hypothesis of denying Euclid's Postulate 5, but preserve the remaining Euclidean assumptions. The fact that they are logically impeccable and therefore of equal mathematical validity represents one of the great revolutions of thought. As we shall see, they destroyed forever the alleged necessity of Euclidean Geometry as a law of nature and thus were a serious blow to the aprioristic thinking of the Kantian philosophers (followers of the 18th century philosopher, Immanuel Kant (§ 1.2)). In Einstein's words, Lobatschewsky 'challenged an axiom' and in doing so his revolutionary thinking abolished the necessary 'truth' of Euclidean Geometry.

Writing of the non-Euclidean Geometry and the way in which Lobatschewsky transformed our ideas about the Universe, William Clifford (1845–1879, English) claimed: 'It is quite simple, merely Euclid without the vicious assumption, but the way things come out of one another is quite lovely.' . . . 'What Vesalius was to Galen, what Copernicus was to Ptolemy, that was Lobachevski to Euclid.'

Some of the chief differences between Euclidean and non-Euclidean Geometries (all of them metrical) may be tabulated thus:

	Euclidean	non-Euclidean	
Geometrical property	Euclid	Lobatschewsky-Bolyai	Riemann
Number of lines through a given point parallel to a given line	1	2	0
Angle-sum of a triangle	π	$< \pi$	$> \pi$
Effect on this angle-sum of increase in area of triangle	unchanged	decreases	increases

Fuller details of the subtleties of parallelism in Lobatschewsky-Bolyai Geometry are given in § 40.2 in Lobatschewsky's own words.

Confronted by this variety of possible geometries, one naturally asks: Which one is 'true'? What significance, if any, have the non-Euclidean geometries for the builder, the architect, the engineer, the navigator, the nuclear physicist, the astronomer? These are important questions which, however, are more appropriately answered later on in Chapter 9, where more detailed mathematical and historical information regarding non-Euclidean Geometry, together with some of its philosophical considerations, is provided. For reasons which will be explained in Chapter 8, the Lobatschewsky-Bolyai and Riemann geometries are called Hyperbolic Geometry and Elliptic Geometry respectively.

Warning: Pay no heed, in all this work, to the nagging and beguiling, but discredited, voice within you which whispers that at the back of your mind you know this is all nonsense. It is the selfsame voice which down the centuries has betrayed us. 'What nonsense to say that this solid table on which I am writing is a collection of electrons moving with prodigious speed in empty space, which relative to electronic dimensions are as wide as the spaces between the planets in the solar system! What nonsense to say that the thin air is trying to crush my body with a load of 14 lb. to the square inch! What nonsense that the star cluster which I see through the telescope, obviously there *now*, is a glimpse into a past age 50,000 years ago!' (Arthur Eddington).

Euclidean models of these non-Euclidean systems have been constructed and in this way the consistency of the non-Euclidean Geometries has been sustained. Eugenio Beltrami (1835–1900, Italian) showed in 1868 that every theorem of plane non-Euclidean Geometry could be interpreted as a theorem holding for a class of surfaces in three-dimensional Euclidean space. A logical flaw in non-Euclidean Geometry could now be interpreted as a logical flaw in classical Euclidean surface geometry.

'It is clear now why all of the efforts to prove Euclid's Parallel Postulate were destined to end in failure. The Postulate can never be proved, for its proof would order the rejection of the parallel postulates of the equally consistent non-Euclidean Geometries' (H. E. Wolfe [121]).

Whenever reference is made to a model (or representation) of a geometrical system, the reader is assumed to have an intuitive understanding of what is meant. Later, in § 4, we explain more precisely the meaning attached to the word.

§ 1.31 A non-Euclidean Model. Quite a simple model exists for the elliptic plane, namely the surface of a sphere with antipodal points identified, e.g. N and S (Fig. V) are regarded as coinciding. Great circles, in this model, represent lines, e.g. the only elliptic lines are the meridians of longitude and the equator, and other diametral plane sections of the sphere.

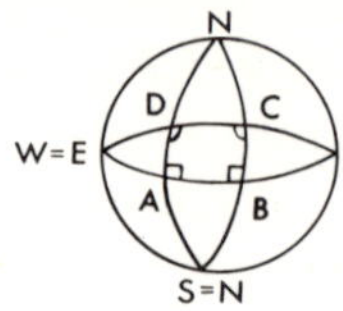

FIG. V. Model of elliptic non-Euclidean geometry

Clearly, as in spherical trigonometry, the angle-sum of a triangle, say *ABN*, is greater than 180°.

Can parallelism exist in the elliptic plane? Unequivocally we answer: No! as the following argument makes plain:

Possible definitions of parallelism are

(i) two lines are parallel if they never meet, however far they are extended;

(ii) the locus of points on one side of a line and at a fixed distance from it is a line parallel to the given line.

Interpreting these ideas on the spherical surface, we see that (i) fails since two great circles meet (in antipodal points) and (ii) fails since the stated locus is a small circle and is therefore, by definition, not a straight line.

Hence parallels do not exist on the surface of the sphere (and in the elliptic plane).

But, you ask, what about railway lines? Are these not parallel? (Engine-drivers may justifiably be excused if they dismiss the mathematicians' answer as a piece of casuistry!) See § 42.21.

Arab and Greek navigators and scientists, such as Ptolemy (2nd century, Egyptian-Greek)—who had also laboured in vain on Euclid's Postulate 5—were aware of the geometry of the sphere and indeed had amassed a considerable deal of information about its trigonometry. Why then did they not discover this model of non-Euclidean Geometry? Simply, it seems, because our model vitiates one of the cardinal Euclidean beliefs, namely that two lines cannot enclose an area such as *NADN* and *NBCN* do. Identification of antipodal points would therefore never have occurred to them.

If $S \neq N$, then we should have two lines *SADN* and *SBCN* meeting in two points, which is an equally unacceptable thought.

§ **1.32 Significance of non-Euclidean Geometry.** Briefly, it

(i) dealt a serious blow to the Kantian philosophy of space and geometry;

(ii) became a useful tool for the geometrical description of the Theory of Relativity;

(iii) showed that Mathematics doesn't deal with absolute truths, but gives freedom to investigate the consequences of any hypothesis one chooses, irrespective of its relevance to the world of sense-perception;

(iv) gave an insight into the working of the human mind, revealing the heights to which human thinking can soar.

§ 1.4 Hilbert's Axioms. Most famous of all the axiomatic systems evolved in modern times as a satisfactory foundation of Euclidean Geometry is that expounded by Hilbert during the winter semester of 1898–1899 at the University of Göttingen. How does he overcome the obstacle of defining point and line? Quite simply, by recognising that they are undefinable and taking them (and planes also) along with an undefined relation of incidence connecting them, as primitive concepts (an idea due in 1899 to Mario Pieri (1860–1904, Italian)). From these, Hilbert constructed five sets of axioms, namely

- Axioms of connection (incidence)
- Axioms of order (largely originating with Pasch in 1882)
- Axiom of parallels (Euclid's Axiom in Playfair's form)
- Axioms of congruence
- Axiom of continuity (*Archimedes' Axiom*).

Full details are set out in Hilbert's classic, *Grundlagen der Geometrie* ('Foundations of Geometry') [50]. Useful comparisons of Hilbert's and Euclid's system may readily be made by consulting Adler [2] or Tuller [148] wherein the basic ideas of Euclid and Hilbert are juxtaposed as Appendices.

While one hopes that the reader will investigate the axioms of Euclid and Hilbert for the plane more thoroughly, an indication of their fundamentally different approaches is thought desirable. Although some of these axioms have been discussed already, we reassemble them here for convenience. Note that Euclid plunges into his definitions immediately whereas Hilbert, with the experience of the centuries to guide him, is more circumspect.

§ 1.41 Euclid's Axioms.

The Definitions of Book 1

1. A point is that which has no part.
2. A line is breadthless length.
3. The extremities of a line are points.
4. A straight line is a line which lies evenly with the points on itself.
5. A surface is that which has length and breadth only.
6. The extremities of a surface are lines.
7. A plane surface is a surface which lies evenly with the straight line on itself.

....

23. Parallel straight lines are straight lines which, being in the same plane and being produced indefinitely in both directions, do not meet one another in either direction.

The Postulates

Let the following be postulated.

....

5. That, if a straight line falling on two straight lines makes the interior angles on the same side less than two right angles, the two straight lines, if produced indefinitely, meet on that side on which are the angles less than the two right angles.

§ 1.42 Hilbert's Axioms.

(For brevity, a summary is given of Hilbert's axioms of incidence and his axiom of parallels—the language and spirit of his approach are preserved.)

Let us consider three distinct systems of things: the things composing the first system, we will call *points*; those of the second, we will call *straight lines*; and those of the third system, we will call *planes*.

We think of these points, straight lines and planes as having certain mutual relations, which we indicate by means of such words as 'are situated', 'are parallel', . . . (Hilbert then explains the use of such phrases relating to incidence as 'lies on', 'goes through', 'is a point of', etc.).

Axioms of Connection (*Incidence*)

1. Two distinct points always completely determine a straight line.
2. Any two distinct points of a straight line completely determine that line.
3. Three points not situated in the same straight line always completely determine a plane.
4. Any three points of a plane, which do not lie in the same straight line, completely determine that plane.
5. If two points of a straight line lie in a plane, then every point of the straight line lies in the plane.

....

7. On every straight line there exist at least two points and in every plane at least three points not lying in the same straight line.

Axiom of Parallels (*Euclid's Axiom*)

In a plane there can be drawn through any point, not lying on a straight line, one and only one straight line which does not intersect the line. This straight line is called the parallel to the given line through the given point.

§ 1.43 Other Systems of Axioms for Euclidean Geometry. Other detailed axiomatic approaches to Euclidean Geometry may be found in, say, Borzuk and Szmielew [15], Forder [41], Moise [77], and Room [146]. Also consult G. D. Birkhoff, 'A Set of Postulates for Plane Geometry, Based on Scale and Protractor', *Annals of Mathematics*, 33 (1932) 329–345. An interesting recent set of axioms for Euclidean Geometry is that devised by the School Mathematics Study Group (SMSG)—reference School Mathematics Study Group: Geometry, Yale University Press (1961) or Tuller [148].

Another approach to a postulational system of Euclidean Geometry (of arbitrary dimension), which is due to Prenowitz, involves the notion of a convex set, a concept occurring widely in Mathematics, e.g. in the theory of games, linear programming and functional analysis. Details are given in W. Prenowitz: 'A Contemporary Approach to Classical Geometry', Slaught Memorial Papers, No. 9, *Amer. Math. Monthly* 68 (1), Part 2 (1961) 1–67.

It is interesting to record in passing that in 1902 Eliakim H. Moore (1862–1932, American) found that some of Hilbert's axioms (of connection and order) were redundant, i.e. that Hilbert's system was not fully independent as some axioms were logically deducible from others.

'Points' and relations connecting them are the raw materials a geometer works on, irrespective of their interpretation, just, as it has been said, as a painter uses his paints without worrying about the chemical composition of the molecules in them. To illustrate this assertion, we turn to a remarkable example from biological science.

§ **1.44 A Biological Example.** Experiments on the fruit fly, Drosophila, in relation to a certain biological mechanism of linked genes in a chromosome, had produced statistical percentages which interested Hilbert. Writing of these percentages in 1930 he exclaimed that they 'satisfy the linear Euclidean axioms of congruence and the axioms concerning the geometrical concept of "betweenness". Thus, the laws of heredity appear as an application of the linear congruence axioms . . .; simple and precise—and at the same time wonderful beyond any stretch of the boldest imagination.'

So our axioms need not necessarily be confined to spatial relations, though, it must be remembered, the correspondence noticed by Hilbert is not exact, however astonishingly approximate. Fuller details of the genetic aspect of this illustration may be found in Colin's *Elementary Genetics*, p. 73.

§ **1.5 Huntington's Axioms.** An interesting system of axioms for three-dimensional Euclidean Geometry was constructed in 1913 by E. V. Huntington (1874–1952, American) who took a *sphere* as an undefined element and *inclusion* as an undefined relation, so that a point (or point-sphere) is defined as 'any sphere which contains no other sphere within it.' Huntington's system necessarily lacks the simplicity of Hilbert's system, which involves three dimensions and not just the plane. For details, consult Huntington ('A set of postulates for abstract geometry, expressed in terms of the simple relation of inclusion', *Mathematische Annalen*, 73 (1913) 522–559; 'The Duplicity of Logic', *Scripta Mathematica*, 5 (1938) 149–157, 233–238), or Robinson [92] which contains a readable summary of the ideas involved.

Chief differences between Huntington's set of axioms and other well-known sets are, quoting Huntington,

(i) the use of the solid body instead of the point as an undefined concept,
(ii) the extreme simplicity of the undefined relation of inclusion;
(iii) the systematic definitions of the straight line, the plane, and the 3-space, which can be readily extended, if desired, to space of n dimensions; and
(iv) the attempt to separate the 'existence postulates' from the postulates expressing 'general laws'.

Let it be noted that there is nothing in Huntington's system which requires 'points' to be *small*! By way of illustration, he gives the example of the set of all spheres where diameters are greater than or equal to one inch. 'Points' are then simply the inch-spheres.

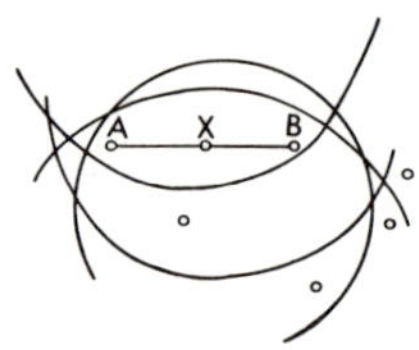

FIG. VI. Huntington's definition of a point X and a segment [AB]

Huntington defines a segment as follows (Fig. VI): Let A and B be any given points. If X is a point such that every sphere which contains A and B also contains X, then X is said to belong to the segment $[AB]$, i.e. the set of points uniquely determined by A and B.

§2 PROJECTIVE GEOMETRY

§2.1 Compass Geometry. Constructions in Euclidean plane geometry rely entirely on the straight-edge (i.e. an unmarked ruler) and the compasses. Independently, Georg Mohr (1640–1697, Danish) [75] in 1672 and Lorenzo Mascheroni (1750–1800, Italian) [68] in 1797, discovered that the fundamental Euclidean constructions could still be carried out using only the compasses, i.e. ignoring the straight-edge altogether. Thus, for instance, the mid-point of the segment joining two given points can be constructed by compasses alone using the method of inversion in a circle (Fig. VII) (Pedoe [84]). However, it should be noted in this compass geometry that no compasses can possibly connect two given points by a straight line. Soviet scholars, among others, have recently shown the possibility of carrying out constructions in the Bolyai-Lobatschewsky plane similar to those of Mascheroni in the Euclidean plane.

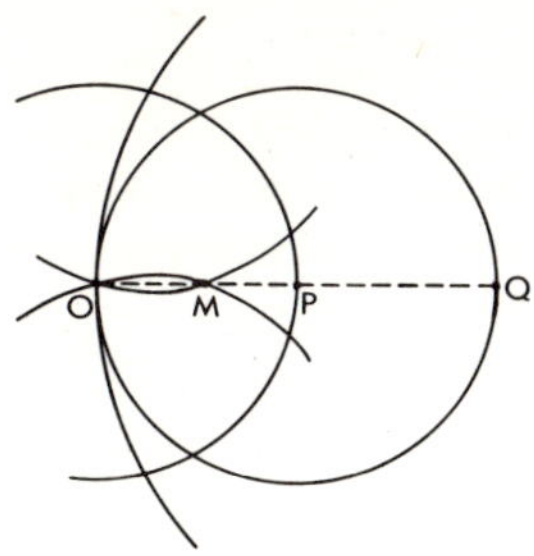

FIG. VII. Mohr-Mascheroni compass geometry (construction of the mid-point M of the segment OP)

§ 2.2 Straight-edge Geometry. On the other hand, supposing that the compasses are dispensed with, what constructions are possible using just the straight-edge? Certainly not Euclid's first construction which involves the intersection of two circles. It transpires that many of the distinguishing features of Euclidean Geometry, such as measurement (angle and length), parallelism and the existence of circles, are lost. Nevertheless, much remains. Credit for developing this type of geometry goes to Abû'l-Wefâ (940–998, Persian), Jean-Victor Poncelet (1788–1867, French) and Jakob Steiner (1796–1863, German). Note that Steiner, in his publication of 1833, required in a plane a given circle with fixed centre and radius, as an aid to his ruler constructions. Whereas in Euclidean Geometry, constructions are made by means of straight-edge and compasses, in Projective Geometry (§ 2.6) things are easier as we require only the straight-edge for our constructions. (See Axioms PP 1–PP 3, Chapter 1.)

An account of the history of the ruler and compasses constructions is given by the author in 'Constructions possible using ruler and compasses', the *Mathematical Gazette*, 44 (350) (1960) 270–276.

§ 2.3 Non-Metrical Theorems of Pappus and Desargues. While Euclidean Geometry is rooted in notions of measurement, there exist some basic theorems of a non-metrical nature which were quite unknown to Euclid. Two of these, one due to Pappus (4th century, Egyptian-Greek), who towered above his contemporaries 'like the peak of Teneriffa above the Atlantic', and the other to the architect Girard Desargues (1593–1662, French), are absolutely fundamental in understanding modern Geometry, as we shall see later. Based on the incidence of point and line, they are dependent only on the straight-edge for their construction.

Pappus (who, incidentally, is recorded as having witnessed a solar eclipse on October 18, 320) states his famous theorem in Book 7 of his *Collectio Mathematica* , ([80] pp. 685–688—French translation), which, in his original lettering (together with modern set theory notation), may be expressed thus:

Pappus' Theorem: If A, E, B are any three points on one line and Γ, Z, Δ are any three points on another line, then the points $H = AZ \cap E\Gamma$, $M = A\Delta \cap B\Gamma$, $K = E\Delta \cap BZ$ are collinear. (Pappus says: '*Je dis que la ligne qui passe par les points H, M, K est droite.*')

Pappus proves his theorem in two parts, (i) (Lemma XII) when AB is parallel to $\Gamma\Delta$, and (ii) (Lemma XIII) when AB is not parallel to $\Gamma\Delta$ (Fig. VIII; cf. [80] p. 687).

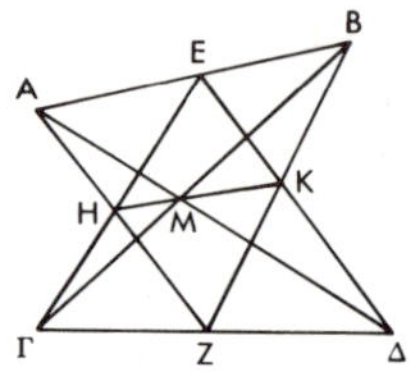

FIG. VIII. Pappus' theorem

Much as it is desirable to reproduce the spirit of Desargues' discovery in a similar way, we find his presentation lacking in simplicity so we content ourselves with a standard statement of his famous result:

Desargues' Theorem: If two triangles ABC, $A'B'C'$ are in perspective (i.e. AA', BB', CC' are concurrent), then the points of intersection of corresponding sides (namely AB, $A'B'$; BC, $B'C'$; CA, $C'A'$) are collinear.

(See Fig. 5(a), § 7, in which the perspective triangles are imagined to exist in either two or three dimensions. Also see Fig. IX(a) for the three-dimensional case.)

It should not be overlooked that Desargues' interest in perspective was allied to his professional work as an engineer and architect. Though Desargues' major work was his *Treatise on Conics* (1639), he is best known for his '*Brouillon proiect d'une atteinte aux éuénemens des rencontres d'un cone avec un plan*' which is characterised by its botanical nomenclature. Worth mentioning is the fact that Desargues, a native of Lyons, often had his name on his publications abbreviated as G.D.L. (G. Desargues Lionais).

§ 2.31 Art and Perspective. Our knowledge of perspective and projection owes much to the artistic studies on perspective of Filippo Brunelleschi, Leone Alberti, Leonardo da Vinci ('Let no one who is not a mathematician read my works'), Paolo Uccello, Albrecht Dürer and other Renaissance artists. Historians usually credit Brunelleschi, architect of the beautiful Dome in Florence, with the discovery of the correct principles of perspective (about 1425). [So obsessed was Uccello with the effects of scientific perspective in his paintings that Donatello upbraided him, saying: 'Ah, Paolo, with this perspective of yours you are sacrificing the substance for the shadow.'] The picture painted by the artist (surrealists excepted!) may be regarded as a projection of the original on the canvas with the eye of the painter as the centre of perspective (Fig. IX(b)). Absurdities which can arise if the basic principles of perspective are ignored by the artist are strikingly depicted in William Hogarth's (deliberately executed) steel engraving, 'False Perspective'.

Systematic development of the sort of geometry (independent of measurement) typified by the Theorems of Pappus and Desargues is less than a century and a half old, as Poncelet's great work on the subject was published in 1822. It suggests that there does exist a worthwhile geometry in which distance is not an inherent property. Is there any simple realisation of the goal we are seeking?

§ 2.4 Conical Projection. Elements at Infinity.

'And Enoch walked with God: and he was not; for God took him.'
(*Genesis* V,24)

In Fig. IX(a) drawn in Euclidean space of three dimensions, $P' \in \mathscr{C}'$ in π' is the *general* (*conical*) *projection* (or *image*, or *map*) of $P \in \mathscr{C}$ in π from V, the *vertex* (*centre*) *of projection*; $\pi \cap \pi'$ is the *axis of projection*. Of course, $\mathscr{C}$ is not necessarily a conic, but is some geometrical curve or figure. As a particular case, $\mathscr{C}$ may be a triangle ABC in which case Desargues' Theorem in three dimensions results.

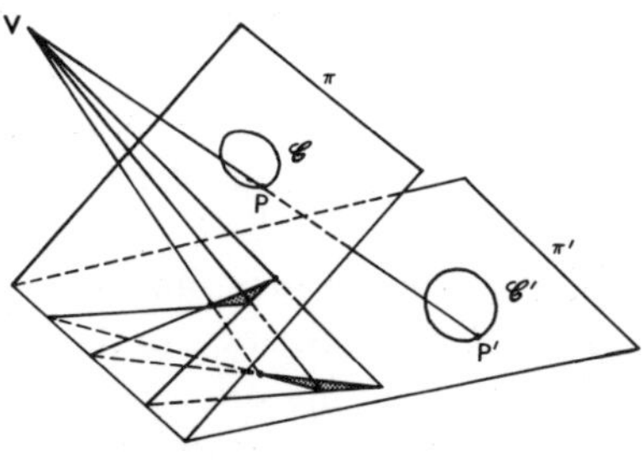

FIG. IX. (a) Conical projection

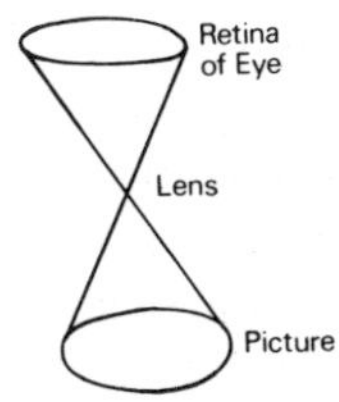

FIG. IX. (b) Optical (conical) projection

Generally, it will happen that VP meets π' in a specific point.

§ 2.41 Point at Infinity. Augmented Line. Suppose, however, that V lies in a plane $\sigma \parallel \pi'$ (Fig. X(a)). Clearly, under projection from V,

(i) line $CQP \to$ line $CQ' \parallel VP$;.
(ii) lines CP, DP in π meeting at P on $HK \to \parallel$ lines CQ', DR' in π'.

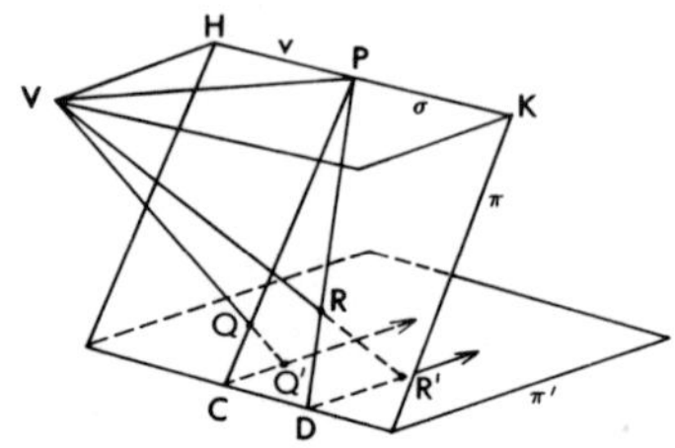

FIG. X. (a) Projecting to infinity

But what about the projected image of P? It is not located at any point of π'. To overcome this difficulty, assume the existence of a physically fictional 'point' P'_∞, the projection in π' of P. This new object is called a *point at infinity*. The intersection of the parallel lines in (ii) may be thus expressed as

(iii) $CQ' \cap DR' = P'_\infty$,

i.e. lines parallel to CQ' have a common point at infinity.

Exactly one point at infinity lies on any ordinary line. Logically, what situation(s) would arise if we postulated the existence of two points at infinity on a line? Either we could say that, through a given point (not on a given line) in a Euclidean plane, *two* parallels could be drawn to the given line, or we could assert (as might appear to be the case if we stood between the rails of a railway track and looked in opposite directions along the track) that two distinct lines have more than one point in common. Both possibilities are inadmissible in Euclidean Geometry.

By adjoining a point at infinity to an ordinary (Euclidean) line we are able to treat parallel lines on the same basis as other lines, i.e. any two lines in the plane, whether parallel or not, intersect in a point.

A Euclidean line extended by the addition of a point at infinity may be called an *augmented Euclidean line*. If we do not wish to distinguish the point at infinity on this augmented line from the other points on it (since any ordinary point can be projected into a point at infinity), we may refer to the augmented line as a *projective line*.

Without attempting to visualise the augmented Euclidean line, we accept the fact that it behaves as though it were closed (like a circle of immense radius), though this does not mean that the line is to be thought of as being 'bent' in any way; see Struik [109] p. 3 and Young [124] p. 15. In Fig. XI(a), which shows how a 1—1 correspondence can be established between points P' of a circle and points P of an augmented line, we see that O corresponds to itself while V corresponds to the point at infinity on the line.

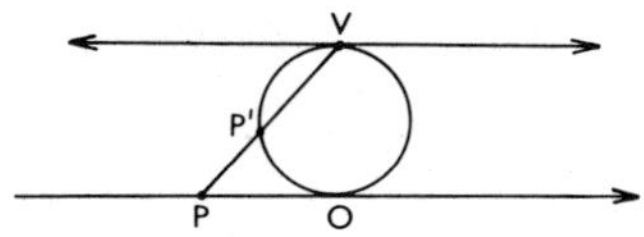

FIG. XI. (a) Correspondence between points on an augmented line and points on a circle

Furthermore, Figs. XI(b), (c) show that on the augmented Euclidean (projective) line the order of points is cyclic and accordingly we cannot say that C lies between A and B (for what then of D?), but only that C and D separate A and B. On the other hand, Fig. XI(d), obtained from Fig. XI(c) by omitting D_∞, shows that the order on the ordinary Euclidean line is serial and that we can assert that C lies between A and B.

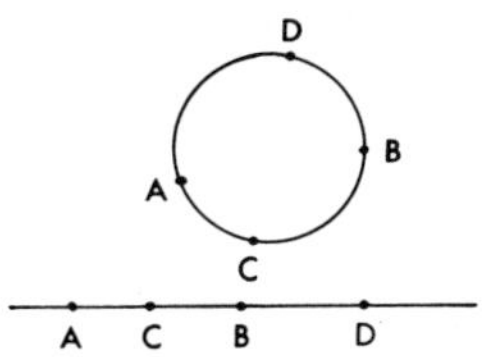

FIG. XI. (b) (closed) Projective line

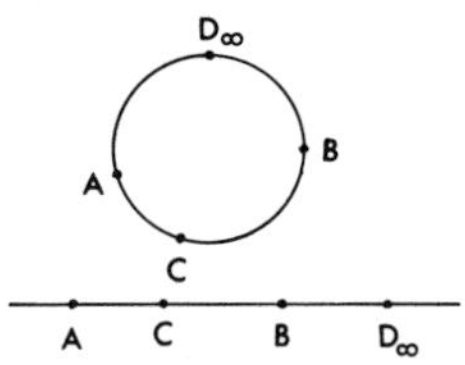

FIG. XI. (c) (closed) Augmented Euclidean (projective) line

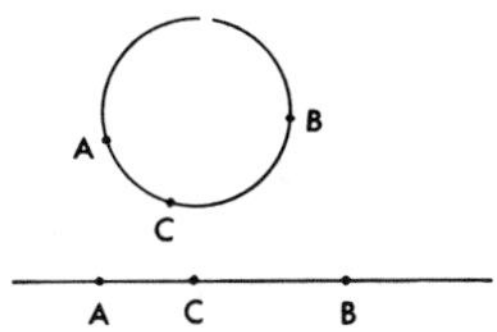

FIG. XI. (d) (open) Euclidean line (non-projective)

§ 2.42 Line at Infinity. Augmented Plane. Revert now to Fig. X(a). Following our line of reasoning in § 2.41, we assume that

$\sigma \cap \pi = HK$ (*vanishing line*) $\rightarrow$ a physically fictional 'line' l'_∞ in π'.

This new object is called the *line at infinity* in π'.

It follows that

(a) l'_∞ is the locus of all points P'_∞; and

(b) $l'_\infty = \sigma \cap \pi'$.

Similarly, l_∞ in π is the projection of $\pi' \cap$ (plane σ' through some vertex $V' \parallel \pi$).

There is just one line at infinity in each plane.

Evidently, given *any* line λ in π, π' may be chosen so that with suitable choice of vertex $\lambda \rightarrow l'_\infty$ in π'. This process is called *projecting to infinity* (Fig. X(a)). An interesting problem occurs when we make a detailed examination of the projection of points on PC (Fig. X(a)). Isolating lines VP, PC, CP'_∞ in Fig. X(a) we obtain Fig. X(b) which provides the details.

Neither direction nor distance is associated with the line at infinity. Indeed, it accounts for all the directions in the plane. Nor is it located in any particular position in the plane, i.e. it is a mathematical idea, not a geographical position. Once we have created the line at infinity, its existence is as 'real' as that of any other line, though it will have some special properties. Later, in our treatment of plane geometry, it may appear that the phrases 'point at infinity' and 'line at infinity' are misleading (and, in fact, there do exist several substitute phrases for them). However, we will retain them in order to preserve something of the historical flavour of the origins of the geometry we intend to develop.

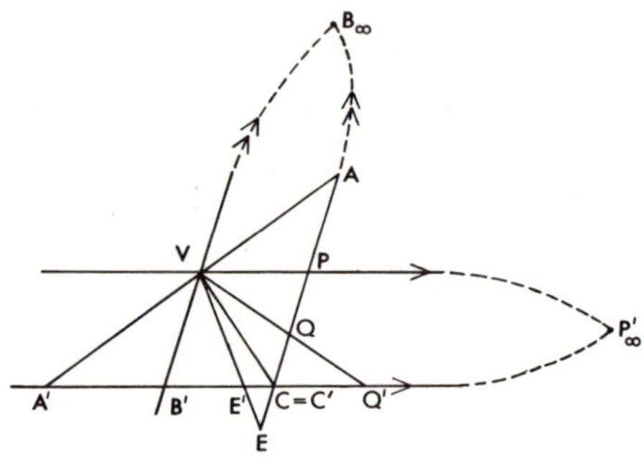

FIG. X. (b) Projecting to infinity (section)

A Euclidean plane extended by the addition of a line at infinity may be called an *augmented Euclidean plane*. As the line at infinity must be treated in the same way as any other ordinary line in the plane (which may be projected into the line at infinity), we also speak of the augmented plane as a *projective plane*.

Symbolically, we may write

$$l = l_E + P_\infty$$
$$\pi = \pi_E + l_\infty,$$

where l_E, π_E refer to an ordinary (= unaugmented) Euclidean line and plane respectively, and l, π to the corresponding augmented Euclidean (projective) line and plane respectively. We may think of π_E as being embedded in π.

As we are aware, the adjective 'Euclidean' carries with it the connotation of distance. But the adjective 'projective' will be seen to involve no metrical ideas whatever, so if we think of augmented Euclidean lines and planes as projective lines and planes we must imagine them as being devoid of metrical content. (Later we will find that the concept of an affine plane is equivalent to that of a non-metrical Euclidean plane.)

§ **2.5 Historical.** Serenus (*c.* 450 A.D., Greek) originated the idea of conical projection but no real use of it was made before Desargues. Though the Greeks such as Apollonius (3rd century B.C.) knew a lot about conic sections, no projective theorems were ever discovered until the time of Pappus. The concept of point at infinity is due to the astronomer, John Kepler (1571–1630, German) and Desargues.

On p. 105 of his Brouillon project [30] 1, Desargues states that, for parallel lines, '. . . *toutes ces droites sont entr'elles d'une mesme ordonnance, dont le but est à distance infinie* . . .', i.e. parallel lines have an infinitely distant point in common.

Says Kepler [55] p. 92, (see Fig. XII): '*In parabole unus D est intra sectionem, alter vel extra vel intra sectionem in axe fingendus est infinito intervallo à priore remotus, adeò ut educta HG vel IG ex illo caeco foco* [blind focus] *in quodcunque functum sectionis G sit axi DK parallelos*',
i.e. *GH*, *GI* meet the axis at one point at infinity (Kepler's 'blind focus').

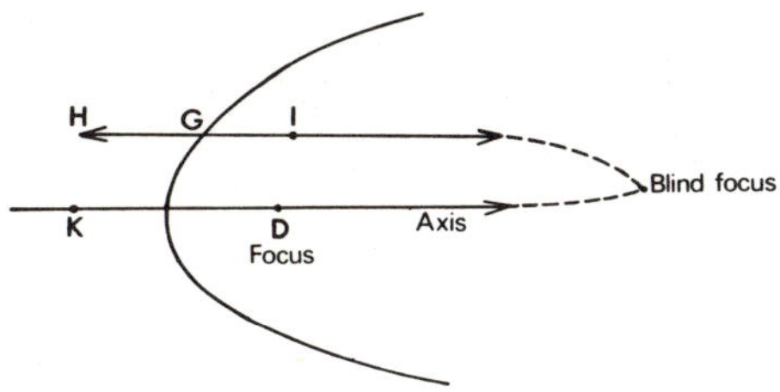

FIG. XII. Kepler's 'blind focus'

The concept of line at infinity is due to Poncelet [87] p. 49: '. . . *tous les points à l'infini d'un plan puissent être considérés idéalement comme distribués sur une droite unique, située elle-même à l'infini sur ce plan.*'

In three-dimensional space, the set of all lines at infinity, one for each plane, (and therefore the set of all points at infinity) constitutes a *plane at infinity* π_∞. (Why *must* this be so?) Poncelet, to whom the concept of plane at infinity is due, expressed his idea in these words: '*Tous les points à l'infini de l'espace peuvent être censés appartenir à un seul et même plan, nécessairement indéterminé de situation.*'

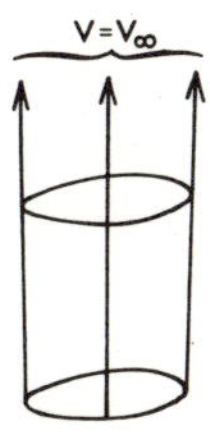

FIG. IX. (c) Cylindrical projection

Notice that if $V\ (= V_\infty)$ is at infinity in Fig. IX(c), the conical projection then becomes *parallel* (*cylindrical*) *projection*. Properties which remain invariant under parallel projection (e.g. bisection, parallelism, betweenness, ratios of segments) form the subject matter of *Affine Geometry*, the word 'affine' (*affinis* [Latin] ≡ 'related') in this sense being due to Euler (1707–1783, Swiss). *Orthogonal projection* occurs if $PP' \perp \pi'$. Observe (Fig. XIII) that, in π, a conic cutting HK in two real points → an hyperbola in π', a conic not cutting HK ($= v$) in two real points → an ellipse in π', a conic touching HK → a parabola in π'. (Refer also to Pedoe [85] pp. 15–17.) Tangents to the conic at the two real points of intersection project into the asymptotes of the hyperbola.

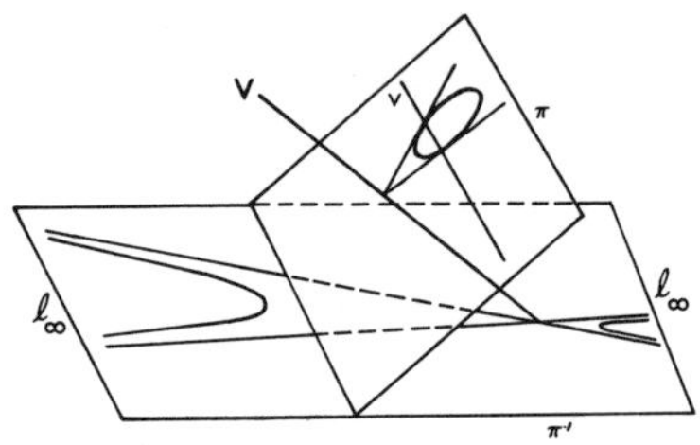

FIG. XIII. Projection of a conic

§ **2.6 Projective Invariants. Projective Geometry.** *A projective property* is one which is invariant under a projection and *Projective Geometry* is the study of such properties.

In Fig. IX(b) supposing V to be 'between' π, π' (so that the diagram of $\mathscr{C} \to \mathscr{C}'$ is a double cone), we observe that the geometric operation of projection is equivalent to taking a photograph of the object $\mathscr{C}$. This is so since the camera collects the rays from $\mathscr{C}$ in its lens (V) and projects them on the sensitised photographic plate π'. From the mathematical point of view, it is manifestly immaterial whether V is between π, π' or is as shown in Fig. IX(a). Thus, in simple language, we could regard a projective property as one which is preserved when a photograph is taken. Continued photographs do not alter these projective properties. (Equivalently, V may be the lens of an eye and $\mathscr{C}'$ the image on the retina of the object $\mathscr{C}$ which the brain subsequently inverts.) An observer looking at a given figure will notice its different appearance as he changes position. Certain features of the figure will, however, remain invariant no matter what the observer's position is. Such invariant properties are the subject matter of Projective Geometry. Viewed from this standpoint, one can appreciate why von Staudt called Projective Geometry the 'Geometry of Position'.

We then ask ourselves: What properties of a figure are preserved by a projection? To help answer our question, we choose some examples from architecture. Consider the two illustrations shown: portion of the Palazzo Ducale (Doge's Palace in Venice, Italy (Fig. XIV(a)); and the Patio de la

Acequia, Generalife Gardens, of the Alhambra in Granada, Spain (Fig. XIV(b)) in which the vertical arches represent patterns in ornamental shrubbery and the central inner section represents a water channel. Other similar architectural examples spring readily to mind, e.g. the Taj Mahal at Agra, India; Angkor Wat in Cambodia; or the Parthenon, Athens, Greece.

FIG. XIV. (a) Doge's Palace (Palazzo Ducale), Venice

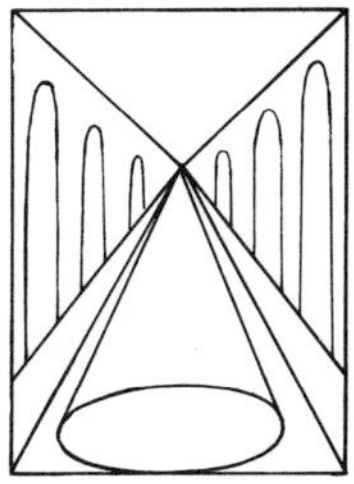

FIG. XIV. (b) Generalife Gardens, Granada

Clearly, the following qualities are preserved: point, straight line, collinearity, concurrency, so these will occur in Projective Geometry. On the other hand, observe that

(α) circles $\rightarrow$ ellipses;
(β) equality of lengths is not preserved;
(γ) parallel lines $\rightarrow$ lines not, in general, parallel;
(δ) an angle $\rightarrow$ an angle not, in general, of the same size.

Ideas such as measure (linear and angular), parallelism, perpendicularity and the distinction between conics are clearly alien to Projective Geometry since they are altered by projection. There will be triangles and quadrangles but not (for example) isosceles triangles or parallelograms, as length is not necessarily invariant under projection. Being plane sections of a circular cone, conics may be projected freely into other conics and hence no distinction is made in Projective Geometry among circles, ellipses, hyperbolas and parabolas.

Furthermore,

no distinction is made in Projective Geometry between points (lines) at infinity and ordinary points (lines) which are not at infinity, since each may be projected into the other at will.

Redraw portion of Fig. X(b) as Fig. X(c) and label the lines l, l' as indicated.

Observe on l that

(i) Q lies between A and C, and
(ii) Q divides the segment AC internally (in the ratio AQ/QC).

Projecting l from V on l', observe that

(i)′ Q' does not lie between A' and C', and
(ii)′ Q' divides the segment $A'C'$ externally (in the ratio $A'Q'/Q'C'$).

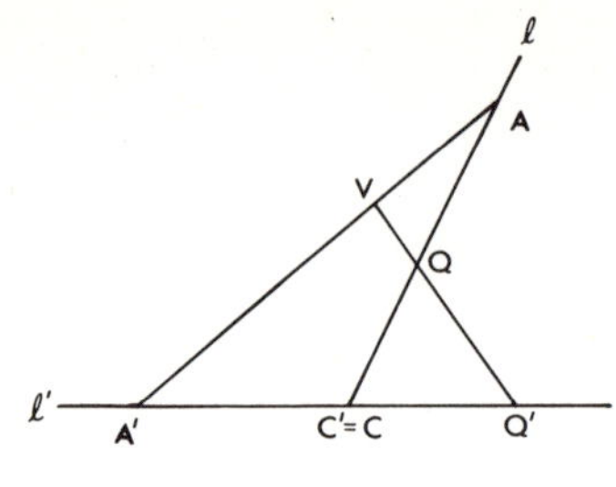

FIG. X. (c) Projective non-invariance of (i) Betweenness, (ii) Segment—division

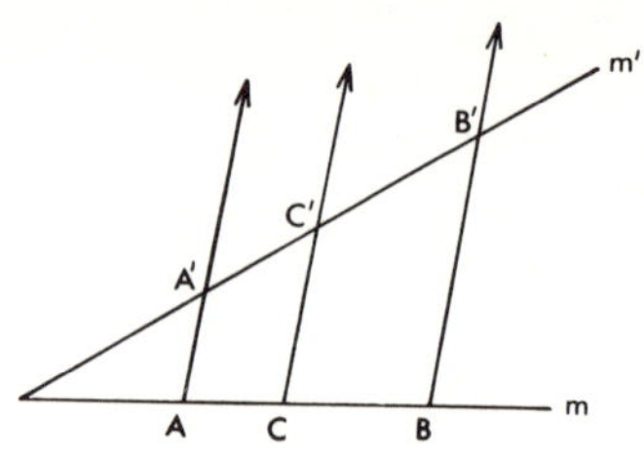

FIG. X. (d) Affine invariance of (i) Betweenness, (ii) Segment—division

Thus, *betweenness and segment-division are not projective invariants.* Consequently, the concepts of betweenness and segment-division (in particular, the concept of mid-point) will not occur in Projective Geometry.

On the other hand, Fig. X(d), which is unrelated to Fig. X(c), reveals that under parallel projection

(iii) C between A and B (on m) projects into C' between A' and B' (on m'), and

(iv) $\dfrac{AC}{CB} = \dfrac{A'C'}{C'B'}$.

Thus, *betweenness and segment-division (e.g. mid-point) are affine invariants* and will therefore occur in Affine Geometry, as we have previously mentioned.

§ **2.7 Cross-Ratio.** Whereas the ordinary ratio of two lengths is altered by projection, cross-ratio, which is the ratio of ratios, is unaltered. As it is defined at this stage, in terms of segments of lines, cross-ratio is a metric invariant though it can be made a non-metric invariant by abstracting distance from the definition. Knowledge of cross-ratio dates back to the early Christian centuries, notably to Pappus, who was aware of its invariance.

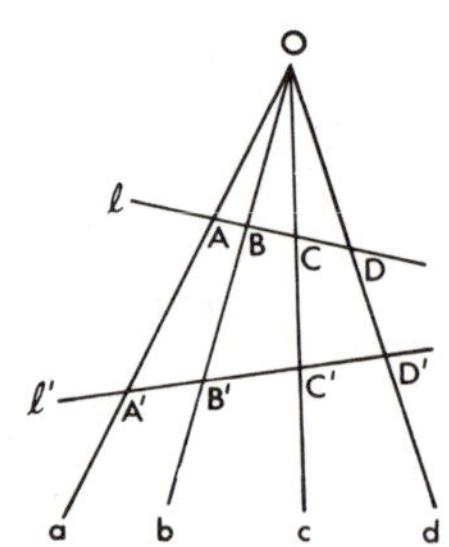

FIG. XV. Invariance of cross-ratio

Consider the four concurrent lines a, b, c, d, meeting at O (Fig. XV). Let the transversals l, l' cut these lines in A, B, C, D and A', B', C', D' respectively. Now

$$\begin{aligned}\frac{AC}{BC} &= \frac{AC/OA}{BC/OB} \cdot \frac{OA}{OB}\\ &= \frac{OA}{OB} \cdot \frac{\sin AOC}{\sin OCA} \Big/ \frac{\sin BOC}{\sin OCB} \quad \text{using the Sine Rule}\\ &= \frac{OA}{OB} \cdot \frac{\sin AOC}{\sin BOC}.\end{aligned}$$

The cross-ratio of the four collinear points A, B, C, D in that order is defined to be the number

$$\begin{aligned}\frac{AC/BC}{AD/BD} &= \frac{OA}{OB} \cdot \frac{\sin AOC}{\sin BOC} \Big/ \left(\frac{OA}{OB} \cdot \frac{\sin AOD}{\sin BOD}\right) \quad \text{by the above result twice}\\ &= \frac{\sin AOC}{\sin BOC} \Big/ \frac{\sin AOD}{\sin BOD}\\ &= \frac{A'C'/B'C'}{A'D'/B'D'} \quad \text{by similar reasoning.}\end{aligned}$$

That is

the value of the cross-ratio is preserved by the projection.

Whereas under an affine transformation it is the ratio of segments which is left invariant, notice that under a projective transformation it is the cross-ratio (or double ratio—*Doppelverhältniss* as it is called in German texts) which remains invariant. French writers use the words *rapport anharmonique* for 'cross-ratio', the term coined by Clifford in 1878.

§ 2.8 Affine Geometry. A little earlier, mention was made of Affine Geometry as being the special case of Projective Geometry when the vertex V $(= V_\infty)$ is at infinity, i.e. when the conical projection becomes cylindrical projection.

Equivalently, plane Affine Geometry results if the line at infinity is deleted from the projective plane, so that lines which met on l_∞ are now parallel. Affine Geometry occupies an intermediate position between Projective Geometry and (metric) Euclidean Geometry, which occurs as a very special case of Affine Geometry.

To illustrate a fundamental difference between Projective Geometry and Affine Geometry, we borrow the following example from the contemporary Russian mathematician, I. M. Yaglom, quoted in Coxeter [28] p. 104: 'I. M. Yaglom, near the beginning of the second volume of his *Geometric Transformations* (in Russian) shows two sketches of his house. The former depicts his living-room on a sunny day. The window panes cast shadows on the floor. The rectangles are distorted into parallelograms, but those that are equal remain equal. Affine Geometry deals with properties that are maintained by such "parallel projection". His other sketch is a night scene outside the house. A lamp inside casts shadows of the same window panes on the lawn. The rectangles are distorted into quadrangles of various sizes, but their sides are still straight. Projective Geometry deals with properties that are maintained by such "central projection".'

Projective Geometry includes all properties of Affine Geometry which are still true after conical projection. Because it is wider in scope than Affine Geometry, Projective Geometry contains statements which are not true in Affine Geometry; for example, in Projective Geometry two coplanar lines always intersect in a point, but in Affine Geometry two coplanar lines either meet in a point or are parallel.

Reverting to Fig. X(a) for a moment, we recall that under the process of projecting to infinity, the purely projective statement that lines CQ, DR meet in P (plane π) is converted into the affine (non-projective) statement that CQ' and DR' are parallel (plane π'). Since the line at infinity is treated like any other line in Projective Geometry, we can get around the process of projecting from one plane on to another by agreeing to say, in the case of Fig. X(a), that the plane π may be made affine by choosing HK as the line at infinity. (See Coxeter [27] pp. 6–7 for the application of this convention to Desargues' Theorem in the plane. Also look at Exercise 8.53.) Statements 'choose l as l'_∞' and 'project l to l'_∞' are thus interchangeable.

Affine Geometry (which is wider in scope than Euclidean Geometry) consists of those properties of Euclidean Geometry which are true after parallel projection. Thus every property of Affine Geometry holds also in Euclidean Geometry, though some Euclidean Geometry properties are without meaning in Affine Geometry, e.g. the distance between two points, which is a basic feature of Euclidean Geometry, is meaningless in Affine Geometry.

Roughly speaking, we may think of Affine Geometry as being Euclidean Geometry with all notions of measurement (linear and angular) omitted. Just as we speak of the augmented Euclidean plane, so we may likewise speak of the *augmented affine plane*, i.e. the affine plane enriched by the adjunction of elements at infinity. Parallel (= non-intersecting) lines in the affine plane become lines meeting on l_∞ in the augmented affine plane.

Summarising the relations among the various planes we have mentioned, we have the following situation:

(i) projective plane π;
(ii) augmented affine plane (a projective plane);
(iii) augmented Euclidean plane (= metrical augmented affine plane; a projective plane if we ignore metrical aspects);
(iv) affine plane $\pi_a = \pi - l_\infty$;
(v) Euclidean plane (= metrical affine plane).

§ 3 ANALYTICAL AND ALGEBRAIC TREATMENT

§ 3.1 Parallelism and Infinite Elements. Inadequacy of Cartesian co-ordinates. Let us now turn to some of the familiar analytical ideas used in two-dimensional geometry; that is to say, a point in the plane is described by a pair of co-ordinates (X, Y) with respect to two (perpendicular) axes of reference. These co-ordinates, called *(rectangular) Cartesian co-ordinates*, after their inventor, René Descartes (1596–1650, French)—cf. Descartes [31]—are usually understood to belong to the field of real numbers.

Consider the two distinct coplanar lines

$$aX + bY + c = 0 \tag{i}$$

$$a'X + b'Y + c' = 0 \tag{ii}$$

These lines are easily found to intersect in the point

$$P \equiv \left(\frac{bc' - b'c}{ab' - a'b}, \frac{ca' - c'a}{ab' - a'b}\right) \tag{iii}$$

Suppose now that $ab' - a'b = 0$, i.e. $-\frac{a}{b} = -\frac{a'}{b'}$, i.e. the lines are parallel with slope $-\frac{a}{b}$. Equations (i) and (ii) then have no solution, and hence P can have no co-ordinate representation. Thus, as we already know, lines in the Euclidean plane are divided into two classes, namely

(i) lines which meet; and

(ii) lines which do not meet, i.e. parallel lines.

Now it would greatly strengthen our position analytically if we were able to eliminate this distinction and to treat all lines on the same basis. In particular, we want to be able to say that

any two lines meet in a point.

From our earlier discussion, we recall that such a possibility involves the notion of infinite elements in the augmented Euclidean plane. We would like to be able to describe infinite elements analytically. Since ordinary Cartesian co-ordinates have been shown to be inadequate to deal fully with parallelism, we must exhibit a more comprehensive system of co-ordinates. We give an outline of the method.

§3.2 Homogeneous Co-ordinates. To overcome the difficulty which arises when the co-ordinates in (iii) become infinite, we might consider writing the co-ordinates (X, Y) of P as ratios, namely

$$X : Y : 1 = bc' - b'c : ca' - c'a : ab' - a'b$$

(always provided that $0 : 0 : 0$ is excluded). At least one of the three quantities occurring in the ratios is not zero, whether the distinct lines (i), (ii) are parallel or not.

These remarks suggest that we might think of a point as being determined by an ordered set of three numbers, in which only the ratios matter.

Writing

$$\left.\begin{aligned} X &= \frac{x}{z} = \frac{kx}{kz} \\ Y &= \frac{y}{z} = \frac{ky}{kz} \end{aligned}\right\} (k \neq 0)$$

we see that any point (X, Y) is expressible in terms of *three* co-ordinates $(x, y, z) \equiv (kx, ky, kz) \equiv k(x, y, z)$ where only the *ratios* of the co-ordinates are important, i.e. from x, y, z there are derived just *two* independent ratios. Thus $(6, -4, 2)$, $(3, -2, 1)$ and $(1, -\frac{2}{3}, \frac{1}{3})$ (say) all represent the same point. Obviously $(0, 0, 0)$ is excluded as a possible point, since it is geometrically meaningless (the ratio $\frac{0}{0}$ being meaningless).

Equations of the lines (i), (ii) now become homogeneous in the variables x, y, z, namely

$$ax + by + cz = 0 \qquad \text{(i)}'$$
$$a'x + b'y + c'z = 0 \qquad \text{(ii)}'$$

and their point of intersection P has co-ordinates

$$P \equiv (bc' - b'c, ca' - c'a, ab' - a'b) \qquad \text{(iii)}'$$

Such co-ordinates x, y, z (not all zero) as we have thus introduced into the analytical plane are called (real) *homogeneous Cartesian co-ordinates* in contradistinction to the non-homogeneous Cartesian co-ordinates X, Y. When $z \neq 0$ equations homogeneous in Cartesian co-ordinates x, y, z, can always be reduced to their familiar non-homogeneous Cartesian equivalent by setting $z = 1$. Non-homogeneous equations in x, y can always be rendered homogeneous by inserting appropriate powers of z. Practice should be gained in moving freely from one form to the other. For a pictorial approach to homogeneous co-ordinates using a geometrical model in ordinary Euclidean space of three dimensions, see Eves [36]2, pp. 70–72.

Homogeneous co-ordinates were first used in the 1820's by the German mathematicians Julius Plücker (1801–1868), Augustus Möbius (1790–1868) and Karl Feuerbach (1800–1834) in the period 1827–1829 and their powers fully exploited by Karl von Staudt (1798–1867, German) in 1847.

When the lines (i)′, (ii)′ are parallel, instead of (iii)′ we obtain

$$P_\infty \equiv (bc' - b'c, ca' - c'a, 0) \qquad \text{(iv)}$$

i.e. the third co-ordinate is $z = 0$. In (iv) putting $a' = \lambda a$, $b' = \lambda b$ $(\lambda \neq 0)$, which is the case if (i)′, (ii)′ are parallel, we see that

$$P_\infty \equiv (b, -a, 0)$$

on ignoring the common factor $c' - \lambda c \neq 0$ (for if $c' = \lambda c$ then the lines (i)′, (ii)′ coincide). That is, P_∞ is the specific point at infinity common to the set of lines parallel to (i)′. Note that the z co-ordinate of P_∞ is zero.

Clearly then

$$z = 0 \textit{ is the equation of the line at infinity,}$$

i.e. the locus of all the points at infinity (one for each system of parallel lines).

The line $z = 0$ is as factual as any other line, though it has some special properties. For a discussion of the infinite in Analysis and Geometry, consult Hardy [133] pp. 502–503.

As $x = 0$, $y = 0$ are the equations of the $y-$ and $x-$ axes respectively in a Cartesian framework (axes possibly oblique), we must expect the triangle $OX_\infty Y_\infty$ to play a special role in the augmented plane, $X_\infty Y_\infty$ being the line at infinity, $z = 0$. (Refer to Fig. XVI(a).) In the projective plane, it is called the *triangle of reference*. Along with the *unit point* U (1, 1, 1), it fixes the framework for the non-metrical system of homogeneous co-ordinates in the projective plane.

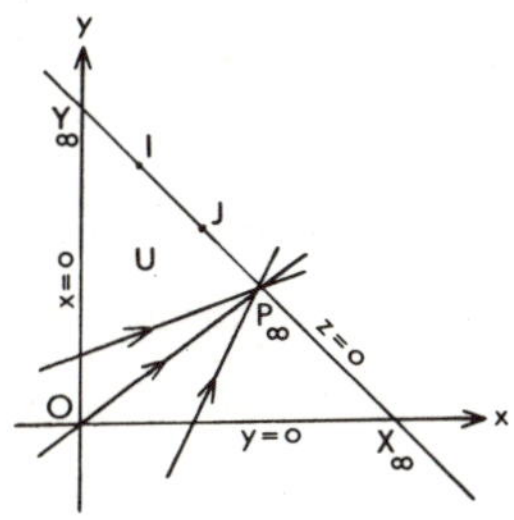

FIG. XVI. (a) Triangle of reference

Besides Cartesian homogeneous co-ordinates, other different (metrical) systems of homogeneous co-ordinates exist, for example, areal and trilinear homogeneous co-ordinates, in each of which the equation of the line at infinity is *not* given by $z = 0$. For details, consult Sommerville [105] pp. 156–158. Also see J. W. Mellor, *Higher Mathematics for Students of Chemistry and Physics*, Dover (1946), where it is shown how trilinear co-ordinates are used in the graphic representation of the composition of certain ternary alloys, and the mixtures of salts.

As we already know (though it is worth repeating)

any line can be projected into the line at infinity

and

in the projective plane there is no line at infinity

since any line could be so chosen. Thus, in the projective plane all lines are on an equal footing, as are all points.

Since the projective plane is non-metrical, the co-ordinates used in relation to it must have no metrical significance. This means that the homogeneous Cartesian co-ordinates used in the augmented plane are unsuitable, and so we seek a more general system of co-ordinates. These new co-ordinates will be called *(general) homogeneous co-ordinates* and their introduction by von Staudt was classed by Whitehead as 'one of the triumphs of modern mathematical thought'.

Point and line are defined abstractly and non-metrically in general homogeneous co-ordinates in a way similar to that for point and line in homogeneous Cartesian co-ordinates. That is to say, a point is an ordered triple (x, y, z) of numbers, real or complex, but not all zero, in which only ratios matter, and a line is defined by the equation $lx + my + nz = 0$. However, note that

in general homogeneous co-ordinates $z = 0$ is not a special line

since there are no points at infinity in a (general) projective plane.

If we wish to specialise a projective plane as an augmented plane (so that then $z = 0$ would play a special role), we find that

homogeneous Cartesian co-ordinates are a special case of general homogeneous co-ordinates.

§ 3.3 Circular Points at Infinity. Consider any circle with equation

$$x^2 + y^2 + 2gxz + 2fyz + cz^2 = 0 \tag{v}$$

in its homogeneous form. It cuts the line at infinity $z = 0$ in the two points given by

$$x^2 + y^2 = 0 \tag{vi}$$

and this is true of any circle in the plane, i.e. of a circle with respect to any orthogonal co-ordinate framework.

Solutions of (vi) are the points

$$\begin{aligned} I &\equiv (1, i, 0) \\ J &\equiv (1, -i, 0) \end{aligned} \tag{vii}$$

where $i^2 = -1$. Of course, we could write I_∞, J_∞ for I, J.

These points are naturally called *the circular points at infinity* (since all circles pass through them and they are on the line at infinity $z = 0$). Thus (Fig. XVI(b))

all circles pass through the circular points at infinity I, J.

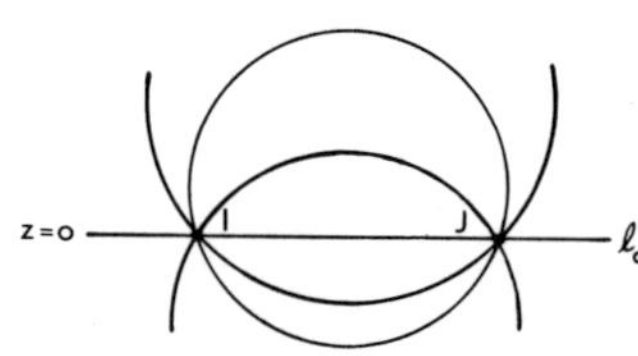

FIG. XVI. (b) Circular points at infinity

From time immemorial the special qualities of the circle—its symmetry, its simplicity, its beauty—have made a forceful appeal to the human mind. An appreciation of its deeper and hidden qualities, as evidenced by its relation to I, J, makes this appeal even stronger.

Regarding I, J we comment that, w.r.t. (vii),

(i) I, J, have the same co-ordinates in any rectangular c-ordinate system;
(ii) I, J are actual (i.e. non-fictitious) points; and
(iii) metrical properties of plane geometry are closely linked with the unique nature of $z = 0$ and with I, J.

We have already seen that any line may be chosen as the line at infinity. It is also true that

any two points may be chosen as I, J

or, as we may also say (see Walker [116] p. 304 for details),

any two points can be projected into I, J.

The mathematical existence of the circular points at infinity was discovered by Poncelet, an officer of engineers in Napoleon's army, when he was a prisoner-of-war (1812–1814) of the Russians at Saratov on the Volga after the French defeat at Krasnoi. Left for dead on the frozen battlefield at Krasnoi, Poncelet survived the rigours of the Russian winter to create his masterpiece, [87], which resulted from his researches in the spring of 1813 while in the prisons of Russia, a veritable mathematician's 'Pilgrim's Progress'. Writing in the Preface of [87], which was ultimately published in 1822, Poncelet expressed his tremendous difficulties, mental and material, in these memorable words: '. . . *privé de toute espèce de livres et de secours, surtout distrait par les malheurs de ma patrie et les miens propres. . . .*'

§ 3.31 Circle at Infinity. We conclude this discussion on elements at infinity in two dimensions with brief mention of some corresponding ideas in three dimensions. As these have no bearing whatever on subsequent work they may be omitted, though they give completeness and have an intrinsic appeal.

Recall (§ 2.5) that the set of lines at infinity in three dimensions is the plane at infinity, π_∞. It follows that the locus of all pairs of points I, J on the lines is a conic (why?), called the *circle at infinity*, Ω_∞, i.e. for any plane α, $\alpha \cap \pi_\infty = (l_\infty)_\alpha$, $(l_\infty)_\alpha \cap \Omega_\infty = I_\alpha, J_\alpha$ where I_α, J_α, $(l_\infty)_\alpha$ refer to infinite elements in α. Draw a diagram to illustrate these ideas.

Take care not to confuse the circle at infinity in π_∞ with a circle in α through I_α, J_α.

§ 3.4 Group Theory. Klein's Erlangen Programme. We now change the direction of our thinking and exhibit the striking connection between geometries and groups. Already we have seen something of the hierarchical structure of Geometry:

Projective Geometry → Affine Geometry → Euclidean Geometry

where, at this stage, we may think of Euclidean Geometry as a special case of Affine Geometry in which measurement is fundamental. Further ramifications will be discussed when the occasion warrants it.

Consider now the non-singular linear transformation (also called *collineation*)

$$\begin{cases} \rho x' = a_{11}x + a_{12}y + a_{13}z \\ \rho y' = a_{21}x + a_{22}y + a_{23}z \\ \rho z' = a_{31}x + a_{32}y + a_{33}z \end{cases}$$

where $\rho(\neq 0)$ is a factor of proportionality and

$$A = \begin{vmatrix} a_{11} & a_{12} & a_{13} \\ a_{21} & a_{22} & a_{23} \\ a_{31} & a_{32} & a_{33} \end{vmatrix} \neq 0.$$

Usually we take $\rho = 1$, since two transformations whose coefficients are proportional may be considered to be identical.

Now, it may be shown that one aspect of this transformation, its projective aspect, is to permute the points and lines of the plane among themselves.

It may also be shown (Semple and Kneebone [99] pp. 29–31) that

every such transformation defines a projective transformation of the plane into itself,

i.e. these transformations characterise Projective Geometry. One does not have to be psychic to recognise that this is a crucial step, and it is with reluctance that we omit the lengthy details. An essential feature of the proof given by Semple and Kneebone is the derivation of the result:

There is a projective transformation which transforms four given points, no three of which are collinear, into four other given points, no three of which are collinear.

Generally, points of the two coplanar tetrads are distinct, though this is not necessary. In the general case, this transformation of quadrangles into quadrangles is achieved by a sequence of four conical projections on to different planes, beginning and ending with the given plane.

Next, it is not difficult to appreciate that these projective transformations form a group under the operation of one transformation followed by another. Otherwise expressed

Projective Geometry is characterised by a certain (projective) group.

As we impose successive restrictions on the nature of the geometry, corresponding restrictions are forced on the equations of the linear transformations and, consequently, on the nature of the group. Accordingly, the geometrical hierarchy with its sub-geometries is reflected in the group-theoretic hierarchy with its sub-group structure:

Projective Group → *Affine Group* → *Euclidean Group.*

Non-Euclidean geometries are also included in this scheme, as we shall see later (§ 44.4, Chapter 9).

Transformations which form the Euclidean group will later be seen to consist of ordinary rotations and reflections which have the fundamental property that they leave I and J invariant. Otherwise stated

the Euclidean group leaves I, J invariant

and

Euclidean Geometry is characterised by the Euclidean group.

One of the most significant advances in mathematical thinking was the recognition by Klein of the intimate relationship between geometries and groups. Accepting a Chair of Mathematics at the University of Erlangen in 1872, Klein, in his inaugural public address, enunciated his famous principle for the codification of Geometry, called the *Erlangen Programme* [in German: *Erlanger Programm*]:

A geometry is the set of properties of figures which are left invariant under the non-singular linear transformations of some group.

So impressed was Arthur Cayley (1821–1895, English) by the marvellous beauty, generality and comprehensiveness of Projective Geometry that he exclaimed: 'Projective Geometry is *all* geometry!' (though he used the word 'descriptive' instead of 'projective'). Cayley's exuberant claim must be balanced against modern geometrical advances which have taken place since his time. Topology, for example, which deals with more general transformations than does Projective Geometry, is certainly not a sub-geometry of Projective Geometry. Nor can Projective Geometry encompass say, the modern developments of classical Riemannian Geometry in relation to Relativity Theory. Nevertheless, Klein's synthesis, though inadequate in terms of 20th century Mathematics, still provides us with a panoramic view of the geometry to which it is applicable.

§ 3.5 Central Collineations. Among the plane collineations are the perspective (*central*) *collineations* which (unlike other collineations) are collineations expressible as a product of two projections in three dimensions. Such a central collineation may be constructed as follows:

Suppose a given plane π contains a line l and a point C. Let π_1 ($\neq \pi$) be a plane through l. Further, let O_1, O_2 be any two points collinear with C and not lying in π or π_1. Under the projection from O_1, P (in π) is carried into P_1 (in π_1). Under the projection from O_2, P_1 (in π_1) is carried into P' (in π). These two successive projections (= *perspectivities*) define a central collineation in π with centre C ($= O_1O_2 \cap \pi$) and axis l, which transforms points like P into points like P' (Fig. XVII).

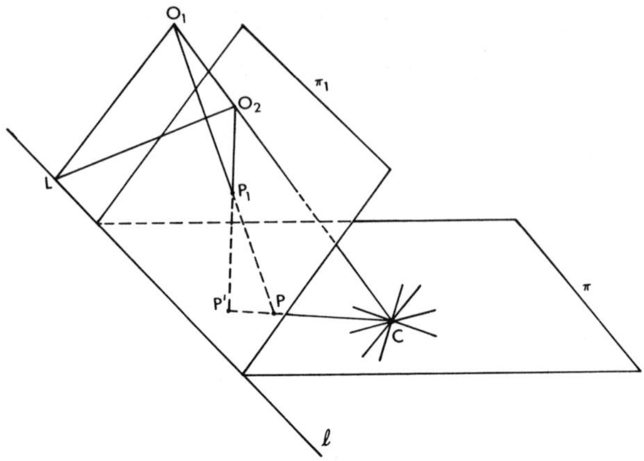

FIG. XVII. Central collineation

Salient features of the central collineation in π are that

(i) C
(ii) l } are invariant;
(iii) any line through C
(iv) C, P, P' are collinear.

Readers should verify these statements. (See Hall [45], Pedoe [85].)

Important cases occur when C lies on l (Chapter 11).

If the plane π_1, containing the figure $\mathscr{C}_1$, of which P_1 is a representative point, is rotated about l to coincide with π, then $\mathscr{C}_1$ becomes a figure $\mathscr{C}'$, of which P' is a representative point. Such a process by which $\mathscr{C}_1 \rightarrow \mathscr{C}'$ by rotation is called a *rabatment* and it may be proved (Filon [39] p. 13) that this rabatment is in fact equivalent to a parallel (cylindrical) projection. Of course, the two figures $\mathscr{C}'$ and $\mathscr{C}$ (of which P is representative) are still related by a central collineation with the properties (i)–(iv) above. $\mathscr{C}$ is the projection of $\mathscr{C}_1$.

Pedoe [85] p. 10 shows how

the existence of the central collineation by rabatment involves Desargues' Theorem in the plane,

a very basic relationship.

§ **3.6 Historical Development of Projective Geometry.** Historically, the development of the body of theory we call Projective Geometry falls into four main periods:

(1) The Greek Period (Euclid, Apollonius, Archimedes, Pappus): 3rd century B.C.–4th century A.D., during which time elementary plane geometry was cast into a logical mould, and conics, curves and solid figures were investigated;

(2) The Analytic Period (Descartes): 17th century, when great impetus was given to Geometry by its union with Algebra, by means of co-ordinates;

(3) The non-Euclidean and Projective Period (Saccheri, Gauss, Bolyai, Lobatschewsky, Riemann, Cayley, Klein, Poncelet, Steiner, von Staudt): 19th century—except for Saccheri—an era of soaring geometrical achievement during which Euclidean authority was overthrown and Projective Geometry arose as a generalisation of the older ideas;

(4) The Axiomatic Period (Hilbert, Pieri, Pasch, Giuseppe Peano (1858–1932, Italian), Giuseppe Veronese (1857–1917, Italian), Oswald Veblen (1880–1960, American): From about 1890, when the logical foundations of the subject were thoroughly examined. [See, for example, O. Veblen: 'A system of axioms of Geometry', *Transactions Amer. Math. Soc.* 5 (1904) 383–384.]

While stressing the development of Projective Geometry, we should not overlook the achievements elsewhere in Geometry, e.g. in Differential Geometry, where the techniques of Calculus and Geometry commingle.

§ 3.7 Applications of Projective Geometry. Reflect for a moment on the two following Euclidean results:

Conics which circumscribe a triangle and also pass through its orthocentre are rectangular hyperbolas;

and

Given two fixed points A, B, then any line cuts AB at O and an arbitrary circle through A,B at P,Q such that $OP.OQ$ = constant.

What hidden connection, if any, is there between these two seemingly distinct results? Strangely enough, they are found to be two aspects of the same theorem in Projective Geometry. Actually they are the special cases of Desargues' Theorem (Theorem 23) covered in Example 37(§ 34) and Exercise 8.5(iii) (in its metrical form) respectively.

Were Projective Geometry restricted merely to exposing such unexpected and illuminating relationships it would undoubtedly be a 'fine, rare experience' (as Wolfe claimed for non-Euclidean Geometry) and worthy of study as an aesthetic and intellectual discipline in its own right. But it is much, very much, more. In its various forms, it has considerable applications to the physical world in which we live. Omitting Euclidean Geometry, the use of which is too well known to elaborate, we shall see in the course of the book how (say) the Physical Sciences, Cosmology, Relativity, Astronomy, Statistics, Philosophy and Art are all enriched by contact with the diverse facets of Projective Geometry.

Certain biological phenomena should also be mentioned. In his scholarly book, *On Growth and Form* , in which, for instance, he revealed how transverse sections of fishes are related to each other by Möbius transformations or by linear transformations, D'Arcy Thompson had this to say (p. 7):

'Cell and tissue, shell and bone, leaf and flowers, are so many portions of matter, and it is in obedience to the laws of physics that their particles have been moved, moulded and conformed. They are no exception to the rule that Θεὸς ἀεὶ γεωμετρεῖ (God does geometry continually). Their problems of form are in the first instance mathematical problems. . . .'

Recently, Hyperbolic (non-Euclidean) Geometry was used to describe space as observed by people of normal vision. In the March 9, 1962, issue of *Science*, Vol. 135, Number 3506) pp. 763–770, Kenneth N. Ogle in his article, 'The Visual Space Sense', reported the following experiment of A. A. Blank (which appeared in *Jour. Opt. Soc. America* 48 (1958)). Subjects were asked to place light sources *a*, *b*, *c* at what they perceived to be the mid-points of the sides of an isosceles triangle formed by three fixed light sources *A*, *B*, *C* in a horizontal plane at eye level. Generally speaking, the results indicated triangles with sides which curved inwards, thus affording further evidence that binocular visual space could be described by Hyperbolic Geometry. (See Fig. XVIII.)

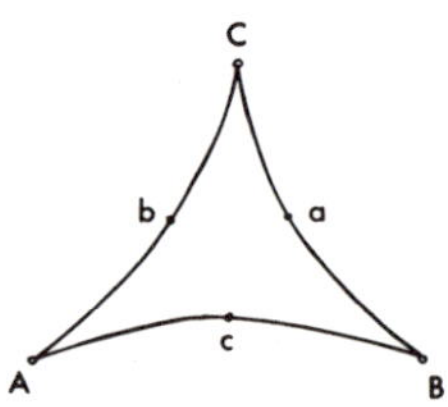

FIG. XVIII. Hyperbolic visual triangle

§4 THE AXIOMATIC APPROACH

'If you would converse with me, define your terms.'
(Voltaire)

'I hate definitions.'
(Disraeli)

§4.1 Nature of an Axiom. Nowadays an axiom is merely a working hypothesis, an initial unproved (and unprovable) statement which we assume. Whether or not it is self-evident or even simple is quite irrelevant (i.e. we are not dealing with Euclid's [Greek] conception of an axiom). No concern is necessarily felt about any physical interpretation of the abstract system, geometrical or otherwise, built up on these foundations (e.g. PP 2 and PS 3 in Chapter 1, § 5.2 and § 6.1, are not always valid for phenomena of sense-perception). Our Projective Geometry has been emancipated from the external world—we are dealing with an abstract mathematical space constructed in our minds.

Says Robinson [92] p. 9: ' "Point" and "line" are objects of thought and our statements concerning them must be exact, not subject to those experimental errors, however small, which are familiar to us in dealing with the external world.' (Compare these words with those of Robert Recorde, § 1.21.) Such an abstract, conceptualised system deduced from the initial assumptions is a framework within which we may obtain mathematical certainty,

> 'an independent world,
> Created out of pure intelligence.'
> (Wordsworth)

On account of the arbitrariness of our choice of assumptions, the number of possible geometrical systems is infinite.

Choosing some elements and relations as more fundamental than others, because of their greater simplicity, necessarily involves having undefined ideas, as each definition depends on previously defined objects and terms and we have to start somewhere. Otherwise we should have a vicious circle. Similarly, some propositions (axioms) must be assumed, since each proposition is deducible from the undefined ideas and previously proved propositions. Undefined elements must initially be regarded only as symbols, devoid of any particular context except that inherent in the basic propositions.

Difficulties which arise through the desire to define everything are amusingly depicted by J. L. Synge (quoted in Coxeter [28] p. 5) in describing the game he calls Vish (short for 'vicious circle'), which he discovered when he followed the hopeless attempt of the *Concise Oxford Dictionary* to define 'point'. Play Vish with 'line' using the same source.

Many different systems of axioms have been devised to provide a logical foundation for Projective Geometry, the most famous in all probability being that due to the American mathematicians, Veblen and Young, in 1910. Others will be mentioned in § 5.2. The system of axioms we shall use is one originating a couple of decades ago with the American, Marshall Hall. Any system of axioms for Projective Geometry should be wide enough in scope to account for the great variety of important models (§ 5.5, § 6.11) which are possible.

While our concern is with a geometrical system of axioms, let us examine the qualities which are desirable in any axiomatic system.

§ 4.2 Qualities desirable in a System of Axioms.

> (Occam's Razor) 'Things not known to exist should not, unless it is absolutely necessary, be postulated as existing.'
> (William of Occam *c.* 1300–1349, English)

Ordinarily one would not postulate a set of axioms from which no interesting or useful consequences could be deduced. Granted that our system of axioms is to have some relevance, we expect it to possess some of the following properties, not all of which are necessary:

$$\begin{cases} \text{consistency} \\ \text{completeness} \\ \text{independence} \\ \text{categoricalness} \\ \text{simplicity.} \end{cases}$$

Let us say briefly what we understand by these properties in the axiomatic context.

Concerning a system of axioms, the most important problem is that of consistency. A system of axioms is *consistent* if it does not lead to contradictions. Usually, consistency is tested by the construction of a representation or *model* of the axioms. To construct a model, we assign specific meanings to the undefined elements and relation and then establish its validity by proving the axioms. (Students familiar with the axioms of group theory will recognise as a model the set [say] of reals $\neq 0$ under multiplication.) Clearly, it is necessary that the axioms be consistent (otherwise no models of the system could possibly exist).

A system of axioms is *complete* if no essential axiom has been omitted. Completeness guarantees that the system tells us enough. Euclid's system of axioms, as we have seen, does not satisfy the criterion of completeness. If a set of axioms is incomplete, it is possible that two contradictory statements may both be consistent with it. Axioms common to Euclidean and Hyperbolic Geometry, for instance, form an incomplete system in which both the Euclidean and Hyperbolic parallel axioms are consistent. (Theorems which are deducible from just these common axioms will be valid in both geometries.)

On the other hand, the system may tell us more than we require, that is, the axioms may not be independent. An axiom system is *independent* if each of its axioms is independent, i.e. is not a logical consequence of the other axioms of the system. Unless the superfluous non-independent axioms serve some specific purpose (e.g. pedagogic), it is generally desirable to restrict the number of axioms to a minimum. A standard procedure for testing the independence of a given axiom is to exhibit a model satisfying all the other axioms of the system as well as the negation of the one given, e.g. the existence of non-Euclidean Geometry showed that Euclid's Postulate V was independent. (For a revealing verification of this method of testing the independence of a set of axioms see Seidenberg [98].) Independence is thus not a necessary criterion for a set of axioms. (Refer to Moore, § 1.43.)

A system of axioms is said to be *categorical* if any two of its models are isomorphic, i.e. if corresponding elements and relations of the two models are in 1–1 correspondence. Essentially, a categorical system of axioms describes just one unique mathematical system. An example of a non-categorical (though extremely valuable) system of axioms is that for a field, because two models of the field system are the set of real numbers and the set of rational numbers between which there certainly cannot be established a 1–1 correspondence. Categoricalness is thus not a necessary property of an axiom system.

In passing, it is worth mentioning that the problems of consistency and categoricalness have been the subject of much philosophical investigation, e.g. by the Austrian logician, Kurt Gödel.

Finally, it adds elegance to the axiomatic theory if the axioms can be stated in simple language, using simple ideas. While we are concerned with a scientific (mathematical) exposition of Geometry, we must always appreciate the aesthetic values which are inherent features of it.

Perhaps it is worth including a simple illustration of an axiom system embodying some of these ideas. Borrowing from Annita Tuller [148] p. 26, consider a set of undefined elements a, b, c, . . . and an undefined relation R between pairs of these elements satisfying

Axiom 1. If $a\ R\ b$ then $a \neq b$.

Axiom 2. If $a\ R\ b$ and $b\ R\ c$, then $a\ R\ c$.

(For example, if the undefined elements are positive integers and $a\ R\ b$ means $a > b$, we have a familiar model of the axiom system.) The reader is asked to show that the system is consistent and independent, but not categorical.

§ 4.3 Value of the Abstract Approach. Virtues of the abstract approach are

(i) logical rigour, and

(ii) applications in a variety of interpretations.

Situations may arise, for example, in Economics, Physics or Botany, in which the description is given in terms of geometrical language. In short, the Geometry which is described abstractly is not one geometry, but many geometries.

§ 4.4 Summary of the Chapter. Concluding remarks. Based on the physical process of projection and, historically, deeply rooted in metrical geometry, Projective Geometry deals with properties of figures in a space (say a plane) which remain unaltered by a certain type of transformation. It is a construction of the mind, and must not be confused with a physical science; it is an ideal conception, suggested by our visual and tactile experiences of the physical world, 'an etherealisation of common sense', as Kelvin remarked of Mathematics generally. It is an independent discipline whose validity is tested by its own self-consistency and not by its interpretation in terms of related geometries. Projective Geometry requires its own analytical apparatus and its own set of abstract axioms. *The Projective Geometry we develop axio-*

matically in Part 2 will depend on those axioms alone, and not on any motivational and background material in Part 1. Occasionally, however, we will use isolated fragments of previous experience (e.g. the augmented Euclidean plane) to illustrate aspects of the projective theory.

Embarking on this new adventure, the reader must be prepared for further surprises: Herein, within the axiomatic framework of Projective Geometry, he will encounter, in addition to the Euclidean, non-Euclidean and affine geometries, geometries consisting of only a finite number of points, and geometries in which the fundamental theorems of Desargues and Pappus are invalid. At times he will leave the relative familiarity of the plane and explore higher dimensions.

Suggested topics for further study

1. Neolithic geometry and architecture (see A. Thom: 'The Geometry of Megalithic Man', *Mathematical Gazette*, Vol. 45, Number 352 (1961) pp. 88–93; A. Thom: 'Megalithic Astronomy: Indications in Standing Stones', *Vistas in Astronomy*, 7 (1966) 1–57).
2. Euclid's *Elements* (Euclid [35]).
3. Logical weaknesses in Euclidean Geometry, Postulate 5 (Bonola [14], Euclid [35], Prenowitz and Jordan [88]).
4. Hilbert's axioms (Hilbert [50]).
5. Inversion Geometry (Adler [2], Pedoe [84]). {*Peaucellier's linkage*, a mechanical device of pivoted rods for converting circular motion into rectilinear motion, is based on geometrical inversion (A. Peaucellier 1832–1913, French)}.
6. Invariants under conical and orthogonal projection (Hopkins and Hails [52]).
7. Compass Geometry (Kostovskii [59], Pedoe [84]).
8. Geometrical constructions with a ruler (Smogorzhevskii [104], Steiner [108]).
9. Contributions to Projective Geometry from Art (Ivins [53], Kline [58]).
10. Evolution of geometrical ideas (G. H. Hardy: 'What is Geometry?', *Mathematical Gazette* 12 (175) (1925) 309–316; W. V. D. Hodge: 'Changing Views of Geometry', *Mathematical Gazette* 39 (329) (1955) 177–183; A. F. Horadam: 'Euclid to Einstein: Changing Concepts of Geometry', *Australian Mathematics Teacher*, 17 (1) (1961) 7–15; Whittaker [119]).

    For the philosophically minded:

(a) Kant's conception of space and geometry (I. Kant: 'Critique of Pure Reason', Macmillan (1952));

(b) Bertrand Russell: 'An Essay on the Foundations of Geometry', Dover (1956);

(c) Kurt Gödel: 'On Formally Undecidable Propositions of Principia Mathematica and Related Systems', Oliver and Boyd (1962).

EXERCISES ON THE INTRODUCTION

1. Euclid's definition of a point implies that it has position but no size. Comment on this. Does it not involve concepts of line and distance in relation to some given framework?

2. Another ancient definition of a point (probably due to Pythagoras [6th century B.C., Greek]) is 'unit having position'. What do you think is meant?

3. Euclid says (Definition 3, Book 1): 'The extremities of a line are points'. Comment on Aristotle's criticism which argued that a point is indivisible and points therefore cannot constitute a line, which is continuous.

Criticise the following Greek concepts (4)–(6) of a straight line. What presuppositions are inherent in each statement?

4. Plato: (429–348 B.C.) 'that the middle of which covers the ends'.

5. Hero: (*c.* 100 B.C.) 'A line stretched to the utmost' (like a taut string).

6. Archimedes: (287–212 B.C.) 'the shortest distance between two points'.

7. Try to invent your own definitions of point and line.

8. Get hold of a 'proof' (Maxwell [71], Prenowitz and Jordan [88]) of the nonsensical result: 'All triangles are isosceles'. Analyse its logical defect.

[Bearing in mind Exercises (1)–(8) above, and Suggested Topics 2, 3 and 4, the interested and adventurous reader is encouraged to attempt to develop his own system of axioms for Euclidean Geometry.]

9. In the spherical model of the elliptic plane, prove: The sum of the angles of a quadrilateral is $> 360°$.

10. Draw diagrams to illustrate I. M. Yaglom's examples of figures in Affine and Projective geometries.

Establish the truth of (11)–(14):

11. With a suitable choice of vertex of projection, any line in a plane π can be projected to infinity and at the same time any two given angles in π can be projected into angles of given magnitude in a suitably chosen plane π'. (See Hopkins and Hails [52] pp. 22–23, Narayan and Kochhar [78] p. 35, Walker [116] p. 303.)

12. Any quadrilateral can be projected into a square.

13. Any triangle can be projected into an equilateral triangle.

14. Ratios of segments are unaltered under an affine transformation.

15. Two angles whose arms intersect in two points on the vanishing line project into angles which are equal to each other. Demonstrate the truth or falsity of this statement.

Deduce each of the following assumptions (16–18) from Postulate 5:

16. Playfair's Axiom (§ 1.21),

17. Posidonius' assumption

Two parallel lines are equidistant.

18. Proclus' assumption (in his 'Commentary on Euclid', Book 1)

If a line intersects one of two parallels, it also intersects the other.

[To establish the equivalence of Postulate 5 to each of these assumptions we may use any of Euclid's first 28 Propositions of Book 1, since Postulate 5 is not used by Euclid until Book 1, Proposition 29.]

19. Try to find out (cf. Heath's Euclid [35] pp. 204–206, 208–210) the unstated assumptions used in the 'proofs' of Postulate 5 by

(i) Ptolemy;

(ii) Nasīraddīn.

20. Endeavouring to prove his assertion (Exercise 18, above), Proclus proceeded thus (Euclid [35] p. 207), with reference to Fig. XIX:

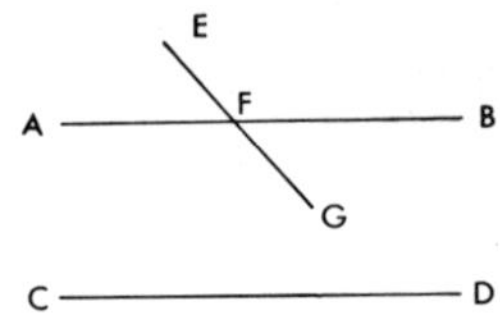

FIG. XIX. Proclus and parallelism

'I say that, if any straight line cuts one of two parallels, it will cut the other also.

For let AB, CD be parallel, and let EFG cut AB; I say that it will cut CD also.

For, since BF, FG are two straight lines from one point F, they have, when produced indefinitely, a distance greater than any magnitude, so that it will also be greater than the interval between the parallels. Whenever therefore they are at a distance from one another greater than the distance between the parallels, FG will cut CD.'

Proclus concluded that he has established the validity of his assertion. Can you detect the hidden assumption used in his reasoning?

THE NATURE OF GEOMETRY: SOME VIEWPOINTS

Comments on the following points of view:

1. Beyond the limits of the stars exist pure ideas, without shape or colour, intangible and invisible, not fixed in sensible particulars but free and independent.'
(Socrates) '. . . . the knowledge at which geometry aims is knowledge of the eternal and not of anything perishing or transient.'

—Plato

2. 'Mathematical objects we consider as the results of abstraction; physical objects have further properties.'
'He [the mathematician] investigates things after eliminating all sensible qualities such as weight, lightness, hardness and softness, also heat and cold . . . leaving only the quantitative and continuous. . . . He investigates them in relation to nothing else.'

—Aristotle

3. 'Our notions of space are rooted in our physiological organism. Geometric concepts are the product of the idealisation of physical experience of space. Systems of geometry, finally, originate in the logical classification of the conceptual material so obtained. All three factors have left their indubitable traces in modern [i.e. 20th century] geometry.'

—Ernst Mach (1838–1916, Austrian)

4. 'That all our knowledge begins with experience there is indeed no doubt . . . but although all our knowledge originates *with* experience, it does not arise *out* of experience.'

—Immanuel Kant

5. 'The axioms of geometry . . . are neither synthetic *à priori* judgments nor experimental facts. They are conventions; our choice among all possible conventions is guided by experimental facts, but it remains free and is limited only by the necessity of avoiding all contradiction. Thus it is that the postulates can remain rigorously true even though the experimental laws which have determined their adoption are only approximate.'

—Henri Poincaré

6. 'The axioms of geometry are—according to my way of thinking—not arbitrary, but sensible statements, which are, in general, induced by space perception and are determined as to their precise content by expediency.'

—Felix Klein

7. 'Projective Geometry . . . will be found, if I am not mistaken, to be wholly *à priori*, to take nothing from experience, and to have, like Arithmetic, a creature of the pure intellect for its object.'

—Bertrand Russell

PART 2

CHAPTER 1

AXIOMS OF INCIDENCE

> 'Our Geometry is an abstract Geometry. The reasoning could be followed by a disembodied spirit who had no idea of a physical point; just as a man blind from birth could understand the Electro-magnetic Theory of Light.'
>
> (H. G. Forder (1889– , English))

> '. . . every deductive science, and especially geometry, must rest on a certain number of axioms which cannot be demonstrated. All treatises on geometry therefore commence with the enunciation of these axioms.'
>
> (Poincaré)

THE object of this Chapter is to lay the axiomatic foundation for a study of the projective plane. From a simple set of axioms we derive many basic properties of the projective plane and then indicate some of the essential features of higher dimensional space.

Unless specifically defined, words have their usual meanings. Concepts not specifically explained are taken for granted, for example, the notion of belonging to a set.

§ 5 AXIOMS OF INCIDENCE

§ 5.1 Fundamental Concepts and Assumptions. In the beginning, we assume the existence of a set of *undefined elements*, or objects, called *points*, of which there may be a finite or infinite number. Points are completely unrestricted in character and must not be imagined as necessarily having any geometrical meaning. A point may, for instance, be a star in the Milky Way, a cell in a micro-organism, a point in the sense of Euclidean Geometry, an astronaut, an electron, a function, an element of a group—in fact, almost anything we choose it to be. Points take their meaning from the assumptions we make concerning them.

Besides points, there is assumed to exist another set of undefined elements called *lines*, also unrestricted in character. For convenience, a line may be thought of as a certain subset of the set of points (although this is not necessary—Forder [41] p. 159). Like 'point', the word 'line' has no meaning other than what is implied by the assumptions. Eventually, we wish to identify these points and lines with the points and lines of Euclidean Geometry, so we have a guide for the assumptions we wish to make about them. Later, we introduce the idea of a *plane* which contains points and lines.

As a notational convention, we will represent

points by capital Roman letters $A, B, C, \ldots, P, \ldots$, lines by non-capital Roman letters $a, b, c, \ldots, p, \ldots$, and planes by non-capital Greek letters $\alpha, \beta, \gamma, \ldots, \pi, \ldots$

Two points A, B are *distinct* (or *coincident*) if they are (or are not) represented by different letters, i.e. $A \neq B$ (or $A = B$). Similarly for distinct or coincident lines, and planes.

Between points and lines there is assumed to be an *undefined relation* which we call *on*. If this relation exists for a point P and a line l (symbolically, $P \in l$), we say:

(i) If P is on l, then l is on P;

(ii) If l is on P, then P is on l.

If the relation does not exist for P and l, we say that P is *not on* l and l is not on P (symbolically, $P \notin l$).

Points on the same line are said to be *collinear*.

Lines on the same point are said to be *concurrent*.

Rephrasing an earlier statement, we see that a line is the set of points on it.

Statements (i) and (ii) above are included by some authors among the axioms for a projective plane which we give in § 5.2.

§ **5.11 Alternative Terminology.** The relation *on* is called an *incidence relation*. A phrase synonymous with 'on' is 'incident with', though we will generally use the simpler expression. The word 'incidence' comes from two Latin words: *in* ≡ 'in' or 'on', (*cidere* =) *cadere* ≡ 'to fall'.

Use of the word 'on' is convenient for stating axioms, and particularly for stressing the duality (§§ 5.4, 6.2) which is basic in Projective Geometry. Other equivalent phrases tend to obscure the simplicity of the ideas we wish to communicate. Borrowing from the familiar language of geometry and the theory of sets of objects, we exhibit the following table of alternative expressions for the incidence relation between a point P and a line l:

P is (lies) on l	*l is (lies) on P*
P is incident with l	l is incident with P
P is contained in l	l contains P
	l passes through P.

Having introduced the basic language, symbolism and ideas, we are now in a position to construct the sort of object with which we are largely concerned, namely the projective plane. Assumptions we make about this object will be given in terms of the incidence relation and consequently may be called the *axioms* (or *propositions*) of incidence. On the basis of these axioms, it will be possible for us to derive the main properties of the projective plane in a systematic way. Symbols '*PP*' in the following definition refer to 'projective plane'. The definition is essentially that of the American mathematician, Hall [45] p. 346, appearing in 1959, though the Axioms *PP*1–*PP*3 were given by Hall as early as 1943—see the reference at the end of this Chapter under Topic (4).

§ 5.2 Definition of a Projective Plane.

Definition: A projective plane is a set of points and a set of lines (certain subsets of points) together with a relation, on, between points and lines, satisfying the following axioms of incidence:

*PP*1. Any two distinct points are on exactly one line.
*PP*2. Any two distinct lines are on exactly one point.
*PP*3. There are four points, no three of which are collinear.

Axioms *PP*1–*PP*3 tell us all that we know about the projective plane at this stage. Points and lines which are (lie) *on* (or 'are incident with' or 'are contained in') the same plane are called *coplanar*; the plane is said to be (lie) on (or to 'be incident with', or 'to contain' or 'to pass through') the points and lines.

Properties of the projective plane deducible from these axioms constitute the subject matter of synthetic *plane Projective Geometry*. Synthetic geometry means geometry in which analytical and algebraic techniques do not occur, i.e. in which co-ordinates are not used. The content of this Chapter is entirely synthetic in nature.

Notational Convention. In *PP*1, the unique line l on two given distinct points A, B, i.e. the *join* of A, B, will be denoted equally well by $l = AB$ or $l = BA$. Likewise, the plane containing a given line m and a given point $P \notin m$ (as allowed by *PP*3) may be denoted symbolically by $\pi = Pm$ $(= mP)$ or, if the points Q,R lie on m, by the notation PQR (or by any permutation of these three letters).

By *PP*2, there is a unique point L on two distinct lines a,b. It is called the *intersection* of a and b and may be written, in set theory notation, $L = a \cap b$ or just $L = ab$. Lines a, b are said to *intersect* (or 'meet') at L, or 'to have L in common'.

§ 5.21 Other Systems of Axioms. Many other systems of axioms exist. Best known of them is that postulated in 1910 by Veblen and Young [113]. Our axioms may be shown to be equivalent to those stated on pp. 16–18, Vol. 1, Veblen and Young.

Garrett Birkhoff ('Combinatorial relations in projective geometries', *Annals of Mathematics* 36 (1935) 743–748) formulated a system of axioms based on the algebraic concept of a lattice and showed the equivalence of his axioms to those of Veblen and Young. Independently, and at much the same time, Karl Menger ('New foundations of projective and affine geometry', *Annals of Mathematics* 37 (1936) 456–482) deduced from his postulates those of Veblen and Young and interpreted his undefined concepts and undefined relation in terms of those given by Veblen and Young. See also Menger's article 'Self-dual postulates in projective geometry', *Amer. Math. Monthly* 55 (1948) 195 in which he expresses his axioms in self-dual form.

Almost every book dealing with Projective Geometry contains at least a section on axiomatics and the interested reader is urged to consult as wide a variety of approaches as possible.

Let us now examine some of the significant features of the Axioms of Incidence.

§ 5.22 Remarks on the Axioms.

(1) *Fundamental Nature of PP*2. Axiom *PP*2 is a characteristic feature of the projective plane. (Certainly, in the ordinary Euclidean plane it is not generally valid, though it is always valid in the augmented Euclidean plane.)
(2) *Significance of PP*3. *Degenerate Planes.* If we exclude *PP*3 from the list of axioms retaining only *PP*1 and *PP*2, we may deduce the existence of certain trivial planes which Hall refers to as *degenerate* (or *improper*) planes. Hall comments that

- (a) every plane in which there are enough points on each line contains these systems of degenerate planes, and
- (b) in defining proper planes, *PP*3 may be replaced by many other axioms which exclude the degenerate cases.

If *PP*3 is omitted, there is nothing in the definition of a projective plane to prevent a set of points being empty, or to consist of just one point.
(3) *Class of Projective Planes.* Axioms *PP*1–*PP*3 describe a class (set) of objects of great variety, distinguished from each other by further properties. Addition of further axioms (e.g. of order or of continuity) and of other restrictions on the plane, enables us to single out from the class of all projective planes a certain subclass. Indeed, it is on this subclass exemplified by the Euclidean plane that our interest concentrates as the theory unfolds.
(4) *Non-degenerate Planes.* Nothing in *PP*1–*PP*3 guarantees an unlimited number of points and lines in the projective plane (as one is accustomed to in the Euclidean plane). Indeed, *PP*3 strongly orientates us towards the possible existence of a plane (or planes) in which there is only a finite number of points and lines.
(5) *Consistency of the Axioms.* Are the Axioms *PP*1–*PP*3 consistent, i.e. free from self-contradiction? One method of establishing the consistency of

our system of axioms is to construct a *model* of the axioms. This means that we assign meanings to the undefined terms (point, line, on) and then demonstrate that the axioms become verifiable theorems for the model. Models of the projective plane are themselves projective planes, of course. A model which must come readily to mind is the augmented Euclidean plane (more precisely, the augmented affine plane, unless we ignore measurement) in which the following identification of terms is made:

Projective plane	*Augmented Euclidean plane*
points	ordinary points and points at infinity
lines	ordinary lines and line at infinity
on	on

Indeed, this is the most natural model for the Axioms *PP*1–*PP*3 and, historically, was the plane from which the projective plane was abstracted. Details of the verification are left to the reader.

Presently, we will exhibit some other models of the projective plane, but at this stage it is deemed wiser to investigate some of the implications of the axioms.

§ **5.23 Logic and Diagrams.** So far, in the projective plane no diagrams whatever have been drawn, or even suggested. This is because our reasoning has been concerned with logic. To introduce diagrams into the theory, and so to give physical representations to the objects of our mathematical thought, we adopt the convention:

Convention: Diagrams are drawn (or imagined) in terms of Euclidean Geometry with which we are all familiar, that is, a point is represented by a dot on the paper (or blackboard).

Ordinarily, a projective line is represented in the usual way as the physical image of the edge of a straight-edge but, as we shall see, this is not always so. Instances will be found in which for aesthetic and other reasons a curve best represents a line in the projective plane.

Inevitably, physical drawings bring with them our intuitions and acquired knowledge of space and geometry. We must set out minds against this source of danger to which our thinking is exposed.

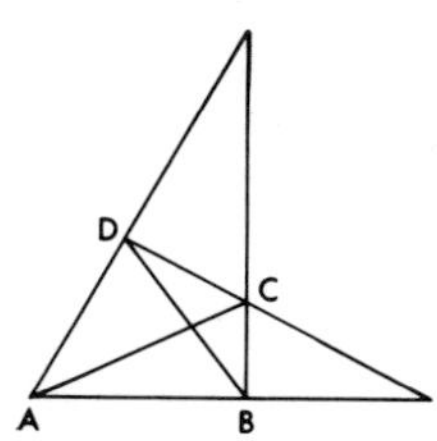

FIG. 1.

Axioms *PP*1–*PP*3 allow the construction of the basic diagram (Fig. 1) which will constantly recur in the progressive development of the theory.

§ 5.3 Applications of the Axioms of Incidence. Next we obtain the most elementary results which are deducible from the initial assumptions.

Theorem 1: There exist four lines, no three of which are concurrent.

Proof: Let *A*, *B*, *C*, *D* be the four points whose existence is guaranteed by *PP*3. Then *AB*, *BC*, *CD* and *DA* are four lines which possess the desired property. [Theorem 1 states a minimum property of the plane.]

Theorem 2: Every line contains at least three points.

Proof: Again, let *A*, *B*, *C*, *D* be the points defined by *PP*3.
Consider the three possible cases (based on the points A):
(i) a line not passing through *A*;
(ii) a line passing through *A*, but excluding *AB*;
(iii) *AB*.
Then a line of (i) is, by *PP*2, intersected by *AB*, *AC* and *AD* in three ($= 1 \times 3$) distinct points;
a line of (ii) is, by *PP*2, intersected by BC and BD in two distinct points and hence contains three ($= 2 + 1$) distinct points;
AB contains the three ($= 1 + 1 + 1$) distinct points *A*, *B*, $AB \cap CD$, by *PP*2.
[Though we have concentrated on *A* in the proof, the Theorem is obviously proved equally well by concentrating on *B* or *C* or *D*.]

Theorem 3: Every point lies on at least three lines.

Proof: Construction of the proof is left as an exercise (Exercise 1.1). Model the proof on that in Theorem 2. See also the ensuing paragraph.

§ 5.4 Principle of Duality in the Plane. Perusing the above deductions from the axioms, one cannot fail to be struck by a certain mutual relationship existing between points and lines on the one hand and between lines and points on the other. Suppose we pair off Axioms *PP*1–*PP*3 and Theorems 1–3 in two columns, thus

*PP*1	*PP*2
*PP*3	Theorem 1
Theorem 2	Theorem 3.

Inspecting the statements for the pairs of corresponding results in the two columns, we readily appreciate that the roles played by point and line, and by line and point in each pair are interchangeable. This observation leads us to the very valuable and elegant *Principle of Duality*, discovered at about the same time by Joseph-Diaz Gergonne (1771–1859, French) and Poncelet, though in a limited way it seems to have been understood earlier by the Renaissance geometer, Franciscus Maurolycus (1494–1575, Italian).

Proofs of theorems deducible from the left column become proofs of dual theorems deducible from the right column. Moreover, as Theorems 2 and 3 are consequences of the Axioms *PP*1–*PP*3, it follows that the dual theorems are consequences of the Axioms *PP*1–*PP*3 also. Thus, the Principle of Duality, which we proceed to quote, is a built-in feature of the projective plane, a consequence of the axioms, and its truth is guaranteed by this fact.

> *Principle of Duality in the Plane:* Any theorems deducible from Axioms *PP*1–*PP*3 concerning the points and lines of a plane remain valid if stated in the 'on' terminology when the words 'point' and 'line' are interchanged.

Thus, in the projective plane, point and line are dual concepts.

Notationally, it is usually convenient to represent the dual of a point P (line l) by the line p (point L).

For practice in application of the Principle of Duality, it is recommended that you draw up a short lexicon, i.e. dictionary, whereby simple projective ideas relating to point and line are 'translated' into their duals. Examples of dual concepts in the plane are

plane duals	
point on line ⟷	line on point
collinear points	concurrent lines
join of two points	intersection of two lines.

Formal usage of duality, in theory, in illustrations and in exercises, will occur from time to time, but the reader should never lose the opportunity to examine for himself the dual of any geometrical statement or figure.

Duality has the great (and economical) value that the dual of a given result does not have to be proved, since its existence and validity are guaranteed by the truth of the Principle. Automatic doubling of the number of theorems and other geometrical results is thus ensured. Duality endows the structure of the projective plane with a strength, a symmetry and a balance which are pleasing to the mind and to the eye. Truly, it is a rich treasure-house for the geometer.

[Can duality apply in (i) the augmented Euclidean plane; (ii) the ordinary Euclidean plane? Why?]

Says Cremona [29] p. vii, quoting from Reye's '*Geometrie der Lage*' (1866): 'Geometry affords nothing so stirring to a beginner, nothing so likely to stimulate him to original work, as the principle of duality ...' (Luigi Cremona 1830–1903, Italian).

One may regard duality as a transformation by means of 1–1 correspondences of points and lines of the projective plane π into the corresponding lines and points of the dual projective plane π^*. Then, a careful analysis (Hall [45] p. 347) reveals that the dual of π^* is indeed π, that is

$$(\pi^*)^* = \pi.$$

(Inherent in this argument is a demonstration that the system π^* *is*, in fact, a projective plane. Try this as an exercise.)

Historically, it is interesting to note that the discovery of the Principle of Duality unhappily provoked an unbecoming quarrel between Poncelet and Gergonne, each of whom claimed the idea as his own. Poncelet, an army officer, was left in no doubt that Gergonne regarded him as anything but a gentleman. Quarrels such as this (and one is reminded of the acrimonious Newton-Leibniz dispute over the invention of the Calculus) serve to convince us of the all-too-human weaknesses of some of the intellectual giants of Mathematics. (See Poncelet [87] 2, towards the end, for his correspondence with Gergonne.) However, let us progress to more wholesome topics!

Triangle. A *triangle* is the figure formed from three non-collinear coplanar points P, Q, R (*vertices*) and the three joining lines QR, RP, PQ (*sides*). Such a triangle may be described symbolically by PQR or by any permutation of these letters. By our remark in § 5.2, this notation serves also to name the plane in which the triangle lies, though no confusion should arise. The dual of a triangle (a *trilateral*) is essentially the same figure. A description of it is left to the reader.

§ 5.41 Complete Quadrangle and Complete Quadrilateral (dual figures). Another simple, yet very important, geometrical figure is formed by a slight extension of Fig. 1. Take four points A, B, C, D. Writing $AB \cap CD = E$, $AD \cap BC = F$, $AC \cap BD = G$ where E, F, G are distinct from A, B, C, D and from each other we obtain the *complete quadrangle* formed from A, B, C, D (Fig. 2(a)). Starting from four given lines a, b, c, d instead of the four given points A, B, C, D, we arrive at the dual construction, the *complete quadrilateral* formed from a, b, c, d (Fig. 2(b)). Dual concepts may be set out in columnar fashion, thus

Complete Quadrangle		*Complete Quadrilateral*	
A, B, C, D	: vertices	a, b, c, d	: sides
AB	: side	$a \cap b$	: vertex
x, y	: opposite sides	X, Y	: opposite vertices
E, F, G	: diagonal points	e, f, g	: diagonal lines
EFG	: diagonal (point) triangle	efg	: diagonal (line) triangle.

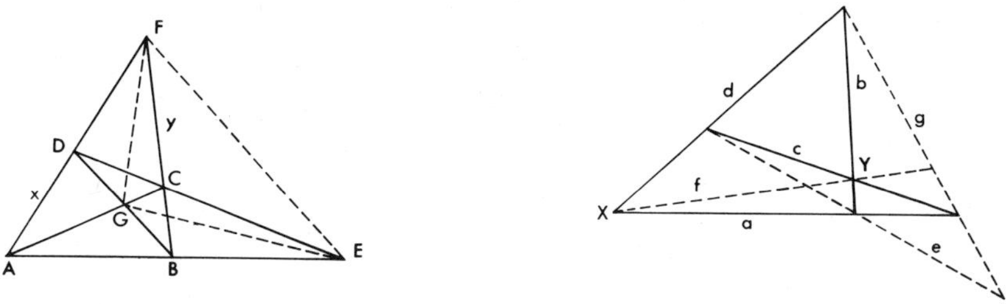

FIG. 2. (a) Complete quadrangle FIG. 2. (b) Complete quadrilateral

Although both figures look rather alike, notice the different ways in which their diagonal triangles (dotted) are formed, that is, neither figure is self-dual. Where, in Fig. 2(a), is the diagonal triangle of (say) the quadrangle $AEFG$?

In Fig. 2(*a*), *note particularly that the diagonal points are not collinear.*

Given a quadrangle, it is possible to construct a quadrilateral whose diagonal triangle coincides with that of the quadrangle. For a proof of this statement, consult Veblen and Young [113] 1, pp. 45–46. Also see Exercise 2.18.

§ 5.42 Quadrangular Set. Associated with a complete quadrangle is a set of six collinear points called a *quadrangular set*. These are the six points in which a line in the plane, not passing through a vertex of the quadrangle, cuts the six sides of the quadrangle, that is, it is a section of the complete quadrangle by a given line.

In Fig. 3(a), we represent the fact that the six points A, B, C, A', B', C' form a quadrangular set based on the complete quadrangle $PQRS$ by the symbol

$$Q\,(ABC,\ A'B'C').$$

This signifies that the pairs A,A'; B,B'; C,C' are on opposite sides of the quadrangle. Alternatively

$$Q\begin{pmatrix} A & B & C \\ A' & B' & C' \end{pmatrix}$$

may be preferred as a meaningful symbol to represent the same situation.

A little thought will make it clear that the following symbols are equivalent to $Q(ABC,\ A'B'C')$ (that is, express the same relation): $Q(BAC,\ B'A'C')$, $Q(AB'C',\ A'BC)$, $Q(A'BC',\ AB'C)$, as well as other similar arrangements of the paired letters.

§ 5.43 Harmonic Set. When A,A' coincide and B,B' coincide, the quadrangular set based on the complete quadrangle $PQRS$ then becomes a special set of four collinear points, called a *harmonic set* (Fig. 3(b)). For simplicity, we write

$$H(AB,\ CC') \equiv Q(ABC,\ ABC').$$

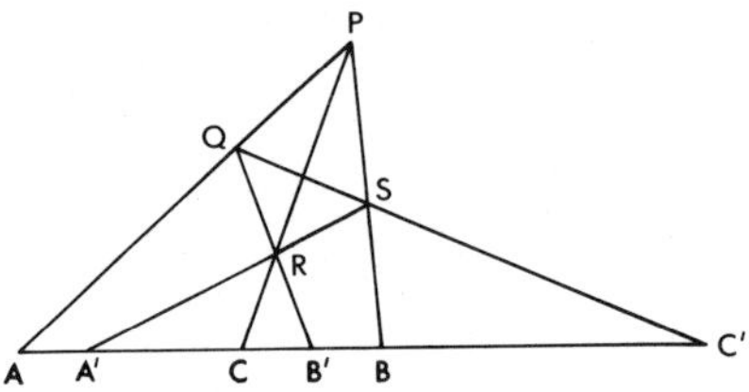

FIG. 3. (a) Quadrangular set

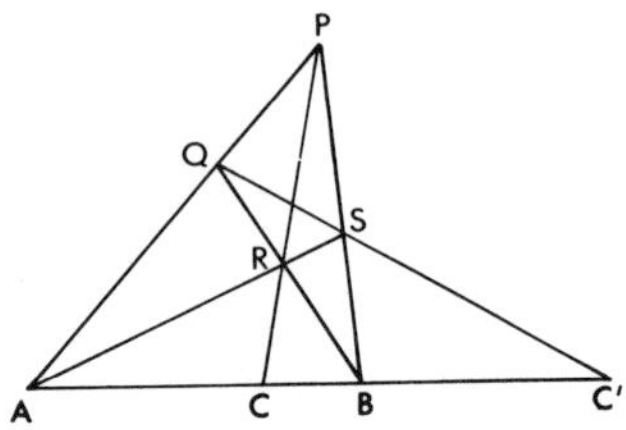

FIG. 3. (b) Harmonic set

With respect to the quadrangle *PQRS*, *A* and *B* are two vertices of its diagonal triangle, and *C,C'* are the points of intersection of the line *AB* with the remaining sides of the diagonal triangle.

Equivalence of $H(AB,CC')$ with $H(BA,CC')$, $H(AB,C'C)$ and $H(BA,C'C)$ clearly exists. Thus the harmonic relation among a set of four collinear points remains harmonic if the order of just one pair of points is altered.

Quadrangular sets and harmonic sets, which were familiar to Pappus, will later on (Chapter 5) be seen to be extremely important ideas.

§ 5.5 Models of the Projective Plane. Earlier, in § 5.22, the notion of a model of a projective plane was introduced. Now we give some well-known examples of this notion.

(1) *Augmented Euclidean (affine) plane.* This has already been discussed under Remark 5 of the Axioms.

(2) *Non-Desarguesian planes.* In these models, Desargues' Theorem, which has been considered in Part 1, does not hold. A particular example of a non-Desarguesian plane, namely the Moulton plane, will be examined in detail in § 9. Other non-Desarguesian planes will be met in Chapter 11.

(3) *Analytical planes.* When points and lines are specified by homogeneous co-ordinates belonging to some field, we obtain an analytical model of the projective plane. Indeed, Chapter 3 is devoted to the study of these planes.

(4) *Finite planes.* Axiom *PP*3 foreshadowed the possible existence of a plane containing only a finite number of points and a finite number of lines. In fact, there exist very many finite planes, the simplest of which we proceed to examine. Because of their intrinsic interest and practical value, a whole Chapter (Chapter 10) is reserved for these finite planes.

Now, we know that each line of a projective plane contains three points (Theorem 2)—this is the minimum number of points on a line in any projective plane. Dually, we know that there is a minimum of three lines through each point (Theorem 3). Curiosity suggests that it might be a worthwhile venture to apply our Axioms to these minimum conditions. Otherwise expressed, we desire to see if there exists a 'miniature' model of the projective plane. Indeed, it is not hard to show that the minimum number of points in a projective plane is seven (Exercise 1.3), and that the minimum number of lines in a projective plane is seven, a fact guaranteed by the Principle of Duality.

Next consider a student organisation in which the duties of the students are arranged according to their membership of (sub-)committees. Make the following specification of our undefined terms (point, line, on) with regard to this organisation:

Projective plane	*Model (Organisation)*
point	student
line	committee
point on line	student on committee

Let there be seven students S_i ($i = 1, 2, \ldots, 7$) and seven committees c_i ($i = 1, 2, \ldots, 7$) selected so that three students are on each committee and each committee consists of three students. Omitting the S in S_i for convenience, and retaining only the suffixes $1, \ldots, 7$, we may describe the organisation by means of the following array

c_1	c_2	c_3	c_4	c_5	c_6	c_7
1	2	3	4	5	6	7
2	3	4	5	6	7	1
4	5	6	7	1	2	3

in which each set of three numbers under c_i ($i = 1, 2, \ldots, 7$) signifies that the corresponding students are on committee c_i. (Other arrays are possible but this pattern will be used again in Chapter 10.) Check now on the Axioms.

*Proof of Axiom PP*1: There are $C_2^7 = 21$ pairs of points chosen from $S_1, \ldots, S_7$, namely, $S_1S_2, S_1S_3, \ldots, S_1S_7, S_2S_3, \ldots, S_6S_7$. From the array, it is easy to see that each pair of points occurs just once. This proves Axiom *PP*1.

*Proof of Axiom PP*2: We have to check that each pair of columns in the array has one element in common. This proof requires the checking of $C_2^7 = 21$ pairs of lines.

*Proof of Axiom PP*3: We are required to verify that there exist selections of four of the S_i, such that no three of them are collinear. An example is S_2, S_3, S_6, S_7. Altogether, 28 verifications are required here.

These proofs of Axioms *PP*1–*PP*3 guarantee that we have produced a model of the projective plane which we naturally call a *7-point plane*. Replacing 'student' and 'committee' by, respectively, 'committee' and 'student', we obtain the dual model. In this dual, how is the incidence relation 'on' specified?

Representing point (student) in the usual way, we obtain a recognisably geometrical picture of this finite projective plane as in Fig. 2(c), which is the same as Figs. 109(a), (b). However, it must be remembered that in this diagram, a line is a certain set of three points, i.e. the physical lines (shown as dotted to emphasise this fact) are merely part of the skeletal framework to support the figure.

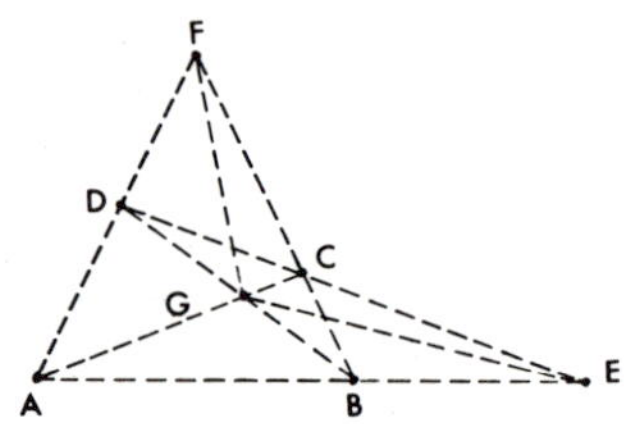

FIG. 2. (c) Finite Fano plane

Variations of this model are easy to imagine. A commonly used model, and a physically meaningful one, arises if we consider an ornament consisting of seven beads attached by threes to seven wires. Making the identifications given in the table below, we obtain the model and its dual:

Projective plane	Ornament	←(Dual) *Models*→	Ornament
point	bead		wire
line	wire		bead
point on line	bead on wire		wire through bead

Although this example is perhaps illuminating, it could be misleading as the physically continuous wire might suggest continuity in the line, whereas it is a mere support for the beads (points).

Another model of this 7-point finite projective plane is the analytical model in which homogeneous co-ordinates are taken over a finite Galois field mod. 2.

Readers are invited to invent other models (concrete representations) of this 7-point plane. See, for instance, the example given in Artzy [4] where, 'point', 'line' and 'on' are interpreted as 'colour', 'ball' and 'ball has colour' respectively, so that the resulting 'geometry' deals with statements concerning coloured balls.

The existence of these widely differing models of the projective plane shows that our Projective Geometry, as defined by the Axioms *PP*1–*PP*3, is not categorical, i.e. there is not merely one Projective Geometry but (infinitely) many geometries. When co-ordinates are introduced and their field is specified, a particular Projective Geometry is singled out.

§ 5.51 Comments on the Existence of Finite Projective Plane Geometries.

(i) The existence of finite planes shows that order and continuity are not implied by the Axioms *PP*1–*PP*3.

(ii) Observe that *in the 7-point plane* (*Fig.* 2(*c*)), *the diagonal points of a complete quadrilateral are collinear.* This is contrary to the non-collinear situation in the general projective plane (Fig. 2(a)). If we wished our projective plane only to resemble our spatial conception of a plane, we would have to add some axiom (Fano's Axiom, § 5.6) which would have the effect of eliminating the 7-point plane from our consideration.

(iii) Each line of the 7-point plane contains three points. This suggests that there may exist finite planes with four, or more, points on each line. As we shall see in Chapter 10, this is indeed so. However, the reader is advised even at this stage to try to construct this finite projective plane with four points on each line.

§ 5.52 Isomorphism of Models. Two projective plane models π,π' are *isomorphic* if there is a 1–1 mapping m_1 of the points $P \in \pi$ onto the points $P' \in \pi'$ (i.e. $P\, m_1 = P'$), and a 1–1 mapping m_2 of the lines $l \in \pi$ onto the lines $l' \in \pi'$ (i.e. $l\, m_2 = l'$) such that $P \in l$ if and only if $P' \in l'$. That is, m_1, m_2 preserve incidence relations. In other words, two models are isomorphic if they are structurally identical and differ only in the material of their structure. An example of two isomorphic planes (models) is the pair of models of the 7-point plane discussed above (student-committee and bead-wire models). Of course, the 7-point plane and the augmented Euclidean plane constitute a simple example of two non-isomorphic models.

§ 5.6 Fano's Axiom. Emphasis has already been given to the fact that the diagonal points of a complete quadrangle are not collinear. This feature distinguishes the complete quadrangle in the general projective plane from the analogous figure in the 7-point finite plane in which the diagonal points are collinear. To eliminate consideration of this possibility, Fano postulated that the diagonal points are not collinear. We always wish to allow for the existence of the 7-point plane in our theory and therefore *we shall not use Fano's postulate as an axiom*. It will nevertheless often be convenient to have it stated for reference purposes:

Fano's Axiom: The diagonal points of a complete quadrangle are not collinear.

One immediate implication of Fano's Axiom, if used as such, would be that every line has at least four points.

Fano's Axiom, due to the Italian mathematician, Gino Fano, may be found in Fano's article, '*Sui postulati fondamentali della geometria proiettiva in uno spazio lineare a un numero qualunque di dimensioni*' which appeared in the journal, *Giornale di Matematiche*, 30 (1892).

In the 7-point (and 7-line) finite plane of Fig. 2(c), there are three points on each line and three lines through each point. We may speak of this symmetrical arrangement of points and lines as a 7_3 configuration and, more especially, as a *Fano configuration.*

A plane is called a *Fano plane* (Fig. 2(c)) if the diagonal points of every quadrilateral are collinear, or, equivalently, if every quadrilateral generates a Fano configuration. It is one of the strange ironies of Mathematics that such a plane, which violates Fano's Axiom, should be called after Fano. More accurately, it ought to be called an 'anti-Fano plane.'

Fano planes are the subject of a nice theorem discovered in 1956 by the American mathematician, A. M. Gleason. An enunciation of Gleason's theorem will be given in Chapter 10 (§ 46), in which finite planes are discussed at some length.

What is the dual statement of Fano's Axiom?

§ 5.7 Partial planes. Free planes. Next, let us look briefly at a system of points and lines of which less is demanded than the satisfaction of Axioms *PP*1–*PP*3 by a projective plane. Indeed, only a weakened form of *PP*1 will be required.

Definition: A *partial plane* is a set of objects called points and a set of subsets of points called lines, with a relation 'on' between points and lines satisfying the postulate,

Partial Plane Axiom: Two distinct points are on, at most, one line.

It follows immediately that

Two distinct lines are on, at most, one point.

Examples of a partial plane are
(i) the affine plane, and
(ii) subsets of points and lines of a projective plane.

Concepts such as isomorphism and duality can be extended to partial planes. Degenerate partial planes also exist. A partial plane may be complete.

A *configuration* is a finite partial plane.

Example of configurations are
(i) finite planes (e.g. the 7-point Fano plane of Fig. 2(c)),
(ii) the figure (Fig. 5(a)) for Desargues' Theorem in the plane, and
(iii) the figure (Fig. 7) the Pappus' Theorem.

Suppose now we are given a partial plane π_0. Further, suppose we restrict π_0 to contain at least four points, no three of which are collinear. By constructing further points and lines and thus extending π_0—see the literature under Topic 4 at the end of this Chapter for precise details—we create a sequence of partial planes $\pi_0, \pi_1, \pi_2, \ldots, \pi_n, \ldots$, each being embedded in the subsequent one. Eventually, we have a projective plane called the *plane freely generated* by π_0. In other words, every partial plane can be embedded in a projective plane.

Under certain conditions on π_0, the plane freely generated by it is what is called a *free projective plane*. This is in particular the case when π_0 consists of four points (no three of which are collinear) and no lines. An interesting feature of such a plane is that it is non-Desarguesian and therefore non-Pappian (§ 8), while Fano's Axiom is valid in it.

§ 6 PROJECTIVE SPACE OF N DIMENSIONS ($N \geqslant 3$)

Occasionally, in the development of plane Projective Geometry, we need more room to manoeuvre and so we have to postulate the existence of points 'outside' the plane; i.e. points not incident with (on) the plane. Indeed, this is a vital ingredient of the study of Desargues' Theorem (§ 7). Taking this particular objective as our motivation, we proceed to a brief, but rigorous, discussion of projective space of three dimensions.

As this is the first mention of dimensions, we make the following definitions:

The *dimension* of a $\begin{cases} \text{point is } 0; \\ \text{line} \quad 1; \\ \text{plane} \quad 2. \end{cases}$

In speaking of projective objects of greater complexity than point, line and plane, we will refer to them as *projective spaces of n dimensions* ($n > 2$), or more simply as *projective n-spaces* ($n > 2$) where n is an integer. Sometimes, in the case $n = 3$, we refer to the space as a *solid*. An alternative notation is to represent a point, a line, a plane and a 3-space by the symbols [0], [1], [2] and [3] respectively. Thus, the symbol for a projective space of n dimensions is [n]. (No possible ambiguity can arise from the two uses of a square bracket to enclose a number, namely, a dimension number or a reference number as the case may be, for the context should make the intention clear.)

One interesting feature of our axioms for space may be noted in advance. In ordinary Euclidean space, two distinct lines may fail to meet because they are either

(i) parallel (i.e. are coplanar), or
(ii) skew (i.e. are non-coplanar).

Possibility (i) was covered when axiomatising the projective plane by insisting in *PP*2 that any two distinct lines shall meet in a point. Possibility (ii) cannot be covered in this way, because skew lines are non-coplanar so we must reconsider our approach a little. Two skew lines, i.e. lines which have no point in common, may be said to intersect in a [-1]('a space of dimension -1').

When building the projective 3-space axiomatically, it is customary to begin afresh, i.e. not to use *PP*1–*PP*3. There are many acceptable systems of axioms available. However, we choose to follow the contemporary Dutch mathematician, A. Heyting [49] (1963), whose system of axioms has the advantage of highlighting the duality of 3-space without too much trouble. Our *undefined* elements are *points*, *lines* and *planes*. Between pairs of these elements there is postulated to exist an undefined symmetrical incidence relation which, again, we call *on*.

§ 6.1 Axioms of Incidence of Projective Space of 3-Dimensions.

*PS*1. Any two distinct points are on exactly one line.
*PS*2. A line and a point, not on the line, are on exactly one plane.
*PS*3. A plane and a line, not on the plane, are on exactly one point.
*PS*4. If a point is on a line and the line is on a plane, then the point is on the plane.
*PS*5. There are five points which are not all on one line, and no four of which are on a plane.

Definition: A projective space of three dimensions is a set of points, a set of lines and a set of planes, together with a relation 'on' between points and lines, lines and planes, points and planes, satisfying the Axioms of Incidence *PS*1–*PS*5.

Properties of the projective space deducible from these axioms form the content of synthetic *Projective Geometry of three dimensions.*

§ 6.11 Remarks on the Axioms.

1. Notational and terminological conventions, as stated for § 5, will be

used here also, e.g. *PS*3 may be rephrased in more familiar language as 'a plane and a line, not in the plane, meet in exactly one point'. If this line is l and the plane α, we indicate their unique point of intersection by $\alpha \cap l$ (or $l \cap \alpha$). [In Deduction IV below the line common to two planes α,β may be represented by $\alpha \cap \beta$.]

2. *PS*5 is necessary in order to avoid trivial cases.
3. Models of projective 3-space are
   (i) *Augmented Euclidean* (*affine*) *space*, i.e. ordinary Euclidean (affine) space extended by the adjunction of a *plane at infinity*, containing all the points at infinity and all the lines at infinity in space;
   (ii) *Analytical projective* 3-*spaces* in which the elements are described in terms of homogeneous co-ordinates. These models are analogous to the analytical models for the projective plane;
   (iii) *Finite spaces*, about which we say a little more in Chapter 10.

Details of the proofs of the following results are left to the reader.

§ 6.12 Deductions from the Axioms.

I. A line which is on two distinct points of a plane is on that plane.
II. Three non-collinear points are on exactly one plane.
III. Every plane is on four points, no three of which are collinear.
IV. Two distinct planes are on exactly one line.
V. There are five planes which are not all on one line, and no four of which are on a point.

Notice that in Axioms *PP*1–*PP*3 and *PS*1–*PS*5 we have attached what appear to be two different meanings to the word 'plane'. It is of fundamental importance to show that they are, in fact, identical. We shall state this as a deduction from the Axioms, and prove it.

*Deduction VI: Axioms PP*1–*PP*3 *are valid in every plane of a projective 3-space.*

Proof: *PP*1 and *PP*3 are immediate consequences of *PS* 1–*PS* 3
Now let α be a plane of the projective 3-space.
Let l,m $(l \neq m)$ be two lines in α.
Choose a point A not on α. Represent the plane Al by $\beta(\neq \alpha)$.
Now $m \notin \beta$
$\therefore\ m \cap \beta = B$ (say) by *PS*3.
Suppose $B \notin l$
$\therefore\ \alpha,\beta$ both contain B,l
$\therefore\ \alpha = \beta$ (by *PS*2) which is a contradiction $\because\ \alpha \neq \beta$ by hypothesis
$\therefore\ B \in l$
$\therefore\ l \cap m = B \in \alpha$
$\therefore$ *PP*2 is valid for α.
Hence, the deduction is established.

Thus we have shown that every plane in projective 3-space is a plane as we already know it and so its previously established properties may be exploited in three dimensions.

As consequences of the above theory, it may be noted that any two distinct lines in 3-space have at most one point in common, that is to say, either the lines have exactly one point in common (in which case they are coplanar), or they have no point in common (i.e. they are skew).

There are many other interesting incidence properties in [2] and [3] which we do not pursue here, but which may be consulted in the literature.

§ 6.2 Principle of Duality in three Dimensions. Interchange point and plane in the Axioms *PS*1–*PS*5 and we observe duality according to the following correspondence:

*PS*1	IV
*PS*2	*PS*3
*PS*3	*PS*2
*PS*4	*PS*4
*PS*5	V

i.e. *PS*4 is a self-dual axiom.

As in the case of the plane, duality is a built-in feature of projective space, being a necessary consequence of the choice of axioms.

We state the Principle of Duality in projective space as follows:

Principle of Duality in [3]*:* Any theorems deducible from Axioms *PS*1–*PS*5 concerning points, lines, and planes of [3] remain valid if stated in the 'on' terminology, when the words 'point' and 'plane' are interchanged, and the word 'line' remains unchanged.

That is to say, in [3]

$$\begin{cases}\text{point is the dual of plane,}\\ \text{plane is the dual of point, and}\\ \text{line is the dual of line.}\end{cases}$$

Expressed otherwise

$$\begin{cases}\text{point and plane are dual elements, while}\\ \text{line is a self-dual element.}\end{cases}$$

The reader is advised to set out pairs of dual statements for [3] as was done for the plane in § 5.4.

To emphasise the difference between the plane and space duals of a given plane figure, we remark that the plane dual of a set of collinear points is a set of concurrent (coplanar) lines, whereas the space dual of the same set of collinear points is a set of planes (in [3]) with a common axis.

To illustrate further the axiomatic method in [3], we prove the following important result:

Theorem 4: There is a unique line passing through a given point and meeting each of two given skew lines, neither of which contains the point.

Proof: Take the given skew lines and the given point as a, b, P respectively.
Let $Pa = \alpha$ (say), $Pb = \beta$ by $PS2$
Now α, β are distinct since $\alpha = \beta$ implies that a,b intersect, contrary to hypothesis.
$\therefore \alpha \cap \beta = d$ (say) by IV, § 6.12
$\therefore d,a$ intersect, and d,b intersect. Also, $P \in d$ ($\because P \in \alpha, P \in \beta$).
$\therefore d$ is the required line (Fig. 4(a)).

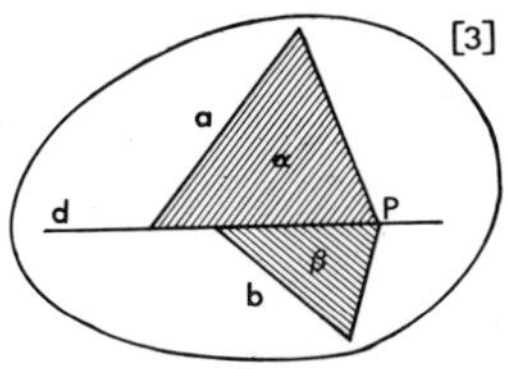

FIG. 4. (a) Unique transversal of two lines (and through a point) in [3]

Observe that this is a partial proof only; the complete solution, which requires the establishing of the uniqueness of the transversal d, is left as an exercise (Exercise 1.25).

§ 6.3 Projective Space of n Dimensions. When axiomatising the projective plane, we saw that we could manage with just three axioms ($PP1$–$PP3$). However, in setting up an axiom system for three dimensions, we found that we needed more than this number, namely five axioms ($PS1$–$PS5$). This suggests that as the dimension number increases, so too does the difficulty in formulating a suitable set of axioms.

As we are not going to use n-dimensional space in any essential way, but include it for its own interest and as a natural extension of lower dimensional theory, we shall be content merely to describe some of its salient features.

Building up a picture of projective n-space can be painlessly achieved by beginning with a definition of a projective plane used by some writers: ($n = 2$) If $P_i (i = 0, 1, 2)$ are three points not on the same line and if $l = P_1 P_2$, then the set of all points on the lines joining P_0 to the points of l, and the set of lines joining pairs of these points, is called the *projective plane* determined by P_0 and l.

[It follows that the plane is also determined by P_1 and the line P_0P_2 and also by P_2 and the line P_0P_1. Axioms *PP*1–*PP*3 may be shown to be valid for the plane so defined.]

Extending this idea to higher space, we define a [3] and a [4] as follows:
($n = 3$) If $P_i(i = 0, 1, 2, 3)$ are four points not all coplanar and if $P_1P_2P_3 = \pi$, then the set of all points on the lines joining P_0 to the points of π, and the set of lines joining pairs of these points, is called the *projective* 3-*space* determined by P_0 and π.
($n = 4$) If $P_i(i = 0, 1, 2, 3, 4)$ are five points, not all on the same 3-space, and if $P_1P_2P_3P_4 = \sigma$ (a 3-space), then the set of all points on the lines joining P_0 to the points of σ, and the set of lines joining pairs of these points, is called the *projective* 4-*space* determined by P_0 and σ.

And so on for $n = 5, 6, 7, \ldots$

Generally, if we have constructed an $[n - 1]$ in this fashion, we define an $[n]$ as follows:

If $P_i(i = 0, 1, \ldots, n)$ are $n + 1$ points, not all on the same $[n - 1]$, and if $P_1P_2 \ldots P_n = \nu$ (an $[n - 1]$), then the set of all points on the lines joining P_0 to the points of ν, and the set of lines joining pairs of these points, is called the *projective n-space* determined by P_0 and ν.

Elements of an $[n]$ are thus
points, lines, planes, solids, [4]s, . . ., $[n - 1]$s.

A space $[n - 1]$ in $[n]$ is called a *hyperplane* (or a *prime* or, more suggestively, *a flat space*), since it is an extension in higher space of the familiar object, the plane, in [3]. (Thus, in [4], a solid is a hyperplane.)

Bearing in mind that the dual of a point is, in [2] a line, in [3] a plane (i.e. the corresponding hyperplane), we are led to expect that, generally,

in [n], point and hyperplane are dual elements,

i.e. . . . [0] . . . $[n - 1]$ . . .

(Note that the sum of the dimensions of two dual spaces in $[n]$ is always $n - 1$. See Exercise 1.40.)

Consequently, in [4], pairs of dual spaces are

$$\begin{cases} \text{point, solid;} \\ \text{line, plane.} \end{cases}$$

What elements are dual in [5], [6], . . ., $[n]$? Under what circumstances can an element be self-dual? Are there any dual elements in [1]? Answers to these questions should be discernible from the duality pattern emerging in [2] and [3] and are left to the reader as exercises.

§ 6.31 Three Important Theorems in $[n]$. While it is not our wish to digress too much from our main object of study, the plane, it is felt that the properties or intersections and joins of spaces in $[n]$ are too intriguing to be overlooked entirely. By the *intersection* of spaces we mean the space of greatest dimension which is common to them. By the *join* of spaces we mean the space of minimum dimension which contains them all. Assume, without proof (refer to Semple and Roth [100] pp. 3–4 for the proofs), the following three theorems in $[n]$ about a pair of spaces, a $[p]$ and a $[q]$, neither of which is contained in the other ($p < n, q < n$).

Theorem A: If $p + q \geqslant n$, then $[p] \cap [q] = [p + q - n]$ ($\geqslant [0]$).

Theorem B: The join of $[p]$ and $[q]$ ($p + q \leqslant n - 2$) is $[p + q + 1]$ ($\leqslant [n - 1]$).

Theorem C: The join of $[p]$ and $[q]$ ($p + q \leqslant n - 2$), having $[r]$ in common, is $[p + q - r]$ ($\leqslant [n - 1]$).

Using the formal symbol $[-1]$ (§ 6) to indicate that two spaces have no points in common whatsoever, we see that Theorem C becomes Theorem B when $r = -1$.

Examples of these Theorems:

Theorem A:

in [2], two distinct lines meet in a point;
. . . [3], two distinct planes meet in a line;
. . . [3], a plane and a line not in the plane meet in a point;
. . . [4], two distinct planes meet in a point;

Theorem B:

. . . [2], two distinct points lie in a line;
. . . [3], a line and a point not on it lie in a plane;
. . . [4], two skew lines lie in a [3];

Theorem C:

in [3], two intersecting lines lie in a plane;
. . . [4], a line and a plane meeting in a point lie in a [3].

It may be noted that the examples relating to Theorems B and C are true in $[n]$ and not just in the spaces specified, which are the spaces of minimum dimension for which the examples are true. Many of the above examples will be recognised as axioms used for [2] and [3].

Employing Theorem A, we construct the following intersection table for spaces in [4] (see also Veblen and Young [113] 1, pp. 25–26):

$\cap$	line	plane	solid
line	$[-1]$	$[-1]$	point
plane	$[-1]$	point	line
solid	point	line	plane

Intersection table for spaces in [4]

A similar intersection table for spaces in [5] is worth constructing, e.g. in [5] two planes do not generally meet, but if they meet in a point they must lie in a [4]; in [5], two solids meet in a line.

Generally, a $[k]$ and a $[n-k]$ meet in a point while a $[k]$ and a $[n-k-1]$ do not meet.

Particular attention should be directed to the fact that nothing in the axioms precludes the existence of a finite line, plane, space, . . ., $[n]$, in each of which there is only a finite number of points. This type of Projective Geometry, naturally called *Finite Geometry*, will be dealt with in a separate Chapter (Chapter 10) so as not to disturb the flow of argument relating to the special class of projective plane exemplified by the augmented Euclidean plane. This is at least as special as the finite planes, though distinguished from them by other different properties.

EXAMPLE 1: In [4], three skew lines have a unique transversal (= line intersecting them).

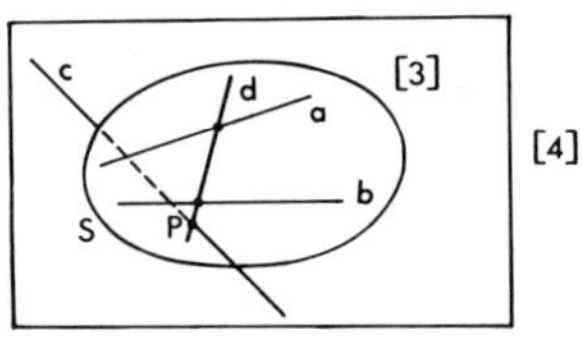

FIG. 4. (b) Unique transversal of three lines in [4]

Solution: Let the three skew lines be a, b, c.
Then a and b lie in a [3]. Call it S.
Also, by Theorem A, $S \cap c =$ a point, P (say).
By Theorem 4, there is a unique transversal, d, drawn in S from P to cut a and b
$\therefore$ in [4], there is a unique transversal, d, of 3 given skew lines a, b, c (Fig. 4(b)).

§ 6.4 Physical Applications of Higher Space. Geometrical language is often a convenient and suggestive medium for expressing problems of a non-geometrical nature. In this respect, the geometry of higher space, quite apart from its intrinsic interest to the mathematician, plays a special role. Once expressed geometrically, the problem may be found to possess interesting facets due to the known geometry of the space.

Relativity theory, as is well known, requires a four-dimensional space to describe it since it deals with sets of four numbers (three for space and one for time) which are consequently non-homogeneous co-ordinates of the four-space. It ought to be remembered, however, that this application and those below refer to spaces which may not closely resemble our projective space, though they may be related to it (e.g. they may be higher-dimensional Euclidean spaces).

Certain dynamical systems, e.g. systems of gas molecules, are described naturally in space of many dimensions; for example, a system consisting of rotations and translations in three dimensions properly requires six dimensions to explain it. Electronics and hydrodynamics are other areas of application of six-dimensional space.

Variables in Statistics may be imagined as co-ordinates in many-dimensional space. Similar notions are applicable in economics theory and in psychology theory, as, for instance, in multiple factor analysis. In linear programming, the simplex method presents itself naturally in a many-dimensional setting.

Phase spaces occurring in theoretical chemistry and physics are many-dimensional while Hilbert space, which is a useful concept for explaining certain physical situations, is a mathematical space of infinitely many dimensions.

Spaces of dimension $n > 3$ have no direct physical counterpart in the experiential world in which we live—they are purely conceptual constructions, creations of the mind without tangible reality.

Suggested topics for further study.

(1) Thorough, detailed axiomatic treatment of the projective plane (Goodstein and Primrose [42]; Robinson [92]; Veblen and Young [113]).

(2) Complete quadrangle and quadrilateral (J. Clawson: 'The complete quadrilateral', *Annals of Math.* (Ser. 2) 20 (1918–19) 232–261; N. A. Court: 'Pencils of conics', *Bulletin of Math.* 28 (1955) 7–15; M. Srirangamma (Mrs): 'The configuration of complete quadrangle and quadrilateral', *Math. Student* 30 (3,4) (1962) 157–161).

(3) Application of n (Euclidean) dimensions to Statistics (Kendall [54]).

(4) *Partial planes*, *free planes*, and the theorem of the Russian, Lydia Kopejkina (§ 46.5) (M. Hall (Chapter 11 reference); R. Hartshorne [132]; A. Seidenberg [98]; L. A. Skornyakov (Chapter 11 reference)).

EXERCISES 1

1. Every point on a projective plane lies on at least three lines (Theorem 3). Prove.

2. List all the planes satisfying *PP*1 and *PP*2 but not *PP*3.

3. In a projective plane, show that there are at least seven points and at least seven lines.

4. Look at the axioms for Euclidean Geometry given by Hilbert [50]. Compare them with those of Veblen and Young [113] for Projective Geometry.

A set of axioms for a projective plane given by some authors (e.g. Seidenberg [98]), where 'point', 'line' and 'on' are undefined, is:

Axiom 0: There exist at least one point and one line.
Axiom 1: On two distinct points there is at least one line.
Axiom 2: On two distinct points there is at most one line.
Axiom 3: On two distinct lines there is at least one point.
Axiom 4: There are at least three points on any line.
Axiom 5: Not all points are on the same line.

Using these Axioms 0–5, establish the results 5–12:

5. On two distinct lines there is at most one point (Hint: Use Axiom 2).

6. On two distinct lines there is just one point (our Axiom *PP*2).

7. On two distinct points there is just one line (our Axiom *PP*1).

8. If C and D are two distinct points on the line AB, then A and B are on the line CD. (That is, *any two points on a line determine the line.*) Hint: Use Axioms 1, 2.

9. There are at least three lines on any point. (This is our Theorem 3, i.e. Exercise 1.1, derived from a system of axioms different from Axioms *PP*1–*PP*3.)

10. Not all lines are on the same point.

11. Which of the Axioms 0–5 and Exercises 1.5–1.10 are the duals of Axioms 0–5? (Note that Axiom 0 is self-dual.)

12. Axioms 0, 4 and 5 are together equivalent to Axiom *PP*3. Prove. (Consequently, the above axiom system is equivalent to Hall's system and therefore to that of Veblen and Young.)

13. Consider the lines and planes through a given point O in ordinary Euclidean 3-space. Using familiar facts about Euclidean 3-space, verify that the following identification gives a model of a projective plane (defined by Axioms *PP*1–*PP*3):

Euclidean 3-space	*Projective plane*
line through O	point
plane through O	line
line through O in a plane through O	point on a line

In the three dimensional Euclidean model of Exercise 1.13, what Euclidean properties are used to verify Exercises 1.14–1.16?

14. Given that A, B, C are three non-collinear points, and D is a point on BC ($D \neq B$, $D \neq C$), and E is a point on CA ($E \neq C$, $E \neq A$), then there is a point F on AB such that D, E, F are collinear. (This result may be regarded as a projective version of Pasch's Axiom mentioned in § 1.2, Part 1.)

15. Pappus' Theorem (a projective theorem stated in § 2.3, Part 1).

16. The non-collinearity of the diagonal points of a complete quadrangle. (Exclude the the 7-point plane from consideration here.)

17. Given the diagonal triangle and one vertex of a complete quadrangle, construct the complete quadrangle. Is it unique?

18. Demonstrate that $Q(ABC, A'B'C',)$ implies $Q(AB'C', A'BC)$.

19. Establish the existence of the unique point C' such that $Q(ABC, A'B'C')$ where A, B, C, A', B' are five collinear points. (That is, each point of a quadrangular set is uniquely determined by the remaining five points.) (Hint: Use Desargues' Theorem, a projective theorem stated in § 2.3, Part 1.)

20. What are the duals of (i) a quadrangular set of points, (ii) a harmonic set of points? Illustrate by a diagram in each case.

Projective Space of $n(\geqslant 3)$ *Dimensions*

21. Any two planes in [3] have either (i) a common line, or (ii) they coincide. Prove.

22. Prove that, in [3], three planes (i) have a common point, or (ii) they have a common line, or (iii) they coincide.

23. Given that the lines of each pair from three lines have a common point, prove that the three lines are either (i) coplanar, or (ii) concurrent.

24. Any two lines in [3], each of which meets three given skew lines, are skew to each other.

25. Establish the uniqueness of the line d in Theorem 4. (Hint: Assume the existence of another distinct solution d' and then obtain a contradiction to this hypothesis.)

Enunciate the space duals of the statements (26)–(28) in [3]:

26. Two coplanar lines intersect in a point.

27. Two planes intersect in a line.

28. A line and a plane (not containing the line) intersect in a point.

29. What is the space dual of the triangle ABC?

30. State the space dual of Theorem 4. Illustrate with a diagram.

31. Describe the space dual of a *tetrahedron*, i.e. the figure formed from four distinct planes and thus consisting of four vertices, six lines (edges) each joining two vertices and four planes (triangular faces) each containing three vertices.

32. Two figures α, β in [3] are described thus: α has eight vertices, 12 edges and six quadrangular faces; β has 12 vertices, 30 edges, 20 triangular faces. How would we describe the duals of α, β in [3]? (What are α, β and their duals called in Euclidean Geometry if, in each case, all edges are equal in length? What are the groups of rotational symmetries for α, for β, for their duals, and for the tetrahedron?)

33. If A, B, C, D are four non-coplanar points, prove that the lines formed from distinct pairs of A, B, C, D are skew.

34–38. Establish the Deductions I–V in § 6.1, using only Axioms *PS*1–*PS*5.

39. State the n-dimensional dual of a triangle.

40. In $[n]$, what elements are (i) dual, (ii) self-dual? (Distinguish between n even and n odd.)

Write down the duals, in $[n]$, of the statements (41), (42):

41. Three hyperplanes intersect in an $[n - 3]$.

42. Two planes intersect in a line.

43. Determine the general intersection space in $[n]$ of the m spaces $[p_1], [p_2], \ldots, [p_m]$.

(Answer: The dimension is $\sum_{i=1}^{m} p_i - (m - 1)\, n$ assumed $\geqslant 0$).

44. Starting with a [4] and a point not on it, write down a definition of projective 5-space.

45. Construct an intersection table for [5].

46. In [6], two distinct planes α,β meeting in a line lie in a space θ, and two other distinct planes γ,δ ($\gamma,\delta \neq \alpha,\beta$) meeting in a point lie in a space ϕ.

(i) What are the dimensions of θ,ϕ?

(ii) What is the dimension of $\theta \cap \phi$?

47. Extend the Euclidean model of the plane (Exercise 1.13) to produce a Euclidean model of $[n]$. (This is the 'very intuitive model' of $[n]$ of which Skornyakov (p. 53 of the reference given in Chapter 11) speaks.)

We have seen how two points determine a unique line, three non-collinear points determine a unique plane, and so on. This leads to the concept of *independent points*; k points are defined to be *independent* if there is no $[k - 2]$ containing them all. Then it may be demonstrated that $n + 1$ independent points determine a unique $[n]$. (See § 12.) Using these ideas, do (48), (49):

48. 49. Establish the truth of Theorems B and C.

50. Show that, in [4], there exists in general a unique plane meeting three given skew planes in lines. In [4], two planes which meet only in a point are said to be *skew*. (See Semple and Roth [100] p. 4.)

CHAPTER 2

THEOREMS OF DESARGUES AND PAPPUS

'Logical consequences are the scarecrows of fools and the beacons of wise men.'

(T. H. Huxley (1825–1895, English))

'The theorems relating to incidence are by far the most important theorems of projective geometry.'

(Hilbert)

AXIOMATIC foundations laid in the previous Chapter enable us to construct the essential features of plane Projective Geometry, central among which are the fundamental and fascinating theorems of Desargues and Pappus. Along with the Fundamental Theorem of Projective Geometry, which we come to in Chapter 4, they constitute the most important results in the projective plane.

§7 DESARGUES' THEOREM

Definition: Two triangles are *in (central) perspective* if the lines joining corresponding vertices of the triangles are concurrent (at the *centre*, or *vertex*, *of perspective*).

For instance, in Figs. 5(a), (b) the two triangles ABC, $A'B'C'$ are in perspective from the centre V, since the joins of corresponding vertices A, A'; B, B'; C, C' are concurrent at V. Symbolically, this fact is designated by

$$ABC \overset{V}{\barwedge} A'B'C'.$$

Equally well, this perspective relationship could be expressed as $CAB \overset{V}{\barwedge} C'A'B'$, or as any other permutations of the letters so long as the correspondence of vertices is preserved. Other examples of two perspective triangles already occur in Fig. 2(a) where, in particular, $EFG \overset{V}{\barwedge} BDC$. Pick out some additional pairs of perspective triangles in this figure.

Bearing the above definition in mind, we turn now to study the crucial theorem which Desargues discovered in 1636, undoubtedly the most significant advance in pure (non-algebraic) Geometry since the Greeks. Among other things, this theorem has important repercussions for the Axioms of Incidence.

Desargues proved it as his '*Première proposition géométrique*' [30] pp. 413–415, an English translation of which appears in Smith [103] pp. 307–309.

Theorem 5 (*Desargues' Theorem*): If two triangles in space are in perspective from a point, then the three pairs of corresponding sides of the triangles intersect in three collinear points (the *Desargues line*, or *axis of perspective*).

There are two cases to consider, depending on whether the two triangles are in (a) distinct planes (hence, in space), or (b) the same plane (i.e. coincident planes).

Proof: (a) *In space* (i.e. planes distinct):

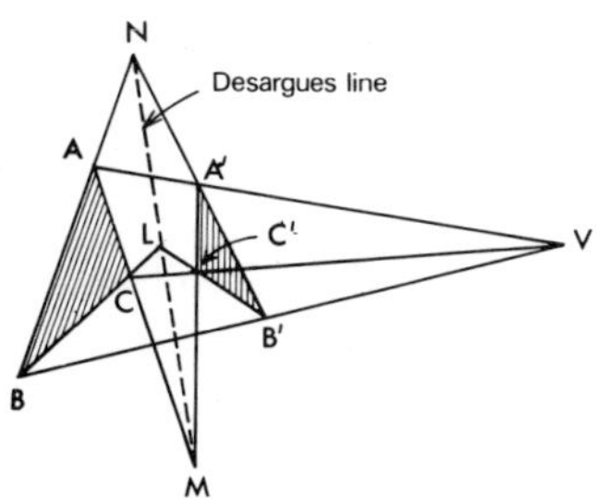

FIG. 5. (a) Desargues' theorem in [3]

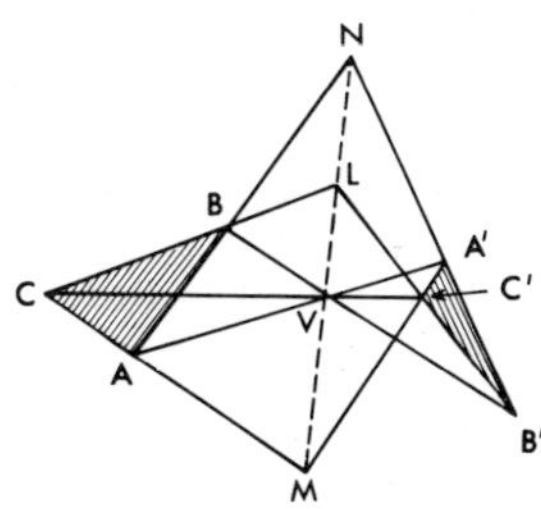

FIG. 5. (b) (minor) Desargues' theorem: Centre on axis

Let $ABC \overset{V}{\underset{\wedge}{=}} A'B'C'$ (Figs. 5(a), (b)).

Since $BB' \cap CC' = V$, then B, B', C, C' are coplanar (by $PP2$)

$\therefore$ $BC \cap B'C' =$ a point, say L (by $PP2$).

Similarly, $CA \cap C'A' = M$, $AB \cap A'B' = N$.

But $ABC \cap A'B'C' =$ a line (by IV, § 6.12) since the planes are distinct.

Also, $L, M, N \in$ both planes

$\therefore$ L, M, N are collinear (Desargues line).

(Whereas in Fig. 5(b), $V \in LMN$, in Fig. 5(a)—the general case—$V \notin LMN$.)

(b) *In the plane*, i.e. planes coincident (proof due to von Staudt):

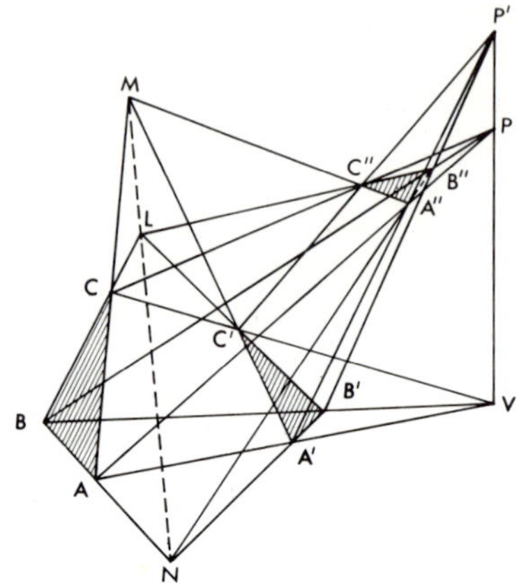

FIG. 5. (c) Desargues' theorem in [2]

Let $ABC \overset{V}{\barwedge} A'B'C'$ in the same plane.

Define L, M, N as in (a).

Through V, construct an arbitrary line VPP' not lying in the coincident planes of ABC, $A'B'C'$, so that the setting is now three-dimensional.

Let $A'' = PA \cap P'A'$, $B'' = PB \cap P'B'$, $C'' = PC \cap P'C'$. (These points exist by $PP2$ since, for example, P, P', A, A' are coplanar (because $PP' \cap AA' = V$).)

Now $ABC \overset{P}{\barwedge} A''B''C''$ and their planes are distinct (Fig. 5(c)). (Here we have used II, § 6.12.)

$\therefore$ by Desargues' Theorem in space,

each of $BC \cap B''C''$, $CA \cap C''A''$, $AB \cap A''B'' \in$ plane $ABC \cap$ plane $A''B''C''$, and likewise, each of $B'C' \cap B''C''$, $C'A' \cap C''A''$, $A'B' \cap A''B'' \in$ plane $A'B'C' \cap$ plane $A''B''C''$.

$\therefore$ BC, $B'C'$, $B''C''$ (lines of intersection of the planes PBC, $P'B'C'$, ABC in pairs) are coplanar or concurrent. Verify this as an exercise (Exercise 1.23).

But BC, PP' are skew, i.e. not coplanar

$\therefore$ they are concurrent

$\therefore$ $L = BC \cap B'C' \cap B''C'' \in$ plane $ABC \cap$ plane $A''B''C''$.

Likewise $M = CA \cap C'A' \cap C''A'' \in$ plane $ABC \cap$ plane $A''B''C''$,

$N = AB \cap A'B' \cap A''B'' \in$ plane $ABC \cap$ plane $A''B''C''$

$\therefore$ L, M, N are collinear (Desargues line).

Theorem 5′ (*Converse of Desargues' Theorem*): If the corresponding sides of two triangles in space intersect in collinear points, then the triangles are in perspective.

Proof: Now $AA'N \overset{M}{\overline{\wedge}} CC'L$

$\therefore$ B, B', $AA' \cap CC'$ are collinear (by Desargues' Theorem)
i.e. BB', CC', AA' are concurrent
i.e. the triangles ABC, $A'B'C'$ are in perspective.

(Variations of this proof, by choices of different triangles at the first step, readily suggest themselves.)

§ 7.1 Remarks on Desargues' Theorem.

(1) Note particularly that *the proof of Desargues' Theorem in the plane made use of three dimensions*, i.e. of points outside (not on) the plane. In fact, if no assumptions other than the Axioms of Incidence in the plane are made, no proof of Desargues' Theorem is possible. (It would have to be assumed as an axiom then.) Geometries in which the axioms of incidence for a plane hold, but for which Desargues' Theorem is not valid, have been constructed, e.g. by Moulton in 1902 (see § 9). Such a system is called *a Non-Desarguesian Geometry*. Peano (1894) and Hilbert (1899) have established the validity of these geometries. Planes in which Desargues' Theorem is not valid are called *non-Desarguesian planes* (§ 5.5). Otherwise they are *Desarguesian.*

Summarising the situation, we assert that *every* [2] *in a* [3] *is Desarguesian.* Proof of the converse statement, namely *every Desarguesian plane is embedded in a* [3] (and hence in [n]), requires methods beyond our present scope. (Hall [45] p. 352 uses collineations, and Seidenberg [98] p. 103 employs analytical techniques.)

Combining these two statements—the validity of the converse statement being assumed—we may restate Theorem 5(b) thus: (Desargues' Theorem in [2]) *If two coplanar triangles are in perspective from a point, then the three pairs of corresponding sides of the triangles intersect in three collinear points if and only if the plane is contained in a* [3].

(2) Fig. 5(c) in the plane without the three-dimensional superstructure may be termed the *configuration of Desargues*, 10_3, meaning that through each of the 10 points A, B, C, A', B', C', L, M, N, V in the configuration there pass three lines, and on every line (10 of them) lie three points. (Note: Figs. 5(a), 5(b), though imagined as being in [3] for the purposes of Theorem 5(a), may equally well be imagined as drawn in the plane. With this in mind, there ought to be no confusion if *we agree to refer to Fig.* 5(a) *as the configuration of Desargues*, rather than the diagram in the more involved Fig. 5(c).)

The configuration of Desargues can be considered in 10 different ways as being generated by two perspective triangles. Expressed otherwise, the diagram can be relettered in 10 different ways without destroying the truth of the theorem; for instance, besides the given pair, there is the pair of perspective triangles $VA'B'$, CML in perspective from C' with Desargues' line ANB. Can you name some other pairs of perspective triangles?

Practice should be gained by solving the following types of problem associated with the symmetry of the Desargues configuration:

(i) Given a point, determine the two triangles which are in perspective from this point, and their Desargues line;

(ii) given a line, determine the two triangles for which this is an axis of perspective, and their centre of perspective.

Detailed below is the complete perspective pattern for the Desargues configuration, which answers these problems (see also Dorwart [32] pp. 103–105):

Pairs of perspective triangles	Centre of perspective	Axis of perspective
ABC $A'B'C'$	V	LMN
MCC' NBB'	L	AVA'
NAA' LCC'	M	BVB'
LBB' MAA'	N	CVC'
$A'NM$ VBC	A	$B'LC'$
ANM $VB'C'$	A'	BLC
$B'LN$ VCA	B	$C'MA'$
BLN $VC'A'$	B'	CMA
$C'ML$ VAB	C	$A'NB'$
CML $VA'B'$	C'	ANB

(3) Certain assumptions which are inherent in the statement of Desargues' Theorem and in the diagrams (Figs. 5(a), (b), (c)) of the Desargues configuration need to be explicitly written down:

(i) A, B, C, A', B', C', V are distinct;

(ii) A, B, C are non-collinear;

(iii) A', B', C' are non-collinear;

(iv) AA', BB', CC' are distinct.

Sometimes it may happen that a vertex of one triangle lies on a side of the other triangle. Nonetheless, Desargues' Theorem is true if the triangles are in perspective. Other special cases are worthy of consideration (Todd [111] pp. 30, 32, 34).

(4) Normally, the centre of perspective does not lie on the axis of perspective, but it may do so, e.g. Fig. 5(b). [Later, (Chapter 11), in dealing with collineations (§ 16), it will be apparent that this distinction has important implications for the structure of the plane.]

(5) *In the plane the Dual of Desargues' Theorem is identical with the Converse of Desargues' Theorem* (Exercise 2.4). However, this situation is entirely different in [3] where the space dual is *not* the same as the space Converse.

[Could you enunciate a space dual of Desargues' Theorem in the plane? It may not prove to be a very rewarding experience!]

(6) In Figs. 5(a), (b), (c) we say that the triangles ABC, $A'B'C'$ are in *axial perspective* with axis LMN. This is the dual concept in [2] of central perspective. Desargues' Theorem and its Converse in [2] may thus be rephrased, respectively

Two triangles which are centrally perspective are axially perspective.
Two triangles which are axially perspective are centrally perspective.

(7) It should be noted that a direct implication of $ABC \overset{V}{\barwedge} A'B'C'$ is the existence of the pairs of perspective triangles $A'BC \overset{V}{\barwedge} AB'C'$, $AB'C \overset{V}{\barwedge} A'BC'$, $ABC' \overset{V}{\barwedge} A'B'C$. (See Exercise 2.9.)

(8) Coxeter, in Chapters 2 and 3 of [28], provides interesting information on some of the more recondite aspects of the Desargues configuration.

(9) There is no uniformity in the literature concerning the name of the theorems we have called Desargues' Theorem and its Converse. (Some authors, e.g. Hilbert, reverse this nomenclature while others combine the two into one theorem which they call Desargues.) A similar confusion exists regarding Pappus' Theorem (which is sometimes called the Pappus-Pascal Theorem and sometimes, e.g. by Hessenberg, Pascal's Theorem). In each case, we have accepted the nomenclature most commonly in use.

A typical problem of medium difficulty relating to the Desargues' Theorem is now discussed in detail. An attempt is made to examine some of the thought-processes operating in the successful solution of the problem. Almost always the question asks us, perhaps indirectly, to detect a pair of perspective triangles.

§ 7.2 Application of Desargues' Theorem.

EXAMPLE 2: P, Q, R, S are four given points and $PQ \cap RS = A$, $QS \cap PR = B$, $PS \cap QR = C$, $AB \cap PS = U$, $AB \cap QR = V$. Prove that VP, UR meet on AC (Fig. 6).

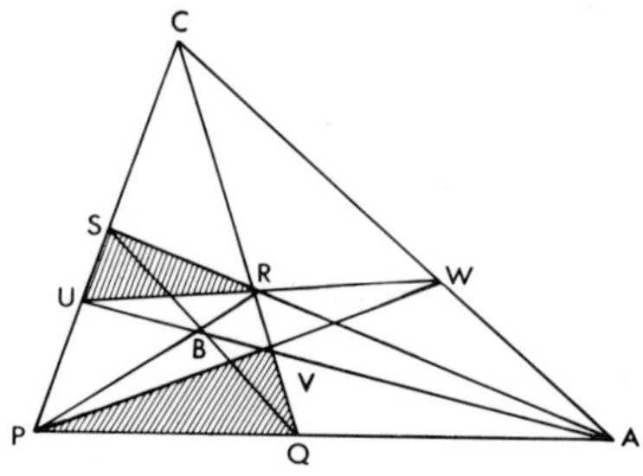

FIG. 6.

Analysis of the Problem: Let $VP \cap UR = W$.

The problem asks us to prove that A, C, W are collinear.

This suggests a direct application of Desargues' Theorem provided we can find two perspective triangles. While the reader may be able to spot these triangles immediately, their discovery will be methodically explained here.

Essentially, we seek a pair of triangles—if they exist—having VP and UR ($= RU$) as corresponding sides. Now $UV \cap PR = B$, so B must be the centre of perspective of two triangles having V,U; P,R as two pairs of corresponding vertices.

To find the third vertex, observe that $A = PQ \cap RS$ suggesting that Q and S must be the third pair of corresponding points since by virtue of the above, P,R already occur. (There are no other relevant points on PQ, RS other than the common point A.)

Then try the triangles VPS, URQ in perspective from B. By Desargues' Theorem, $VP \cap UR$, $PS \cap RQ$, $SV \cap QU$ are collinear

i.e. W, C, $SV \cap QU = W'$ (say) are collinear.

But this result was not required, though it is indeed true. Nor can it readily be used to deduce the required result. Consequently our choice of triangles was wrong.

Therefore, try $VPQ \overset{B}{\barwedge} URS$.

Then, by Desargues' Theorem, $VP \cap UR$, $PQ \cap RS$, $QV \cap SU$ are collinear

i.e. W, A, C are collinear

which was required.

Assembling all these thoughts, we would simply proceed thus:

Proof: Now $VPQ \overset{B}{\barwedge} URS$

$\therefore$ $VP \cap UR$, $PQ \cap RS$, $QV \cap SU$ are collinear, by Desargues' Theorem i.e. W, A, C are collinear

i.e. $VP \cap UR$ lies in AC.

Alternatively: SBQ is the Desargues line of the triangles ARP, CUV

$\therefore$ AC, RU, PV are concurrent, by the Converse of Desargues' Theorem

i.e. $RU \cap PV$ meet on AC.

Can you detect any other approaches? (PBR is the Desargues line of the triangles QSA, UVC implying QU, SV, AC are concurrent (at W', as we have already shown).)

Two solutions are available in this particular problem, though it does not follow that one will always enjoy this good fortune.

§ 8 PAPPUS' THEOREM AND DESARGUES' THEOREM

While more will be said subsequently about Pappus' Theorem, it seems appropriate and natural to discuss its geometrical importance here particularly in view of its significance for Desargues' Theorem. Temporarily then, assume Pappus' Theorem (Fig. 7):

Theorem 6 (Pappus' Theorem): If A, B, C; A', B', C' are two sets of three collinear points, then $AB' \cap A'B$, $BC' \cap B'C$, $CA' \cap C'A$ are collinear.

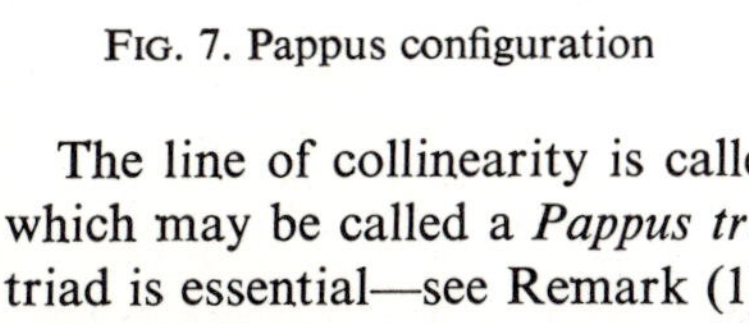

FIG. 7. Pappus configuration

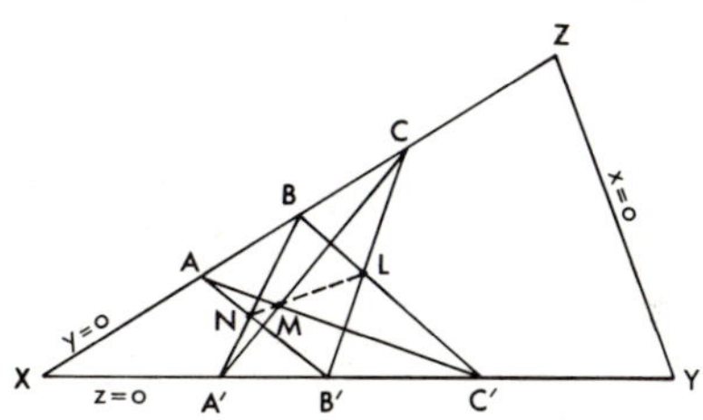

FIG. 7. (a) Pappus' theorem

The line of collinearity is called the *Pappus line* of the two sets, each of which may be called a *Pappus triad*. The order of the elements of a Pappus triad is essential—see Remark (1) below.

On the basis of the Axioms of Incidence $PP1$–$PP3$ alone, Pappus' Theorem cannot be proved. However, if we add Pappus' Theorem to our list of incidence axioms, then, of course, the problem is overcome. A plane in which Pappus' Theorem is valid is called a *Pappian plane*. A plane in which Pappus' Theorem is invalid is *non-Pappian*. Originally, in his *Collectio* (7th book), Pappus deduced his theorem from results already proved by Euclid and used it as a lemma for an understanding of Euclid's lost book, *On Porisms*. Probably, the theorem was known to earlier generations of mathematicians.

§ 8.1 Remarks on Pappus' Theorem.

(1) As A, B, C may be permuted in $3! = 6$ ways, there will be six Pappus lines associated with two unordered Pappus triads.

(2) Pappus' Theorem is later proved (§ 24) to be a special case of Pascal's Theorem for a hexagon inscribed in a conic. Considering Fig. 7 as referring to the hexagon $AB'CA'BC'$, we may state Pappus' Theorem in the following way:

If alternate vertices of a hexagon lie on two lines, then the three pairs of opposite sides meet in three collinear points.

This is a convenient way of conceiving the theorem in relation to Pascal's Theorem (§ 24) (Blaise Pascal 1623–1662, French).

(3) The dual of Pappus' Theorem, establishing the existence of a *Pappus point*, is left as an exercise (Exercise 2.10). By assuming Pappus' Theorem, the dual could be *proved*, if so desired. (Baker [8] pp. 19–20, shows how a special case of the dual of Pappus' Theorem is associated in the Euclidean plane with the physical law known as the parallelogram of forces.)

(4) Pappus' Theorem determines a 9_3 *configuration* (*of Pappus*). Inherent in the configuration and dependent on its symmetry are nine Pappus theorems derived according to the table hereunder (that is, each line in the configuration is a Pappus line for two triads):

Pairs of Pappus triads		Pappus line
ABC	$A'B'C'$	LMN
ABC	LMN	$A'B'C'$
$A'B'C'$	LMN	ABC
$AB'N$	$A'MC$	LBC'
$AB'N$	LBC'	$A'MC$
$A'MC$	LBC'	$AB'N$
$A'BN$	AMC'	$LB'C$
$A'BN$	$LB'C$	AMC'
AMC'	$LB'C$	$A'BN$

The pattern in this table can perhaps be appreciated best by writing down the following schedule (first two triads repeated):

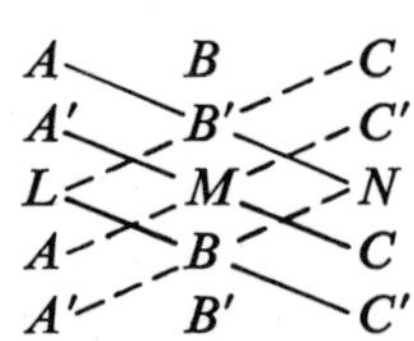

The nine cases are then obtained from the three sets of three lines derived by considering rows (3) and diagonals (6) (but not columns—why?).

(5) The special case of Pappus' Theorem occurring when AA', BB', CC' are concurrent may be proved using Desargues' Theorem. (See Exercise 2.11.)

(6) None of the points A, B, C, A', B', C' lies at $l \cap l'$.

Because of the high degree of symmetry in the Pappus configuration, many other worthwhile geometrical facts may be discovered. The interested reader is encouraged to explore to the utmost the intriguing properties of this configuration. For instance, see Exercise 2.12.

§ 8.2 Applications of Pappus' Theorem.

EXAMPLE 3: If A is any point on PR, and B is any point on QS of a quadrangle $PQRS$, and if $QA \cap PB = U$, $SA \cap RB = V$, then UV passes through $Z = PS \cap QR$ (Fig. 8). Prove.

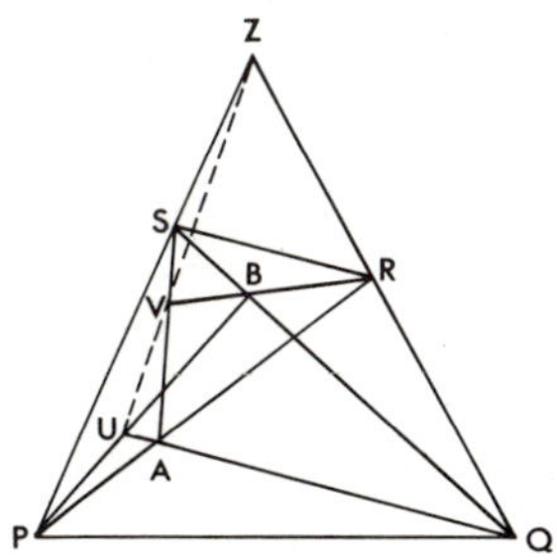

FIG. 8.

Solution: From the Pappus triads $\begin{cases} P & A & R \\ Q & B & S \end{cases}$

it follows that U, V, Z are collinear (Pappus line).

(An analysis, as in Example 2, of some of the thought-processes involved in making the correct choice of Pappus triads is omitted here. Experience should give the student facility in handling this type of problem.)

A striking illustration* of the application of both the Theorems of Pappus and Desargues is now given (Example 2(a)–3(a)).

Suppose we are given only a point Q and two lines a, b whose point of

* From *Configuration Theorems*, by B. I. Argunov and L. A. Skornyakov—translated from the Russian. Copyright © 1963 by the University of Chicago. Published by D. C. Heath & Co., Boston, Massachusetts, U.S.A.

intersection $a \cap b = P$ is 'inaccessible', i.e. cannot be used in the construction. (For example, Q, a, b may be point and lines on a drawing board chosen in such a manner that a, b do not intersect on the drawing board.) Our problem is: How do we construct the line QP? The answer lies in constructing a point R such that P, Q, R are collinear, so that QR is the required line. Essentially, the solution replaces the inaccessible point P by an accessible point R.

Two solutions are offered, the first using Desargues' Theorem, the second using Pappus' Theorem.

EXAMPLE 2(a)–3(a) (Applications of Desargues' and Pappus' Theorems): Given a point Q and an inaccessible point $P = a \cap b$, construct the line PQ. *Solution 1 : (Fig. 6(a)):* Through an arbitrary point $O(\notin a, \notin b, \neq Q)$ draw three arbitrary lines p_1, p_2, p_3.

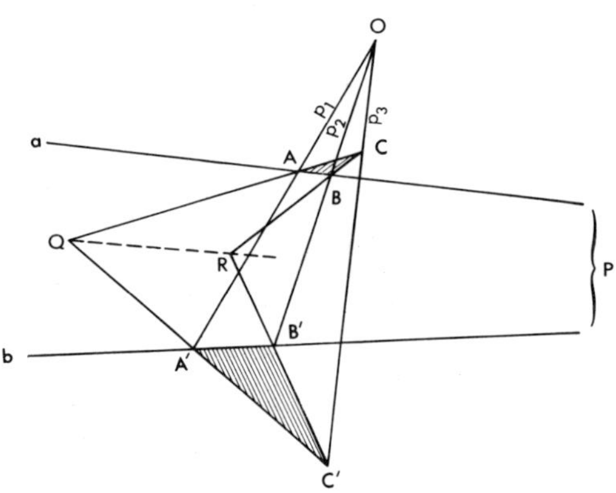

FIG. 6. (a)

Construct the points
$$\begin{cases} a \cap p_1 = A, & b \cap p_1 = A', \\ a \cap p_2 = B, & b \cap p_2 = B', \\ AQ \cap p_3 = C, & A'Q \cap p_3 = C', \\ BC \cap B'C' = R. & \end{cases}$$

Then $ABC \overset{O}{\barwedge} A'B'C'$

$\therefore$ P, Q, R are collinear (Desargues line)

$\therefore$ to construct PQ, merely join Q to R.

Solution 2 (Fig. 8(a)): Take arbitrary points A, $B \in a$ and C, $D \in b$.

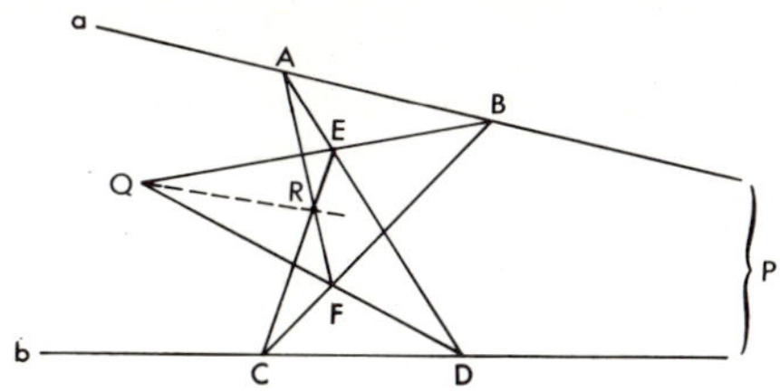

FIG. 8. (a)

Construct the points $\begin{cases} E = AD \cap BQ \\ F = BC \cap DQ \\ R = AF \cap CE. \end{cases}$

Consider the Pappus triads $\begin{cases} ADE \\ CBF. \end{cases}$

$\therefore$ P, Q, R are collinear (Pappus line)
$\therefore$ PQ is constructed by joining Q to R.

§ 8.3 Hessenberg's Theorem. What connection, if any, is there between the Theorems of Desargues and Pappus? This vital question was answered in 1905 by the German mathematician, Gerhard Hessenberg ('*Beweis des Desarguesschen Satzes aus dem Pascalschen*', *Mathematische Annalen* 61 (1905) 161–172).

Theorem 7 (*Hessenberg's Theorem*): Desargues' Theorem is a consequence of Pappus' Theorem.

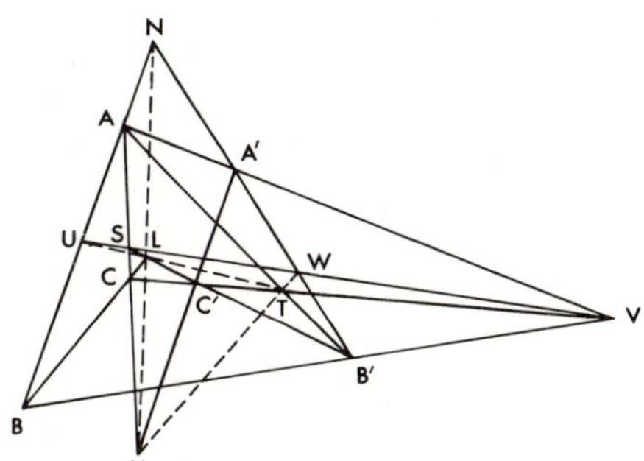

FIG. 9. Hessenberg's theorem

Proof: Construct $ABC \overset{V}{\barwedge} A'B'C'$ with $BC \cap B'C' = L$, $CA \cap C'A' = M$, $AB \cap A'B' = N$. Let $B'L \cap AM = S$, $VS \cap AB = U$, $VS \cap A'B' = W$, $AB' \cap CC' = T$ (Fig. 9).

Then,
by Pappus' Theorem applied to the triads V, B, B'; A, S, C, we have
U, L, T collinear (i)
.... V, A, A'; B', C', S we have
T, W, M collinear (ii)
.... B', T, A; U, S, W we have
L, M, N collinear by (i), (ii)

§ 8.31 Remarks on Hessenberg's Theorem.

(1) *Assuming Pappus' Theorem, Desargues' Theorem in the plane is automatically true.* (But a non-Pappian plane is not necessarily non-Desarguesian.) Diagrammatically, we have the following relationships among the types of planes:

Desarguesian ⟵ Pappian
non-Desarguesian ⟶ non-Pappian

(2) *Cronheim's Theorem.* Certain special possibilities, not covered by Hessenberg, remained unnoticed by anyone for 48 years until, in 1953, *Arno Cronheim produced a more detailed proof of Hessenberg's Theorem.* (Cronheim: 'A proof of Hessenberg's Theorem', *Proceedings Amer. Math. Soc.* 4 (1953) 219–221.) Imperfect though it is in the form presented in Theorem 7, Hessenberg's Theorem is included here as an historically interesting application of Pappus' Theorem. Pickert [86] p. 144, footnote, lists eight defective proofs of Hessenberg's Theorem.

Referring to Hessenberg's Theorem, Cronheim had this to say: 'In all the proofs of this beautiful theorem, only the so-called general case is treated so that some of the points and lines constructed in the course of the proof may actually become indeterminate. In the following lines, we offer a proof of this theorem which takes care of all possible cases.' Cronheim went on to show how the discussion may be reduced to two cases only, the first of which is treated by Hessenberg's argument, while the second one requires a different approach. Despite the care shown by Cronheim, one special case was overlooked by him—this special case is discussed in detail by Pedoe [85] (1963) pp. 39–42.

Geometrically, the importance of Pappus' Theorem lies in the fact that a Desarguesian plane may be derived if we adjoin to the Axioms of Incidence the Theorem of Pappus as a further axiom. Algebraically, Pappus' Theorem has an equal significance as we shall see in the next Chapter.

§ 9 NON-DESARGUESIAN GEOMETRY

Hilbert [50], it seems, was the first to investigate fully the essential nature of Desargues' Theorem though Poncelet [87] had rescued it from oblivion. (Concerning this resurrection of Desargues' work and reputation, Poudra in [30] wrote:
'*M. le général Poncelet dans son Traité des propriétés projectives appela l'attention sur ce véritable et profond géométre et le reconnut . . . comme l'un des fondateurs de la géométrie moderne*'.)
In § 23 of his famous book, Hilbert [50] established the existence of a plane geometry obeying (his) axioms of incidence but in which Desargues' Theorem is not valid. He concluded that 'Desargues' Theorem is not, therefore, a consequence solely of the axioms mentioned: for its demonstration necessitates either the space axioms or all of the axioms of congruence'. To exhibit his non-Desarguesian geometry he chose a rather unsatisfactory and complicated example, however, which involved the use of quadratic equations.

Shortly after Hilbert's work was published, a very simple illustration of a non-Desarguesian geometry was produced by the American astronomer, F. R. Moulton (1872–1952) in 1902. Since then, other types of non-Desarguesian planes have been discovered, for example, by Veblen and Wedderburn, Ruth Moufang, and Hughes. More will be said later about these planes at the appropriate place (Chapter 11). Because of this variety of non-Desarguesian planes, we shall refer to the plane in the Moulton example as the *Moulton plane*, π_M.

§ 9.1 Description of the Moulton plane. It must be remembered that Moulton's setting was the ordinary Euclidean plane geometry which he assumed logically to exist. Explaining the purpose of his work, Moulton wrote: 'The object of this paper is to prove that Desargues's Theorem is not a consequence of Hilbert's axioms I 1–2, II, III, IV 1–5, V by exhibiting a simpler non-desarguesian plane geometry than that given by Hilbert, and one which, by the use of linear equations alone, can be shown to fulfil all of the axioms in question.' In the following exposition, an attempt will be made to recapture some of the spirit of Moulton's ideas.

The following table sets out synthetically Moulton's own description of his geometry against the background of an orthonormal Cartesian co-ordinate system:

Elements of Moulton's plane	Euclidean equivalent
points	points
lines	(i) straight lines with slope $m \leqslant 0$ (ii) (broken) straight line with slope $m > 0$ broken at the x-axis so that the slope above is $c\,m$ ($c \neq 1$; usually $c = \frac{1}{2}$ (Moulton's value) for simplicity) (iii) straight lines with slope $m = \infty$

Elements of the co-ordinate system belong to the field of real numbers.

Thus in Fig. 10(a), the lines A_1A_2KM, B_1B_2V, C_1C_2V, OY and OX are *Moulton* (*non-Desarguesian*) *lines.* One may think of the broken Euclidean line (= Moulton line) as being like a ray of light refracted at the barrier OX.

Analytically, all loci of the Cartesian form

$$y = m(x-a)\,f(y, m)$$

where the function f (not Moulton's symbol) is defined by

(i) $m \leqslant 0 \Rightarrow f(y, m) = 1$

(ii) $m > 0 \Rightarrow \begin{cases} f(y, m) = 1 \text{ if } y \leqslant 0 \\ f(y, m) = \frac{1}{2} \text{ if } y > 0 \end{cases}$

are Moulton lines. Two Moulton lines are parallel if their corresponding portions above or below OX are parallel in the ordinary way.

Extending the Euclidean plane to make it projective by adjoining points at infinity, one for each set of parallel Moulton lines, we derive the *augmented Moulton plane* π'_M ($= \pi_M + l_\infty$) consisting of the Moulton plane and the line at infinity, l_∞.

Properties of the augmented Moulton plane. As an important exercise (Exercise 2.20), it should be verified that

I *The augmented Moulton plane is a projective plane,*
i.e. π_M satisfies *PP*1–*PP*3.

From Fig. 10(a) it is clear that

II *Desargues' Theorem is not valid in the augmented Moulton plane,*
i.e. the Moulton plane *is* non-Desarguesian or, as we might say, Desargues' Theorem is vitiated by the above deformation of the plane; for, if the specially chosen triangles $A_1B_1C_1$, $A_2B_2C_2$ have corresponding sides parallel, then in Euclidean Geometry the lines joining the corresponding vertices are concurrent at V. However, in the Moulton plane, while B_1B_2 and C_1C_2 remain the same, the line A_1A_2 is broken before reaching V. Hence, it will no longer

pass through V. Consequently, Desargues' Theorem is not fulfilled. See also Fishback [40] p. 172.

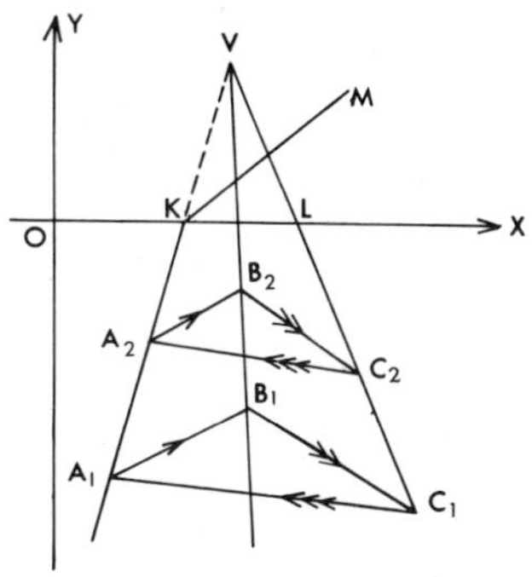

FIG. 10. (a) Invalidity of Desargues' theorem. Non-Desarguesian (Moulton) plane

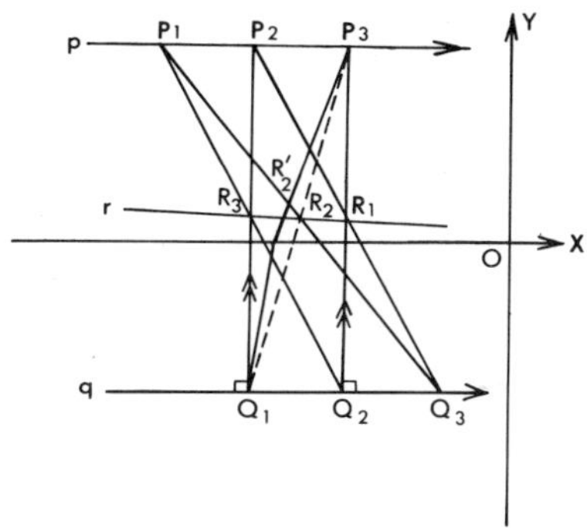

FIG. 10. (b) Invalidity of Pappus' theorem

Similarly, by considering the triads P_1, P_2, P_3 and Q_1, Q_2, Q_3 on p, q respectively, as in Fig. 10(b), it is easy to see that the corresponding Pappus line does not exist. That is,

III *Pappus' Theorem is not true in the augmented Moulton plane,*
i.e. the Moulton plane is non-Pappian.

Of course, result III is really redundant, since Pappus' Theorem cannot possibly hold in a Moulton plane. However, we are reproducing Moulton's argument for historical purposes, bearing in mind that when Moulton made the discovery (in 1902), Hessenberg's Theorem (1905), showing the relation of Desargues' Theorem to that of Pappus, was not known to him.

§ 9.2 Current Research on Moulton planes. A good deal of current research activity is being devoted to the study of non-Desarguesian planes. In the particular case of the Moulton plane, the situation regarding extensions of Moulton's ideas is summarised as follows (where, for convenience, the x-axis is replaced by the y-axis as the line along which 'bending' occurs):

A *Pickert-Moulton plane* over an ordered field F has points (x, y) and lines $x = c$, $y = b + r \circ x$ where

$$r \circ x = \begin{cases} prx & \text{for } r, x < 0 \\ rx & \text{otherwise} \end{cases}$$

$p > 0$, b, c, r, x, y being in F. Moulton's model occurs as a special case when $p = 2$ and F is the field of real numbers. (Günter Pickert is a contemporary German mathematician.)

Generalised Moulton planes defined by Pierce ('Moulton planes', *Canadian J. Math.* 13 (1961) 427–436) have

$$r \circ x = \begin{cases} rx \\ \phi(r).x \end{cases} \quad \text{for all} \begin{cases} x \geqslant 0 \\ x < 0 \text{ and all } r \text{ (in } F), \end{cases}$$

together with certain restrictions imposed on $\phi(r)$ which need not concern us here. Then the Pickert-Moulton plane is a special generalised Moulton plane with F ordered, and

$$\phi(r) = \begin{cases} r \\ pr \end{cases} \qquad \text{for} \begin{cases} r \geqslant 0 \\ r < 0. \end{cases}$$

Suggested topics for further study.

(1) Cronheim's Theorem (Cronheim quoted above, § 8.31(2); Pedoe [85]).
(2) Desargues' '*Première proposition géométrique*' (Desargues [30]).
(3) Life and achievements of Desargues ('*Biographie*' in Desargues [30] 11–45; B. A. Swinden: 'Geometry and Girard Desargues', *Math. Gaz.* 34 (310) (1950) 253–261).
(4) Pappus ([80]; van der Waerden [112]).
(5) Triply perspective triads (W. R. Andress and W. Saddler, with a note by W. Sawyer: 'Perspective triads', *Math. Gaz.* 37 (322) (1953) 247–255; F. Stockton: 'A set of triply perspective triangles associated with projective triads', *Amer. Math. Monthly* 62 (7) Part 2 (1955) 41–51).
(6) Levenberg's generalisation of Moulton's paper (K. Levenberg: 'A class of non-Desarguesian plane Geometries', *Amer. Math. Monthly* 57 (1950) 381–387).
(7) Work of W. A. Pierce on Moulton planes.

EXERCISES 2

1. Points D, E, F are taken in the sides BC, CA, AB respectively of the triangle ABC such that AD, BE, CF are concurrent. Lines EF, FD, DE meet the sides BC, CA, AB respectively in P, Q, R. Prove that P, Q, R are collinear.

2. A, B, C, D are vertices of a quadrangle and $BC \cap AD = X$, $CA \cap BD = Y$, $AB \cap CD = Z$, $BC \cap YZ = L$, $BD \cap ZX = M$, $CD \cap XY = N$. Show that L, M, N are collinear.

3. A, B, C, D are four points; M is any point on BC and N any point on AD. Points P, Q, R, S are taken on AB, AC, BD, CD respectively so that MQP, MRS, PNR are lines. Show that QNS is also a line.

4. In [2], show that the Dual of Desargues' Theorem is precisely the Converse of Desargues' Theorem.

5. A landscape gardener with a flair for Mathematics decides to set out ten shrubs in rows of three so that each shrub lies in three rows. How may he do this?

6. Lines AB, CD intersect in U and AC, BD intersect in V. Define $UV \cap AD = F$, $UV \cap BC = G$, $BF \cap AC = H$. Then prove that GH, CF, AU are concurrent.

7. When three triangles are all in perspective from the same centre, show that the three axes of perspective are concurrent.

8. Given that three triangles are in perspective in pairs with a common axis of perspective, prove that their three centres of perspective are collinear.

9. If ABC, $A'B'C'$ are two coplanar triangles such that AA', BB', CC' are concurrent, and also AB', BC', CA' are concurrent, prove that AC', BA', CB' are concurrent. (Stated as a theorem, this result reads: Two triangles which are *doubly perspective* are also *triply perspective*.)

10. (Dual of Pappus' Theorem.) Through the point P lines a, b, c are drawn; through the point P' lines a', b', c' are drawn. The following points are defined: $U = b \cap a'$, $S = b' \cap a$, $H = c \cap a'$, $K = b' \cap c$, $V = b \cap c'$, $T = c' \cap a$. Prove that US, HT, KV are concurrent. (Use Pappus' Theorem if you wish.)

11. In the Pappus configuration (Fig. 7), suppose AA', BB', CC' are concurrent. Show that then the Pappus line passes through the point of intersection of the lines l, l'. (Hint: use Desargues' Theorem.)

12. Points A_i, B_i, C_i $(i = 1, 2, 3)$ are taken on three lines l_i concurrent at O $(i = 1, 2, 3)$ such that $A_1A_2A_3$, $B_1B_2B_3$, $C_1C_2C_3$ are concurrent at V. Prove that the Pappus lines p_3, p_1, p_2 for the triads on l_1, l_2; l_2, l_3; l_3, l_1 respectively are concurrent at O.

13. Given that two tetrahedra are centrally perspective, show that (i) corresponding faces of the tetrahedra meet in the sides of a complete quadrilateral and (ii) corresponding edges meet in the vertices of this quadrilateral. (The tetrahedra are then said to be *in perspective from the plane* which is common to the quadrangle and the quadrilateral, the quadrangle referring to the vertices in (ii).)

14. Points M, N are taken on the side AB of a triangle ABV, while R is taken on AV. Join RM, RN, VM, VN. Let a line through A cut RM in S and VM in W, and another line through A cut RN in T and VN in Z such that B, Z, W are collinear. Establish that B, S, T are collinear.

15. ABC, $A'B'C'$ are two coplanar triangles in perspective and $BC' \cap B'C = P$; $CA' \cap C'A = Q$, $AB' \cap A'B = R$. Prove that

(i) QR passes through $BC \cap B'C'$,

(ii) (hence) triangles ABC, PQR are in perspective.

16. P, Q, R, S are any four coplanar points. Let X be any point on PR and Y any point on QS. Suppose $QX \cap PY = U$, $SX \cap RY = V$. If $PS \cap QR = L$, prove that L, U, V are collinear.

If, further, $PY \cap SX = A$, $QX \cap RY = B$, prove that PQ, RS, AB are concurrent.

17. Given the Desargues configuration (Fig. 5(a)) let $BC' \cap B'C = P$, $CA' \cap C'A = Q$, $AB' \cap A'B = R$. Prove that

(i) BC, $B'C'$, QR are concurrent,

(ii) AP, BQ, CR are concurrent.

18. The diagonal triangle of a complete quadrangle is in perspective with each of the four triangles formed by a set of three vertices of the quadrangle, the fourth vertex in each case being the centre of perspective. Establish this property.

Hence, show how to construct a quadrilateral having the same diagonal triangle as a given quadrangle. (See Veblen and Young [113] 1, 45–46.)

Exercises 2.19, 2.20 below may be omitted at first reading.

19. Verify, in the augmented Moulton plane, that the two half-lines $y = \frac{1}{2} m_1(x-a_1)$, $y = \frac{1}{2} m_2(x-a_2)$ intersect in the point with Cartesian co-ordinates

$$\left(\frac{m_2a_2-m_1a_1}{m_2-m_1}, \frac{m_1m_2(a_2-a_1)}{2(m_2-m_1)}\right).$$

20. Check that the augmented Moulton plane does indeed satisfy the axioms of a projective plane. (Consult Artzy [4], Pedoe [85] or Whitehead [118] for the details.)

CHAPTER 3

CO-ORDINATES IN THE PLANE

'I hope that posterity will judge me kindly, not only as to the things which I have explained, but also as to those which I have intentionally omitted so as to leave to others the pleasure of discovery.'
(Descartes [31] p. 240)

SYNTHETIC Projective Geometry, the material of the two previous chapters, has a special fascination of its own. However, a continuation along exclusively synthetic lines, while interesting in itself, would describe only a small portion of modern knowledge of the projective plane. With the introduction of analytical and algebraic methods by way of homogeneous co-ordinates, a greater strength is added to our techniques. Moreover, it is hoped that these will furnish the reader with a richer understanding of the nature of Algebraic Geometry and, indeed, of Mathematics.

Just as the raw material of Synthetic Geometry, namely points and lines, were undefined, so it is with the raw material of Analytical Geometry, namely numbers (or co-ordinates). That is 'number' will not be defined, though one may think of real or complex numbers as the occasion warrants. Connecting these undefined numbers (which include a zero 0 and a unit 1) there are two undefined operations, multiplication and addition, represented by the symbols $+$ and $\cdot$ Thus, for elements a, b, of the number system we have the combinations $a + b$, and $a \cdot b$ (or merely ab). Assuming that these co-ordinates belong to an abstract field F, we complete the field operations with subtraction ($-$) and division ($\div$), the inverses of $+$ and $\cdot$ respectively. Nothing is said about order in the abstract field F, but recall that real numbers are ordered, whereas complex numbers are not ordered.

Modern research into the foundations of Geometry has provided us with a variety of methods by which co-ordinates may be introduced into the projective planes. One of these, employing the concept of a ternary ring, will be considered in more detail in a later Chapter (11). Our treatment will use the method of classical homogeneous co-ordinates. These, themselves, may be explained in many different ways, but the most forceful seems to be through the medium of vector spaces.

The symbol ∞. 'We have adroitly defined the infinite in arithmetic by a loveknot, in this manner ∞, but we possess not therefore the clearer notion of it,' said Voltaire. To clarify our use of the symbol, we assume the following properties of the symbol ∞ w.r.t. numbers a, a_1, a_2:

1. $\frac{a}{0} = \infty \qquad (a \neq 0)$
2. $a \,.\, \infty = \infty \qquad (a \neq 0)$
3. $\frac{a}{\infty} = 0$
4. $\infty + a = \infty$
5. $\frac{\infty + a_1}{\infty + a_2} = 1.$

The expressions $\frac{0}{0}$ and $0 \,.\, \infty$ are undefined.

§ 10 GENERAL HOMOGENEOUS CO-ORDINATES

§ 10.1 Homogeneous Co-ordinates of a Point. Experience in analytical techniques suggests to us that a point in a plane can be naturally and adequately described by a set of three numbers (co-ordinates) x, y, z, of which the ratios $x : y : z$ only are of interest. That is, two independent numbers will fix the position of the point. Suppose that these co-ordinates belong to an abstract field F. Concrete instances of F (to which reference will be made when appropriate), are

(i) the field R of real numbers;
(ii) the field C of complex numbers;
(iii) the finite *Galois field* $GF(p^k)$ where p is a prime, k an integer.

[Galois fields are named in honour of the French genius, Évariste Galois (1811–1832), whose tragic and painful death in a duel at the age of twenty was a catastrophe for the world of mathematical ideas.]

Make the following fundamental assumption regarding homogeneous co-ordinates:

*HC*1. In a projective plane a point is uniquely defined by the ratios of three numbers (co-ordinates) x, y, z (not all zero) belonging to some number field. Conversely, these ratios uniquely determine a point in the plane.

That is, the point $P\,(x, y, z)$ is equally well represented by the co-ordinates (kx, ky, kz), provided that the scalar multiplier $k \neq 0$. Thus for example, with $F = R$

$(6, -3, 12) \equiv (2, -1, 4) \equiv (\frac{1}{2}, -\frac{1}{4}, 1) \equiv (1, -\frac{1}{2}, 2) \equiv (-1, \frac{1}{2}, -2) \equiv$ etc.

But especially note that, by definition, (0,0,0) is undefined, so

(0, 0, 0) *is meaningless and represents no point in the plane.*

Numbers (x, y, z) in this context are called *general homogeneous co-ordinates.* Until specified otherwise, the homogeneous co-ordinates will be general, so this adjective may be dropped henceforth. Of course the ordering of x, y, z is important, since, for instance (x, y, z) and (x, z, y) generally represent two different points. A suggestive alternative notation which will be used as appropriate is x_1, x_2, x_3 for x, y, z respectively.

Writing

$$X = \frac{x}{z}\left(= \frac{kx}{kz}\right), \quad Y = \frac{y}{z}\left(= \frac{ky}{kz}\right), \qquad (k \neq 0)$$

we say that (X, Y) are the *non-homogeneous co-ordinates* of P (x, y, z). (Greek letters ξ, η are sometimes preferred to X, Y, especially if ambiguity might otherwise result.)

§ 10.11 Homogeneous Co-ordinates and Vector Spaces. Homogeneous co-ordinates (x, y, z) of P may be thought of as being represented by a column matrix (vector)

$$\mathbf{P} = \begin{bmatrix} x \\ y \\ z \end{bmatrix}$$

with transpose $\mathbf{P}^T = [x, y, z]$. Consequently, $\mathbf{P}$, $k\mathbf{P}$ $(k \neq 0)$ represent the same point. If we wish, we may refer to the point P as the point $\mathbf{P}$.

What we have done in *HC*1 is to associate by means of a many-one mapping f the ordered triples of elements of an abstract field F, i.e. the vectors of a vector $V^3(F)$ over F, with the elements (originally abstract but now taken as geometrical points) of a plane π such that

(a) to every vector $\mathbf{P}$ $(\neq \mathbf{O})$ corresponds a unique point P of π; and

(b) if $\mathbf{Q} = k\mathbf{P}$ $(k \neq 0,\ k \in F)$, then $\mathbf{P}$, $\mathbf{Q}$ correspond to the same point P of π (Fig. 11), and conversely.

This provides us with an analytical model of π (§ 10.2).

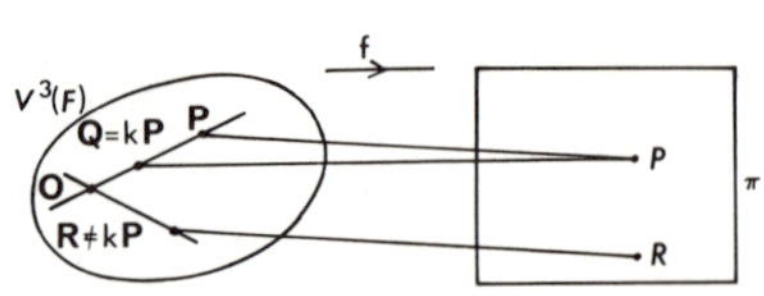

FIG. 11. Mapping of a vector space to a projective plane

Evidently, the mapping f divides the vectors of $V^3(F)$ into equivalence classes, that is, all vectors $k\mathbf{P}$ ($k \neq 0$) equivalent to $\mathbf{P}$ represent the same point P and form an equivalence class. One very special equivalence class consists of just $k\mathbf{O}$, which represents no point.

§ 10.2 The Line. Points of π, as we have seen, are the one-dimensional linear subspaces of $V^3(F)$. Lines of π may similarly be thought of as the two-dimensional linear subspaces of $V^3(F)$. (See also Exercise 1.13.) That is, if $\mathbf{P}_1$, $\mathbf{P}_2$, are two vectors of $V^3(F)$ corresponding to points P_1, P_2 of π respectively, then the set of all $\lambda_1\mathbf{P}_1 + \lambda_2\mathbf{P}_2$ (λ_1, λ_2 scalars) defines a projective line. Systematising these ideas as an axiom, we have:

*HC*2. The line P_1P_2 joining two given distinct points $P_1(x_1, y_1, z_1)$ and $P_2(x_2, y_2, z_2)$ consists of variable points $P(x, y, z)$ for which there are *values of the ratio* $\lambda_1 : \lambda_2$ such that

$$\mathbf{P} = \lambda_1 \mathbf{P}_1 + \lambda_2 \mathbf{P}_2,$$

that is

$$\begin{cases} x = \lambda_1 x_1 + \lambda_2 x_2 \\ y = \lambda_1 y_1 + \lambda_2 y_2 \\ z = \lambda_1 z_1 + \lambda_2 z_2. \end{cases}$$

While the operations $+$, $\cdot$ relate to an abstract vector field, we may conceive of them as behaving like ordinary arithmetical multiplication and addition. Indeed, for general practical purposes, we may imagine F to be the field of complex numbers (or the subfield of real numbers). More detailed comment on Real and Complex Geometry is offered later in this Chapter.

Assumption *HC*2 is similar to incidence Axiom *PP*1, i.e. points P are dependent on P_1, P_2.

Now $\mathbf{P}$ may be equivalently written $-\lambda_3\mathbf{P}_3$. Accordingly, three points $\mathbf{P}_i (i = 1, 2, 3)$ are collinear if there exist scalar multipliers $\lambda_i (i = 1, 2, 3)$ for which

$$\sum_{i=1}^{3} \lambda_i \mathbf{P}_i = \mathbf{O} \tag{i}$$

which is a linear dependence relation.

Next, suppose that four coplanar points $\mathbf{P}_i$ ($i = 1, \ldots, 4$) and four scalars λ_i ($i = 1, \ldots, 4$) can be found satisfying the linear dependence relation

$$\sum_{i=1}^{4} \lambda_i \mathbf{P}_i = \mathbf{O} \tag{ii}$$

From (ii), we have

$$\left.\begin{aligned} \lambda_1\mathbf{P}_1 + \lambda_2\mathbf{P}_2 &= -(\lambda_3\mathbf{P}_3 + \lambda_4\mathbf{P}_4) \\ &= \mathbf{M} \text{ (say)} \end{aligned}\right\} \tag{ii'}$$

This means that M is the point of intersection of the lines P_1P_2 and P_3P_4 (i.e. M is the only point common to the two lines and occurs for certain values of λ_1, λ_2 and λ_3, λ_4). Thus, (ii′) is the analytical expression of Axiom *PP*2.

Verification of Axiom *PP*3 analytically is left to the reader as an exercise.

Accepting this verification as a fact, we see that on the basis of *HC*1 and *HC*2 we have constructed an analytical model of the projective plane, which may be referred to as the *projective number plane* over the field F. (Intuition tells us that this model does indeed describe the familiar (though augmented) Euclidean plane when F is the real field.)

Before proceeding to obtain the equation of the line defined in *HC*2, let us emphasise that, for given vectors **P**, **Q**, the point represented by, say, $\mathbf{P} + 3\mathbf{Q}$, depends entirely on the ratio $\lambda_1 : \lambda_2$ (in *HC*2). For example, if $F = R$, suppose $P \equiv (1, 2, -1)$, $Q \equiv (2, -1, 3)$, then

$$(1, 2, -1) + 3(2, -1, 3) = (7, -1, 8) \quad (\lambda_1 : \lambda_2 = 1 : 3)$$

whereas $4(1, 2, -1) + 3(2, -1, 3) = (10, 5, 5)$ $(\lambda_1 : \lambda_2 = 4 : 3)$
i.e. two different points, yet $4(1, 2, -1) \equiv (1, 2, -1)$.

Eliminate λ_1, λ_2 from the three equations in *HC*2. By elementary determinantal theory, we have the determinantal condition

$$\boxed{\begin{vmatrix} x & y & z \\ x_1 & y_1 & z_1 \\ x_2 & y_2 & z_2 \end{vmatrix} = 0} \tag{1}$$

that is, the linear homogeneous equation in x, y, z,

$$(y_1z_2 - z_2y_1)x + (z_1x_2 - z_2x_1)y + (x_1y_2 - x_2y_1)z = 0$$

which must be the equation of the line P_1P_2.

Conversely, every point with co-ordinates (x, y, z) satisfying (1) lies on the line P_1P_2. Can you prove this statement?

Equation (1) may be more elegantly expressed by replacing the coefficients of x, y, z by l, m, n, thus obtaining the equation of the straight line in the standard form

$$\boxed{lx + my + nz = 0} \tag{1′}$$

or in the matrix form

$$\boxed{\mathbf{L\,X} = \mathbf{O}} \tag{1″}$$

where the row matrix $\mathbf{L} = [l, m, n]$, the transpose $\mathbf{X}^T = [x, y, z]$, and **O** is the 1×1 zero matrix, i.e. in actual fact, the scalar 0.

Observe that the general point with co-ordinates (x, y, z) has been labelled **X** here, rather than **P**, for convenience.

Numbers l, m, n, in (1′) are called *line* (or *tangential*) *co-ordinates* to distinguish them from the *point co-ordinates* x, y, z. Line co-ordinates were the creation of Plücker. Homogeneous co-ordinates in π are then of two types, reflecting the (dual) nature of the two basic elements, point and line, in the plane.

Equation (1′) may be called the *incidence relation* in the plane and can be regarded in two ways:
(a) as the linear *equation of the fixed line* (l, m, n) in the variable point-co-ordinates x, y, z of the *range* of points on it, or
(b) as the linear *equation of the fixed point* (x, y, z) in the variable line-co-ordinates l, m, n of the *pencil* of lines passing through it.

Statement (b) involves a novel concept. Probably the student has not previously imagined that a point could have an equation. What is the equation of the point, say $(1, -2, 3)$? [Answer: $l - 2m + 3n = 0$.] Does the line with co-ordinates $(4, 4, 1)$ pass through the point with co-ordinates $(2, -3, 4)$? [Answer: Yes, since $4 \,.\, 2 + 4 \,.\, -3 + 1 \,.\, 4 = 0$, i.e. (1′) is satisfied.]

Now (a) and (b) are dual statements so that point and line co-ordinates are algebraic duals, i.e. we could call (l, m, n) the *dual co-ordinates* in the plane. Otherwise expressed, equation (1′) is the algebraic statement of the Principle of Duality in [2].

Between the projective plane and its analytical model, the projective number plane over F, the relationship of corresponding ideas may be set out as follows:

Projective plane	*Analytical model*
point P	(x, y, z)
line l	(l, m, n)
P on l	(x, y, z) satisfying $lx + my + nz = 0$
l on P	(l, m, n) satisfying $lx + my + nz = 0$

Observe that if the point with co-ordinates $\mathbf{P}_1 + \lambda \mathbf{P}_2$ lies on the line determined by P_1P_2, then $l(x_1 + \lambda x_2) + m(y_1 + \lambda y_2) + n(z_1 + \lambda z_2) = 0$
i.e. $lx_1 + my_1 + nz_1 + (l\lambda x_2 + m\lambda y_2 + n\lambda z_2) = 0$
i.e. $l\lambda x_2 + m\lambda y_2 + n\lambda z_2 = 0$ (since $lx_1 + my_1 + nz_1 = 0$).
Now, we need to be able to say that this is the same as

$$\lambda(lx_2 + my_2 + nz_2) = 0$$

i.e. $l\lambda = \lambda l$, etc., which is true in a field (and also in a commutative ring). This serves to remind us of the problems which arise if we deal with a less restrictive algebraic system than a field, such as a 'non-commutative field'.

Using *HC*2 along with (1) and (1′), we may deduce the fundamental properties of the line:
(i) A line is determined by *any* two of its points.
(ii) Every linear equation *does* determine a unique line.
(iii) Two lines have a common point.

(iv) Three points (x_i, y_i, z_i) $(i = 1, 2, 3)$ are collinear if

$$\begin{vmatrix} x_1 & y_1 & z_1 \\ x_2 & y_2 & z_2 \\ x_3 & y_3 & z_3 \end{vmatrix} = 0.$$

(v) Three lines $l_i x + m_i y + n_i z = 0$ $(i = 1, 2, 3)$ are concurrent if

$$\begin{vmatrix} l_1 & m_1 & n_1 \\ l_2 & m_2 & n_2 \\ l_3 & m_3 & n_3 \end{vmatrix} = 0.$$

As these present no difficulties, their derivations are left as private exercises. (Analytical methods using homogeneous co-ordinates do not differ in principle from the methods with which the reader is familiar in handling ordinary Cartesian co-ordinates.)

Reverting to (1′), notice that it is satisfied by (0, 0, 0), that is to say, every line would have to pass through (0, 0, 0), and our basic notions of Geometry would be violated, if we did not exclude (0, 0, 0).

§ **10.3 The Triangle of Reference.** (Fig. 12) Among the homogeneous co-ordinates of points in π there are four specially simple sets, namely

$$\begin{cases} X \equiv (1, 0, 0) \\ Y \equiv (0, 1, 0) \\ Z \equiv (0, 0, 1) \\ U \equiv (1, 1, 1) \end{cases}$$

which represent respectively the vertices X, Y, Z of a triangle called the *triangle of reference*, and the *unit point*, U, a characteristic of which is that all its co-ordinates are the same ($U \notin XY, \notin YZ, \notin ZX$); X, Y, Z, U determine a *co-ordinate system*. As line co-ordinates, these four sets of numbers represent the sides $YZ(x = 0)$, $ZX(y = 0)$, $XY(z = 0)$ of the triangle of reference, and the *unit line* $x + y + z = 0$ respectively. That YZ has equation $x = 0$ is easily seen, since by (1) its equation is

$$\begin{vmatrix} x & y & z \\ 0 & 1 & 0 \\ 0 & 0 & 1 \end{vmatrix} = 0$$

that is $1.x + 0.y + 0.z = 0$, that is, its line co-ordinates are (1, 0, 0).

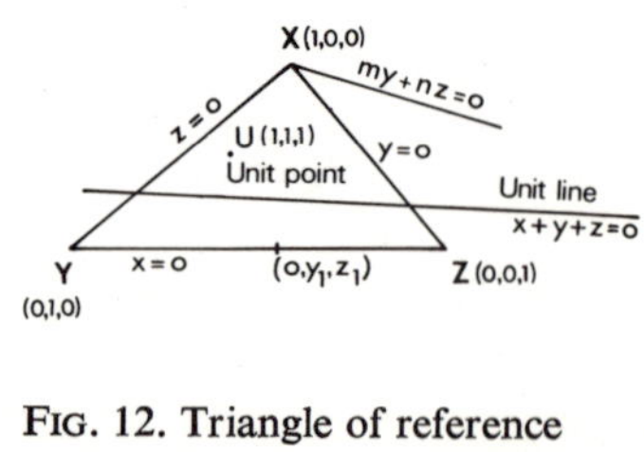

FIG. 12. Triangle of reference

[Some authors prefer to use the square bracket notation [1, 0, 0] to distinguish line co-ordinates from point co-ordinates, but we will always use curved brackets and state their interpretation, where ambiguity may exist. Square brackets are reserved for matrix representation.]

In terms of vector spaces, **X**, **Y**, **Z** form a basis for the vector space $V^3(F)$. Now in a vector space any set of three independent vectors may be taken as a basis. This means that any triangle may be selected as a triangle of reference. Usually, in a problem, it is important to choose a suitable triangle as triangle of reference.

The co-ordinate system can be chosen so that any triangle can be selected as triangle of reference and any point, not lying on a side of the triangle of reference, can be chosen as unit point. Once chosen, they fix the co-ordinate system.

That is, projective co-ordinates can be referred to any quadrangle (no three vertices collinear) as framework of reference. Similarly, a triangle of reference together with any line selected as unit line can be taken as the reference frame for co-ordinates. But the unit point and the unit line cannot, in general, be both simultaneously chosen arbitrarily.

Selection of four points X, Y, Z, U to establish a co-ordinate system does not destroy the projective equivalence of points in [2]. Remember that X, Y, Z, U could be *any* four points in the plane.

Observe that, for the point $P(x, y, z)$ it follows that

$$\begin{aligned}\mathbf{P} &= x\mathbf{X} + y\mathbf{Y} + z\mathbf{Z}\\ &= x\begin{bmatrix}1\\0\\0\end{bmatrix} + y\begin{bmatrix}0\\1\\0\end{bmatrix} + z\begin{bmatrix}0\\0\\1\end{bmatrix}\\ &= \begin{bmatrix}x\\0\\0\end{bmatrix} + \begin{bmatrix}0\\y\\0\end{bmatrix} + \begin{bmatrix}0\\0\\z\end{bmatrix}\\ &= \begin{bmatrix}x\\y\\z\end{bmatrix}.\end{aligned}$$

Properties of π in relation to a given triangle of reference XYZ, which are listed below, ought to be verified as exercises.

(a) Any line through X (1, 0, 0) has an equation of the form $my + nz = 0$, by (1′).

(Note that $m = 0$ and $n = 0$ lead to XY and XZ respectively.)

If $P_1(x_1, y_1, z_1)$ lies on the line, then the equation of XP is $\dfrac{y}{y_1} = \dfrac{z}{z_1}$. (What are the co-ordinates of $XP \cap YZ$? What restrictions are placed on the co-ordinates if (i) $P \in XY$, (ii) $P \in XZ$?) Try particular values of x_1, y_1, z_1 to obtain a variety of lines through X.

(b) Co-ordinates of any point on YZ are expressible as $(0, y_1, z_1)$ which may

be written as, say $(0, 1, \phi) \equiv \left(0, \frac{1}{\phi}, 1\right)$ where $\phi = \frac{z_1}{y_1}$.

Then Y, Z themselves are given by $\phi = 0, \infty$ respectively.

Concerning the difficulties, if any, which students may initially have to overcome in handling homogeneous co-ordinates, Klein [57] p. 87 made the following observation:

> 'Experience shows that operation with homogeneous co-ordinates produces, at least with beginners, something like physical discomfort. I believe that the somewhat indefinite, fluid quality of these quantities, which the arbitrary factor ρ [in our case, k] brings in, is the cause of this feeling, and I hope that such a statement may help to allay this discomfort.'

To Klein's expressed hope I add my own that the presentation given here is fairly painless.

§ **10.4 Subordinate Geometries.** Elements of lower dimension may always be regarded as being embedded in (contained in) higher dimensional spaces. Restricting our discussion to [2], we know that points and lines are embedded *subspaces* of π.

Consider the point $X(1, 0, 0)$. Earlier it was stated that this could be expressed in non-homogeneous form as $\xi = \frac{1}{0} (= \infty)$ and $\eta = \frac{0}{0}$. Now this is meaningless on account of the indeterminate nature of η. But this approach must be faulty since X, being a specific point, is quite determinate. What must be remembered is that X, quite apart from its existence in the wider realm of the plane, has a more restricted existence on $z = 0$ (and on $y = 0$ also).

Considering X as belonging to the subspace $z = 0$, we can ignore this third co-ordinate whose vanishing fixes X on a certain line. Hence, the homogeneous co-ordinates of X on this one-dimensional subspace XY are $(1, 0)$.

Corresponding to the four fundamental points X, Y, Z, U in [2], there are on each line (i.e. in a [1]), three fundamental points $O(0, 1)$, $I(1, 0)$ and $U(1, 1)$, having non-homogeneous co-ordinates (*parameters*) $0, \infty, 1$ respectively, where, for parameter, we have chosen the ratio $\frac{\textit{first co-ordinate}}{\textit{second co-ordinate}}$. For example, on $z = 0$ (i.e. third co-ordinate zero) we would have $O \equiv Y$, $I \equiv X$ and $U \equiv (1, 1, 0)$.

§ 11 APPLICATIONS OF ANALYTICAL METHODS TO BASIC THEOREMS

EXAMPLE 4 (*Desargues' Theorem*):
Solution: From Figs 5, since $\mathbf{V} = \lambda\mathbf{A} + \lambda'\mathbf{A}'$ (say) may be expressed more simply as $\mathbf{V} = \mathbf{A} - \mathbf{A}'$ (co-ordinate vectors of $\mathbf{A}$, $\mathbf{A}'$ being indeterminate to within a scalar multiplier), we have

$$\begin{aligned} \mathbf{V} &= \mathbf{A} - \mathbf{A}' \\ &= \mathbf{B} - \mathbf{B}' \qquad \text{also} \\ &= \mathbf{C} - \mathbf{C}' \qquad \text{also} \end{aligned}$$

$$\therefore \quad \begin{cases} \mathbf{B} - \mathbf{C} = -\mathbf{C}' + \mathbf{B}' = \mathbf{L} \\ \mathbf{C} - \mathbf{A} = -\mathbf{A}' + \mathbf{C}' = \mathbf{M} \\ \mathbf{A} - \mathbf{B} = -\mathbf{B}' + \mathbf{A}' = \mathbf{N} \end{cases}$$

whence $$\mathbf{L} + \mathbf{M} + \mathbf{N} = \mathbf{O}$$

i.e. **L**, **M**, **N** are linearly dependent (refer to (i), § 10.2),
i.e. L, M, N are collinear (Desargues line).

Note: (i) Since this proof is valid in both [2] and [3] (depending on whether the co-ordinates of the points refer to the plane or to space—see § 12), assumptions made for the vectors have wider implications than the Axioms of Incidence have for the plane.

(ii) Reasoning in reverse order, one easily deduces the Converse of Desargues' Theorem.

EXAMPLE 5 (The Harmonic Construction: Construction of the Fourth Harmonic Point): Let A, B, C be three distinct collinear points. We shall construct a unique point D collinear with A, B, C which bears a special relation to A, B, C. Through A, B, C draw arbitrary lines a, b, c and define $a \cap b = P$, $b \cap c = Q$, $c \cap a = R$. Let $AQ \cap BR = S$, $PS \cap AB = D$ (Fig. 13). (We then say that D is the *fourth harmonic point* of A, B, C, or the *harmonic conjugate* of C w.r.t. A, B. Note that A, B, C, D form a harmonic set (Fig. 3(b)), i.e. $H(AB, CD)$ holds, a result known to Pappus [80] pp. 677–678.)

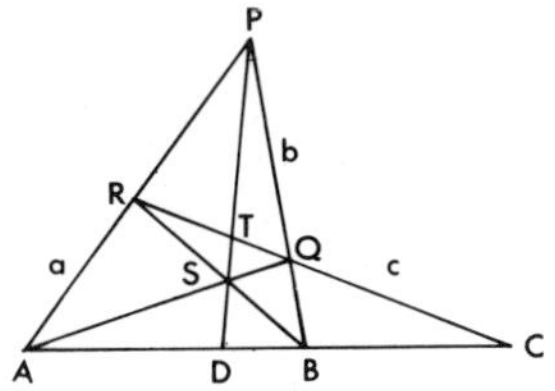

FIG. 13. Fourth harmonic point

Solution: Choose, say, ABQ as triangle of reference with $A \equiv (1, 0, 0)$, $B \equiv (0, 1, 0)$, $Q \equiv (0, 0, 1)$. Let $C \equiv (1, \lambda, 0)$, $S \equiv (1, 0, \mu)$ $(\lambda \neq 0, \mu \neq 0)$.

Then $\mathbf{C} + \mu\,\mathbf{Q} = \mathbf{S} + \lambda\,\mathbf{B} = \begin{bmatrix} 1 \\ \lambda \\ \mu \end{bmatrix} = \mathbf{R}$ $\quad (\because R = CQ \cap SB)$

and $\mathbf{R} - \mathbf{A} = \lambda\mathbf{B} + \mu\mathbf{Q} = \begin{bmatrix} 0 \\ \lambda \\ \mu \end{bmatrix} = \mathbf{P}$ $\quad (\because P = AR \cap BQ$ and BQ is $x = 0)$

$$\therefore \quad \mathbf{S} - \mathbf{P} = \mathbf{A} - \lambda\mathbf{B} = \begin{bmatrix} 1 \\ -\lambda \\ 0 \end{bmatrix} = \mathbf{D}$$

$(\because D = PS \cap AB$ and AB is $z = 0)$

$$\text{i.e.} \quad \mathbf{C} = \begin{bmatrix} 1 \\ \lambda \\ 0 \end{bmatrix}, \mathbf{D} = \begin{bmatrix} 1 \\ -\lambda \\ 0 \end{bmatrix}.$$

Equivalently, $C \equiv (1, \lambda, 0)$, $\mathrm{D} \equiv (1, -\lambda, 0)$.

Remarks: (1) It is a characteristic feature of a pair of harmonic conjugates like C, D that their co-ordinates differ in form by a minus sign only. In terms of **A**, **B** we have $\mathbf{C} = \mathbf{A} + \lambda\mathbf{B}$, $\mathbf{D} = \mathbf{A} - \lambda\mathbf{B}$.

(2) Of course, by Remark 1, C is the harmonic conjugate of D w.r.t. A, B, i.e. a similar construction based on A, B, D yields C. Further, A, B are harmonic conjugates w.r.t. C, D. (See Example 5(a) and § 5.43.)

(3) Fig. 13 is based on the complete quadrangle $ABQR$. The property contained in this construction is known as the *harmonic property of the complete quadrangle ABQR*, which states (cf. § 5.43)
Any two vertices of a complete quadrangle harmonically separate the diagonal point on their side and the point of intersection of their side and the line joining the other two diagonal points.
This property is shared by other sets of four collinear points. For example, suppose $PS \cap QR = T$, then for the quadrangle $PSQR$, $H(PS, TD)$ is valid. Other harmonic sets may be simply constructed in the figure.

(4) It is clearly understood that if A, B, C are distinct, then $H(AB, CD)$ implies that C, D are distinct, i.e. that in every line there must be at least four points. (Fano's finite plane—Fig. 2(c), § 5.6 and Chapter 10—is an exception. In Fano's plane, the co-ordinate field is $GF(2)$ in which $-1 \equiv 1$ (i.e. $-\lambda \equiv \lambda$) so that $D = C$.)

What is the dual statement of that in Example 5?

(5) Occasionally students enquire: Must C lie 'between' A and B? Geometrically, the question is meaningless since axioms of order have been

excluded, i.e. 'betweenness' is a concept alien to the projective plane as defined. Algebraically, the question is meaningless since the field of co-ordinates is abstract and unordered, i.e. it constitutes a wider realm than the field of real numbers (which is ordered).

It is now time for us to prove the statement in Remark (2).

EXAMPLE 5(a): $H(AB, CD) \Rightarrow H(CD, AB)$,
i.e. if C, D are harmonic conjugates w.r.t. A, B
then A, B ,, ,, ,, ,, C, D.

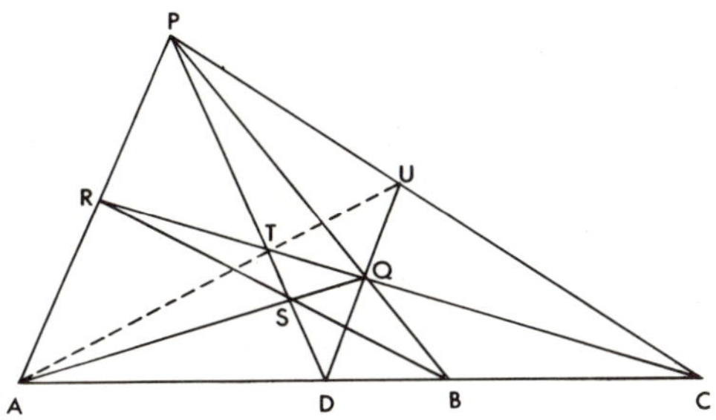

FIG. 13. (a)

Solution 1 (*Geometrical*): In Fig. 13, join CP. Let $U = DQ \cap CP$ (Fig. 13(a)).
Now BCD is the Desargues line for the triangles SRT, QPU.

$\therefore$ $SRT \overset{A}{\barwedge} QPU$ (Converse of Desargues' Theorem)

i.e. A, T, U are collinear
$\therefore$ $H(CD, AB)$ (from the quadrangle $PTQU$).

Solution 2 (*Algebraic*): From Example 5, we have

$$\mathbf{C}+\mathbf{D}=\begin{bmatrix}2\\0\\0\end{bmatrix},\quad \mathbf{C}-\mathbf{D}=\begin{bmatrix}0\\2\lambda\\0\end{bmatrix}.$$

Since we are dealing with homogeneous co-ordinates, the column vectors represent the points A, B respectively. That is to say, in terms of **C**, **D** the representations of A and B differ only by a sign.

Theorem 8 (*Enriques' Uniqueness Theorem*): The fourth harmonic point is unique.

(That is, with the prescribed fourth harmonic point construction of Example 5,

(i) *D is independent of the choice of the arbitrary lines a, b, c,*
(ii) *D is independent of the plane in which the construction is made.*)

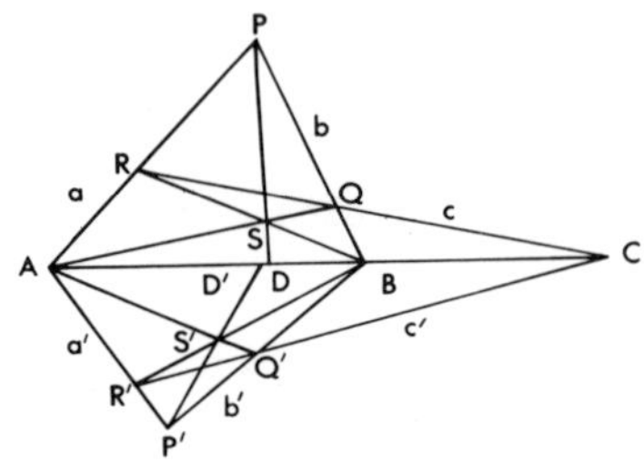

FIG. 14. Enriques' theorem

Proof: Duplicate the fourth harmonic point construction in the same, or another, plane (Fig. 14). By the Converse of Desargues' Theorem applied to the triangles PQR, $P'Q'R'$,

$$PP',\ QQ',\ RR' \text{ are concurrent} \qquad \text{(i)}$$

By the Converse of Desargues' Theorem applied to the triangles RQS, $R'Q'S'$

$$RR',\ QQ',\ SS' \text{ are concurrent} \qquad \text{(ii)}$$

$$\therefore\ PP',\ RR',\ SS' \text{ are concurrent, by (i), (ii).}$$

Apply Desargues' Theorem to the triangles PRS, $P'R'S'$.
$\therefore$ B, A, D'' are collinear, where $D'' = RS \cap R'S'$,

$$\text{i.e. } D' = D = D''$$

$$\therefore\ D \text{ is unique.}$$

Note: Repetition of the construction of the fourth harmonic point leads to the harmonic sequence (§ 13).

EXAMPLE 6 (*Pappus' Theorem*): If A, B, C; A', B', C' are two sets of collinear points, then $AB' \cap A'B$; $BC' \cap B'C$; $CA' \cap C'A$ are collinear (*Pappus line*).

Proof: Extend the Pappus configuration (Fig. 7) to Fig. 7(a) where Y, Z are arbitrary points on $A'B'$, AB respectively and $X = AB \cap A'B'$.
Take XYZ as triangle of reference.

Let $\begin{cases} A \equiv (\alpha, 0, 1);\ B \equiv (\beta, 0, 1);\ C \equiv (\gamma, 0, 1) \\ A' \equiv (\alpha', 1, 0);\ B' \equiv (\beta', 1, 0);\ C' \equiv (\gamma', 1, 0). \end{cases}$

$\therefore$ the equation of BC' is, on using (1) and evaluating the determinant,

$$-x + \gamma' y + \beta z = 0.$$

Similarly, the equation of $B'C$ is $-x + \beta' y + \gamma z = 0$.

Solving, we obtain $L \equiv (\beta\beta' - \gamma\gamma', \beta - \gamma, \beta' - \gamma')$
Likewise, $M \equiv (\gamma\gamma' - \alpha\alpha', \gamma - \alpha, \gamma' - \alpha')$
$N \equiv (\alpha\alpha' - \beta\beta', \alpha - \beta, \alpha' - \beta')$.
Clearly (by adding rows) L, M, N satisfy the determinantal relation (iv), § 10.2.
$\therefore$ L, M, N are collinear (Pappus line).
[Or, $\mathbf{L} + \mathbf{M} + \mathbf{N} = \mathbf{O}$ as in Desargues' Theorem, Example 5.]

Question: What is the equation of the Pappus line?

Note: (i) Eliminating z from the equations of BC, $B'C'$ to obtain the ratio $x : y$ of the co-ordinates for the point of intersection, we derive the term $(\gamma\beta - \beta\gamma)z$ which can only vanish if $\beta\gamma = \gamma\beta$, i.e. if the commutative law of multiplication holds. For Pappus' Theorem to be valid, this law must be valid, i.e. Pappus' Theorem is violated when the co-ordinates belong to a non-commutative field. No such problem arises here, however, since by definition the co-ordinates form a field. Other analytical proofs of Pappus' Theorem, which ought to be consulted, may be found in, say, Baker [8], Faulkner [38] and Fishback [40].

(ii) While on the subject of non-commutative fields, observe that it is incorrect to give the solution of $ax = b$ as $x = \frac{b}{a}$ which could be interpretated as either $a^{-1}b$ or ba^{-1} which, in a non-commutative field, would generally be unequal.

(iii) Other choices of triangle of reference may lead to simpler algebra, though the results may not look so balanced. (See Exercise 3.16.)

EXAMPLE 7 (Co-ordinates of the Vertices of the Complete Quadrangle): Show that the co-ordinates of the vertices of a complete quadrangle may be taken as $(\pm 1, \pm 1, \pm 1)$.

Solution: Take EFG, in Fig. 2(a), as triangle of reference so that $E \equiv (1, 0, 0)$,

$$F \equiv (0, 1, 0),\ G \equiv (0, 0, 1).$$

$$\text{Let } D \equiv (1, 1, 1)$$

$$\therefore \begin{cases} B \equiv (1, 1, 1 + \lambda) \\ A \equiv (1, 1 + \mu, 1) \quad (\lambda, \mu, \nu \neq 0, -1) \\ C \equiv (1 + \nu, 1, 1). \end{cases}$$

Now GCA is given by $\begin{vmatrix} x & y & z \\ 1 & 1+\mu & 1 \\ 0 & 0 & 1 \end{vmatrix} = 0$ or $\begin{vmatrix} x & y & z \\ 1+\nu & 1 & 1 \\ 0 & 0 & 1 \end{vmatrix} = 0$

i.e. $y = (1 + \mu)\, x$ or $y = \dfrac{x}{1 + \nu}$

$\therefore (1 + \mu)(1 + \nu) = 1$ (i)

Similarly $(1 + \nu)(1 + \lambda) = 1$, $(1 + \lambda)(1 + \mu) = 1$ (ii)

Combining (i), (ii), we have $\{(1 + \lambda)(1 + \mu)(1 + \nu)\}^2 = 1$

$\therefore (1 + \lambda)(1 + \mu)(1 + \nu) = \pm 1.$

But, from (i), $(1 + \lambda)(1 + \mu)(1 + \nu) = 1 + \lambda \neq 1$ $(\lambda \neq 0)$

$\therefore 1 + \lambda = -1.$

Similarly $1 + \mu = 1 + \nu = -1$ (i.e., $\lambda = \mu = \nu = -2$)

$\therefore C \equiv (-1, 1, 1)$, $A \equiv (1, -1, 1)$, $B \equiv (1, 1, -1)$, $D \equiv (1, 1, 1)$.

Another way of expressing the result of Example 7 is to say that there exists a system of reference in the plane such that four given points of general position, no three collinear, have co-ordinates $(\pm 1, \pm 1, \pm 1)$.

§ 11.1 Real and Complex Geometry. When the co-ordinates x, y, z of a point belong to the field of real numbers R (or complex numbers C) the points are called *real points* (or *complex points*) and the resulting Geometry is called *Real Projective* (*Plane*) *Geometry* (or *Complex Projective* (*Plane*) *Geometry*), symbolised as $P_2(R)$ (or $P_2(C)$). Similar remarks apply to *real lines* and to *complex lines*. (When the field is $GF(p^k)$, the corresponding finite plane geometry—see Chapter 10—may be given the symbol $P_2(GF(p^k))$). For example $(2, -1, 4)$ is real, whereas $(5, \epsilon, 1)$ is complex (ϵ a complex cube root of unity). However, note that $(ix, iy, iz) = i(x, y, z) \equiv (x, y, z)$, where i is a complex square root of -1, is clearly real for real x, y, z. (Is $(2 + 2i, -1 - i, 4 + 4i)$ a real or complex point? Real, since it is the same point as $(2, -1, 4)$ after taking out $1 + i$ as a factor.)

Since R is a subfield of C, it follows that $P_2(R)$ is obtained from $P_2(C)$ by singling out just the real points and lines, and considering only them. One may think of $P_2(R)$ as being embedded in $P_2(C)$ in such a way that there is a unique correspondence between

real points	and	complex points
(x, y, z)		$(x + 0.i, y + 0.i, z + 0.i) = (z_1, z_2, z_3)$

where $z_i (i = 1, 2, 3)$ are complex numbers.

Points in the Argand (or Gaussian) plane of Analysis (which ought not to be confused with our $P_2(C)$) may be thought of as constituting points of the complex projective line $P_1(C)$ in which the homogeneous co-ordinates are (z_1, z_2) for which $\dfrac{z_1}{z_2} = z = x + iy$. Expressed otherwise, $P_1(C)$ is two-dimensional, whereas the real projective line $P_1(R)$ is one-dimensional. The point at infinity on $P_1(C)$ is the single 'point at infinity' in the Gaussian plane.

Real points and complex points are on the same footing projectively.

Although $P_2(C)$ does not have the same relation to the space of the physical world as does $P_2(R)$, nevertheless it is extremely important because of the added richness of content and technique with which it endows the theory.

Instances of this are that every equation of degree n guarantees us n solutions and, in particular, that any line cuts a conic in two points (both real or both complex, or coincident—see Chapter 7).

Some interesting facts relating to the concepts initiated above are listed herewith:

(i) Every complex line contains at least one complex point and only one real point, e.g. the complex point $(1, \epsilon, \epsilon^2)$ and the unique real point $(1, 1, 1)$ lie on the complex line $x + \epsilon y + \epsilon^2 z = 0$.

(ii) Every real line contains infinitely many pairs of complex points, namely, pairs consisting of the complex point (z_1, z_2, z_3) and its conjugate complex point $(\bar{z}_1, \bar{z}_2, \bar{z}_3)$. Can you prove this statement?

E.g. (a) points $(1, \epsilon, \epsilon^2)$, $(1, \epsilon^2, \epsilon)$ lie on the real (unit) line $x + y + z = 0$,

(b) points $(1, i, 0)$, $(1, -i, 0)$ lie on the real line $z = 0$.

Great significance will subsequently be attached to this latter example. What, in fact, are the duals of (i) and (ii)?

§ 12 CO-ORDINATES IN N DIMENSIONS

Quite clearly, our introduction of homogeneous co-ordinates in [2] by means of a vector space is readily extendible to $[n]$.

Given an $[n]$ as described geometrically in § 6.3, associate with each point P of $[n]$ an infinite set of ordered $(n + 1)$ – tuples $x_1, x_2, \ldots, x_{n+1}$ of elements of a field F(abstract), that is, the vectors in a vector space $V^{n+1}(F)$, such that

(i) to every vector $\mathbf{P}(\neq \mathbf{O})$ corresponds a unique point P in $[n]$;

(ii) if $\mathbf{Q} = k\mathbf{P}$ $(k \neq 0, k \in F)$, then $\mathbf{P}$, $\mathbf{Q}$ represent the same point P in $[n]$.

Excluded by definition as meaningless in this context is the $(n + 1)$ – tuple $(0, 0, \ldots, 0)$ consisting entirely of zeros.

Just as in the plane, so in $[n]$ there exist $P_n(R)$, $P_n(C)$, $P_n(GF(p^k))$ as special cases of $P_n(F)$.

In $[n]$, a linear space is one which is described algebraically by one or more linear equations. The simplest linear space called a *hyperplane* or *prime* (or *flat*) is of dimension $n - 1$ and is defined by the equation

$$\boxed{a_1x_1 + a_2x_2 + \ldots + a_{n+1}x_{n+1} = 0} \tag{1a}$$

where $(x_1, x_2, \ldots, x_{n+1})$ are the homogeneous co-ordinates of a variable point in this $[n - 1]$. (Some authors prefer to represent this point by $(x_0, x_1, \ldots, x_n)$.) Thus, in [2] and [3] a hyperplane is a line and a plane, respectively. Of the $n + 1$ variables in equation (1a), only n are independent.

Substituting the co-ordinates of n distinct points in this equation, we get n equations which can be solved for the ratios $x_1 : x_2 \ldots : x_{n+1}$. Stated otherwise, a hyperplane is uniquely determined by (any) n of its points. Hence, a plane (line) is uniquely determined by any three (two) of its points.

Coefficients $a_1, a_2, \ldots, a_{n+1}$ occurring in (1a) are called *hyperplane* (*prime*) *co-ordinates*. Thus in [2], they are line co-ordinates and in [3] they are *plane co-ordinates*. Equation (1a) establishes a duality between point and hyperplane in $[n]$.

Simplest equations of hyperplanes in $[n]$ are $x_1 = 0, x_2 = 0, \ldots, x_{n+1} = 0$ which determine the 'faces' of a *simplex of reference*. Vertices of this simplex together with a unit point $(1, 1, \ldots, 1)$, consisting of $n + 1$ units, constitute a framework of $n + 2$ points for establishing co-ordinates in $[n]$.

Before leaving this brief outline of the analytical situation in $[n]$, let us consider dual co-ordinates in [3]. Because point and hyperplane are dual elements, it follows that there exist (hyper) plane co-ordinates a_1, a_2, a_3, a_4 forming a dual system to the point co-ordinate system x_1, x_2, x_3, x_4. But we know further that lines are self-dual elements. How should we take this geometrical fact into account when establishing co-ordinate systems? Simply by recalling that a line in [3] may be determined by either (i) two points or (ii) two planes. Accordingly, from the two fixed points, or two fixed planes, whose co-ordinates will be known, we construct new co-ordinates called (*ray*) *line co-ordinates* or (*axial*) *line co-ordinates* respectively in [3]. *Line Geometry*, as this study of properties of lines as basic components in [3] is called, is a stimulating topic but is left for private investigation. Incidentally, it is interesting to learn that Line Geometry, the creation of Plücker (see [143])and Cayley, has been applied to the dynamics of rigid bodies and to the geometry of Einstein's Unified Field Theory.

Case $n = 1$. As a particular case of the above theory, when $n = 1$, we understand how points on a line could be defined in terms of vectors from a two-dimensional vector space $V^2(F)$. Points on the line are given by homogeneous co-ordinates (x_1, x_2) where x_1, x_2 belong to some unspecified field. Non-homogeneous co-ordinates of such points are $x = \dfrac{x_1}{x_2}$. Of course (0, 0) is excluded by definition as meaningless. Base reference points and unit points are, respectively (0, 1), (1, 0) and (1, 1) with non-homogeneous co-ordinates 0, ∞, 1 respectively, as previously remarked at the end of § 10.

§ 13 GEOMETRIC DESCRIPTION OF ADDITION AND MULTIPLICATION

So far, we have *defined* the existence and basic properties of homogeneous co-ordinates in terms of a vector space, and have shown further that these ideas are consistent with the Axioms of Incidence. The co-ordinates are elements of some number field F, generally assumed abstract, in which, of course, division by zero is inadmissible, i.e. ∞ *is not a field element.*

Suppose now we try to establish geometrical constructions corresponding to the algebraic operations connecting elements of the field. Specifically, can we find constructions which give a geometrical meaning to the (abstract) addition and multiplication of elements, namely to $a + b$ and $a.b$ where $a, b \in F$?

Some beautiful and exciting mathematics, which has its origins in the work of von Staudt [115] on 'throws' (in German, *Würfe*), flows from such solutions. Essentially, the technique is to use the points of a line as the raw materials for the constructions. (Though the constructions which follow may appear novel, in the Euclidean plane (Chapter 8) they will be seen to be quite simple.)

Note: In §§ 13.1, 13.3, we take $A \neq I$, $B \neq I$. No point constructed from A and B can coincide with I.

§ 13.1 Addition of Points. Two reference points O, I are chosen on a line l in a given plane π. On l, select the points $\mathbf{A} = \mathbf{O} + a\mathbf{I}$, $\mathbf{B} = \mathbf{O} + b\mathbf{I}$. Then the problem is to construct the point $\mathbf{O} + (a + b)\mathbf{I}$. (See Fig. 15.)

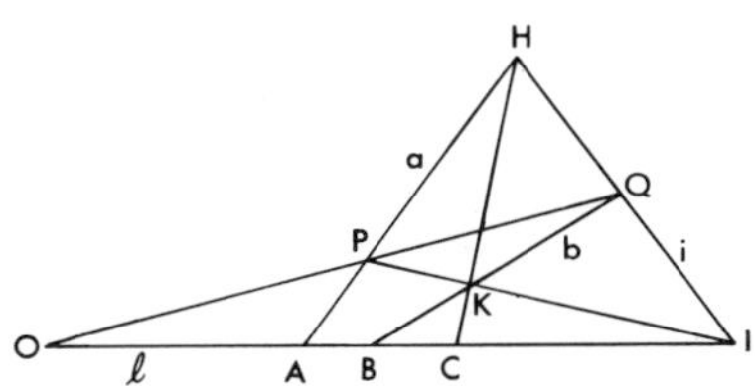

FIG. 15. Addition of points

Choosing the line l to have equation $y = 0$, we have

$$\mathbf{O} = \begin{bmatrix} 0 \\ 0 \\ 1 \end{bmatrix}, \ \mathbf{I} = \begin{bmatrix} 1 \\ 0 \\ 0 \end{bmatrix}, \ \mathbf{A} = \begin{bmatrix} a \\ 0 \\ 1 \end{bmatrix}, \ \mathbf{B} = \begin{bmatrix} b \\ 0 \\ 1 \end{bmatrix}. \text{ Then } \begin{bmatrix} a+b \\ 0 \\ 1 \end{bmatrix}$$

is the point whose construction we desire.

EXAMPLE 8 (Geometric construction of addition):
Solution: Collinear points O, I, A, B are as described above.
Through A, B, I draw coplanar lines a, b, i: $a \cap i = H$, $b \cap i = Q$. Let $OQ \cap AH = P$, $BQ \cap IP = K$, $OI \cap HK = C$.

$$\begin{aligned} \text{Now } \mathbf{P} &= \mathbf{O} + a\mathbf{I} + h\mathbf{H} \text{ since } P \text{ lies on } AH. \\ \therefore\ \mathbf{P} - \mathbf{O} &= a\mathbf{I} + h\mathbf{H}\ (= \mathbf{Q}) \text{ since } PO \cap IH = Q. \\ \text{But } \quad \mathbf{K} &= \mathbf{O} + (a+b)\,\mathbf{I} + h\mathbf{H} \text{ since } PI \cap BQ = K \\ \therefore\ \mathbf{K} - h\mathbf{H} &= \mathbf{O} + (a+b)\mathbf{I} \\ \text{i.e.} \quad \mathbf{C} &= \mathbf{O} + (a+b)\mathbf{I} \text{ since } HK \cap OI = C \\ &= \mathbf{O} + c\mathbf{I} \text{ where } c = a + b. \end{aligned}$$

This geometrical construction gives a pictorial representation of the algebraic operation of addition, i.e. C is the unique *sum* of A and B, as we might say. Symbolically, we may write $C = A + B$ (not vector addition). Since subtraction is the inverse operation to addition, we can obtain B as the unique *difference* of C and A by reversing the construction. Effectively, we have taken OIH as triangle of reference in the Solution.

Remarks:

(i) If $b = -a$, i.e. $c = 0$, then $C = O$, i.e. A, B are harmonic conjugates w.r.t. O, I (Example 5). Symbolically,

$$A + (-A) = O.$$

Students are advised to carry out this and subsequent constructions as an aid to a deeper understanding of the subject matter.

(ii) If $a = b = 1$, then $\mathbf{C} = \mathbf{O} + 2\mathbf{I} = \begin{bmatrix} 2 \\ 0 \\ 1 \end{bmatrix}$ $(2 = 1 + 1$ in the field).

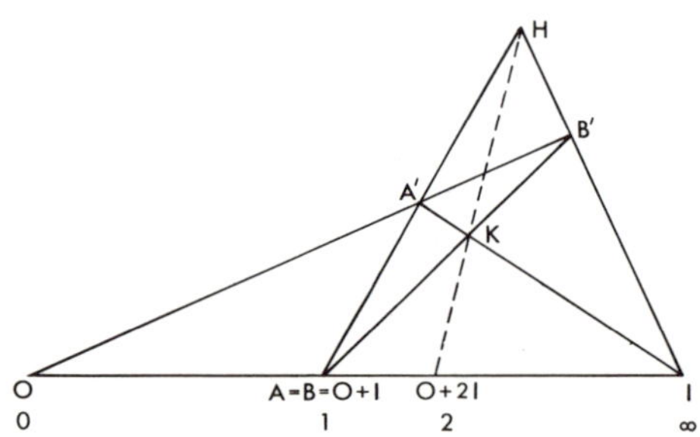

FIG. 15. (a) Construction of P_2

Represent this position of C by P_2. Readers should check that the *construction by which P_2 is derived from O, $A = B$, I is identical with the construction of the harmonic conjugate of O w.r.t. A, I.* That is, $H(OP_2, AI)$ is true (Fig. 15(a). Repetition of the above process eventually leads to the construction of the point $\mathbf{O} + n\mathbf{I} = \begin{bmatrix} n \\ 0 \\ 1 \end{bmatrix}$

How may we construct the point $n\mathbf{O} + \mathbf{I} = \begin{bmatrix} 1 \\ 0 \\ n \end{bmatrix}$, i.e. the point $\begin{bmatrix} \frac{1}{n} \\ 0 \\ 1 \end{bmatrix}$?

(iii) The construction in Fig. 15 is clearly symmetrical w.r.t. A, B, i.e. $\mathbf{O} + (a + b)\,\mathbf{I} = \mathbf{O} + (b + a)\mathbf{I}$. Thus,

$$a + b = b + a.$$

Symbolically.

$$A + B = B + A.$$

Hence, *the commutative law of addition has a geometrical counterpart.*

(iv) Putting $B = O$ (i.e. $b = 0$) in the construction of Fig. 15, we have $C = A$, i.e.,

$$a + 0 = a.$$

Similarly,

$$0 + a = a.$$

In terms of points,

$$A + O = O + A = A.$$

(v) Taken together, (ii) and (iv) imply that on l the point O corresponds to the number 0. Identify the point I with the symbol ∞. Label, for convenience, the points

$$O, A, B, C, I$$

as

$$P_0, P_a, P_b, P_{a+b}, P_\infty.$$

Accordingly, the commutative law of addition, for instance, is expressed geometrically by

$$(P_a + P_b =)\; P_{a+b} = P_{b+a}\; (= P_b + P_a).$$

Recall that ∞ is not an element of the field. Applying this new symbolism to (i), (ii) above, we deduce the validity of $H(P_a, P_{-a}, P_0, P_\infty)$ $(= H(P_1P_{-1}, P_0P_\infty)$ when $a = 1$, Fig. 15(b)) and $H(P_0P_2, P_1P_\infty)$.

(vi) Given three collinear points P_a, P_b, P_c : $\mathbf{P}_a = \mathbf{O} + a\mathbf{I}$, $\mathbf{P}_b = \mathbf{O} + b\mathbf{I}$, $\mathbf{P}_c = \mathbf{O} + c\mathbf{I}$, we can construct, in the manner prescribed in Fig. 15, the points P_{a+b}, P_{b+c} and from these the coincident points $P_{(a+b)+c} = P_{a+(b+c)}$

whence follows the associative law of addition

$$(a + b) + c = a + (b + c).$$

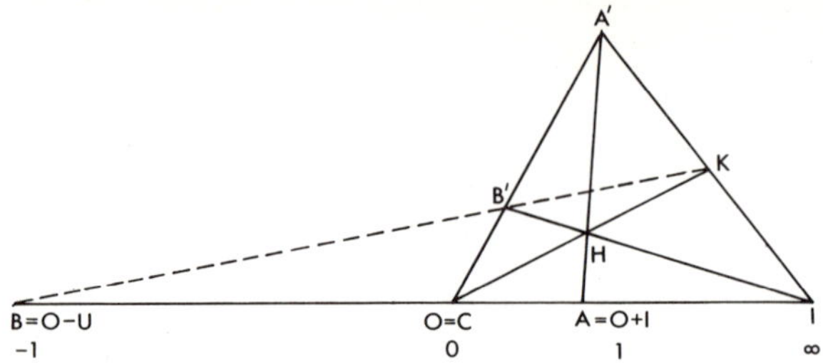

FIG. 15. (b) Construction of P_{-1}

Summarising the situation reached, we see that
the structure of an additive Abelian group is revealed in a fairly simple way in the projective plane, provided that we exclude the symbol ∞ *and the point* P_∞.

Furthermore, notice (Exercise 5.26) that C belongs to a specialised quadrangular set, namely

$$Q\begin{pmatrix} O & A & I \\ C & B & I \end{pmatrix}.$$

Later (Exercise 3.13), it will be seen that the construction of C is independent of the choice of lines through A, B, I.

§ **13.2 Harmonic Sequence.** Anticipating the result of Example 9 that the point U has the number 1 associated with it, we may construct a harmonic sequence (Fig. 15(c)). From the points P_2, P_{-1} just introduced (Figs. 15(a), 15(b)), we construct in succession the harmonic sets

$$H(P_1P_3,\ P_2P_\infty),\ H(P_2P_4,\ P_3P_\infty),\ \ldots,\ H(P_{n-1}P_{n+1},\ P_nP_\infty),\ \ldots$$

and

$$H(P_0P_{-2},\ P_{-1}P_\infty),\ H(P_{-1}P_{-3},\ P_{-2}P_\infty),\ \ldots,\ H(P_{-n+1}P_{-n-1},\ P_{-n}P_\infty),\ \ldots$$

leading to the *harmonic sequence* (based on P_0, P_1, P_∞)

$$\ldots,\ P_{-3},\ P_{-2},\ P_{-1},\ P_0,\ P_1,\ P_2,\ P_3,\ \ldots,\ P_\infty,$$

an idea going back to Möbius in 1827. Plainly, the same harmonic sequence is determined equally well by any three of its points. Taken for granted in these harmonic constructions is the assumption (Fano's Axiom) that the diagonal points of a complete quadrangle are not collinear. Another assumption, due also to Fano ('*Sui postulati fondamentali della geometria proiettiva in uno spazio lineare a un numero qualunque di dimensioni*', *Giornale di Matematiche*, 30 (1892) p. 124), requires the existence of a harmonic sequence with an infinite number of points. (Fano's words: '*Esiste* una *serie armonica non mai rientrante in sè stessa*'. That is to say, if the number of points on a line is finite, then ultimately the harmonic construction will produce no new points, i.e. the harmonic sequence will be *re-entrant*.)

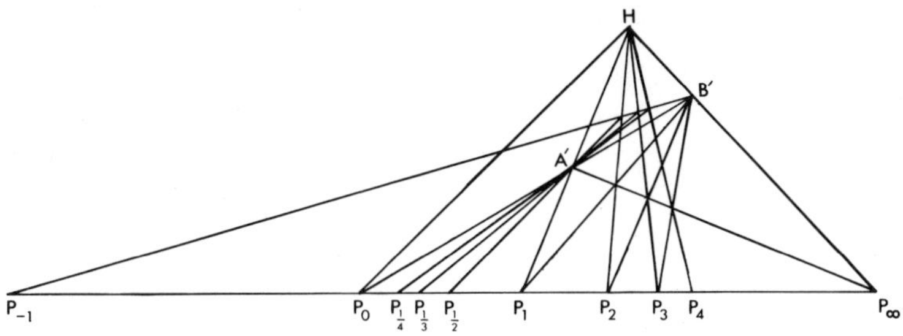

FIG. 15. (c) Harmonic sequence. Möbius net of rationality

§ **13.3 Multiplication of Points.** Next we turn to the problem of constructing the point corresponding to the multiplication $a.b$ (or simply ab), given the points $\mathbf{O} + a\mathbf{I}$, $\mathbf{O} + b\mathbf{I}$. That is, we want to establish the point $\mathbf{O} + ab\mathbf{I}$, i.e. the point $\begin{bmatrix} ab \\ 0 \\ 1 \end{bmatrix}$.

EXAMPLE 9 (Geometrical construction of multiplication):

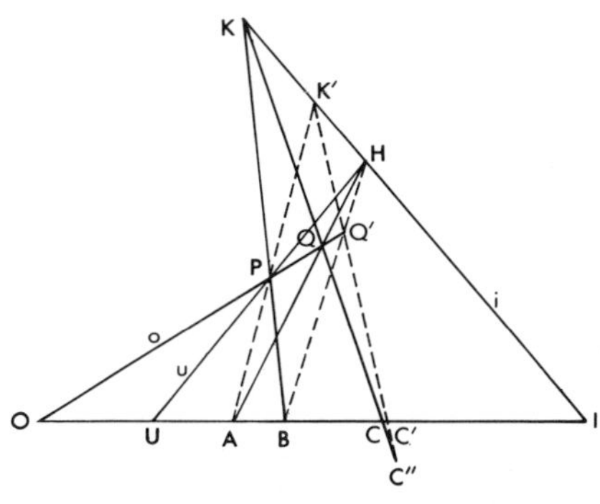

FIG. 16. Multiplication of points. Hilbert's theorem

Solution: Collinear points O, I, A, B, U ($\mathbf{U} = \mathbf{O} + \mathbf{I}$) are given as in Fig. 16. (Recall that addition is independent of U.) Through O, I, U coplanar lines o, i, u are drawn:

$$o \cap u = P,\ u \cap i = H,\ AH \cap o = Q,\ BP \cap i = K,\ KQ \cap OI = C.$$

$$\text{Allow} \quad \begin{cases} \mathbf{A} = \mathbf{O} + a\mathbf{I} \\ \mathbf{B} = \mathbf{O} + b\mathbf{I} \\ \mathbf{Q} = \mathbf{O} + a\mathbf{I} + h\mathbf{H} \end{cases}$$

i.e. effectively OIH is taken as triangle of reference.

$$\therefore \quad \mathbf{P} = a\mathbf{O} + a\mathbf{I} + h\mathbf{H} \qquad \text{since } UH \cap OQ = P$$
$$\therefore \quad \mathbf{K} = (a\mathbf{O} + a\mathbf{I} + h\mathbf{H}) - a(\mathbf{O} + b\mathbf{I}) \text{ since } PB \cap HI = K$$
$$= (a - ab)\mathbf{I} + h\mathbf{H}$$
$$\therefore \quad \mathbf{C} = (\mathbf{O} + a\mathbf{I} + h\mathbf{H}) - \{(a - ab)\mathbf{I} + h\mathbf{H}\}$$
$$\text{since } KQ \cap OI = C$$
$$= \mathbf{O} + ab\mathbf{I}$$
$$\text{i.e.} \quad \mathbf{C} = \mathbf{O} + c\mathbf{I} \qquad \text{where } c = ab.$$

This construction provides us with a geometrical illustration of the operation of multiplication, i.e. C is the unique *product* of A and B, which we may express symbolically as $C = A\,.B$, i.e. $P_{ab} = P_a P_b$.

Reversing the procedure, we obtain a construction (corresponding to the quotient of two numbers) for the unique *quotient* A of C and B, written $A = C/B$, i.e. $P_a = P_{ab}/P_b$.

Remarks:

(i) If $ab = 1$ (i.e. $b = a^{-1}$, $a = b^{-1}$), then $C = U$.
In terms of points, we have
$$AA^{-1} = A^{-1}A = U \text{ where } A \neq O \text{ and } A^{-1} = U/A.$$
See Fig. 16(a) for the construction of $P_{\frac{1}{2}}$ (i.e. $a = 2$, $b = \frac{1}{2}$).

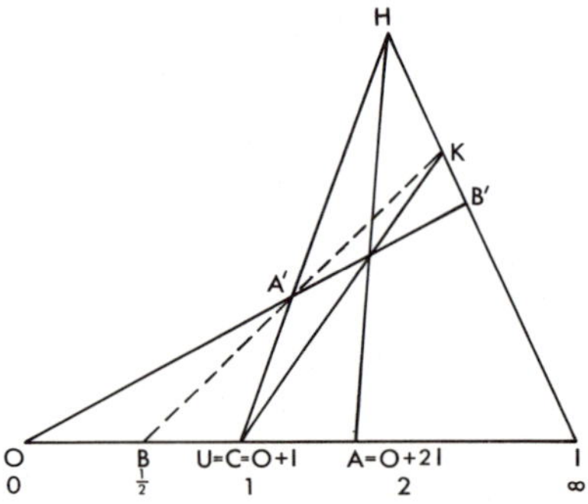

FIG. 16. (a) Construction of $P_{\frac{1}{2}}$

(ii) Setting $B = O$ in Fig. 16 (i.e. $b = 0$), we have $C = O$ also, i.e.,
$$a0 = 0.$$
Similarly,
$$0a = 0.$$
That is, for points, we have
$$AO = OA = O.$$

(iii) With $B = U$ in Fig. 16 (i.e. $b = 1$) we have $C = A$, i.e.

$$a1 = a.$$

Similarly

$$1a = a.$$

In terms of points

$$AU = UA = A.$$

(iv) Thus (iii) implies that U corresponds to the symbol 1 in the number field. In Fig. 16, write

$$P_1, P_{ab}$$

for

$$U, C.$$

(v) Given the collinear points P_a, P_b, P_c, we may construct the points P_{ab}, P_{bc} and hence the coincident points

$$P_{a(bc)} = P_{(ab)c},$$

corresponding to the associative law of multiplication

$$a(bc) = (ab)c.$$

Likewise, we may establish geometrical interpretations of the distributive laws, namely

$$a(b + c) = ab + ac$$
$$(b + c)a = ba + ca.$$

Summarising the position, then, we see that (with P_∞ omitted) constructions have been found to illustrate all the laws of algebra for a field, except the commutative law of multiplication. In carrying out the proofs of some of the above results, Desargues' Theorem is necessary. Consequently, we may say that

a division ring manifests itself in the geometry of a Desarguesian plane.

In a moment, we will pursue the question of the commutative law of multiplication. Before doing that, however, observe (Exercise 5.27) that C in Fig. 16 is a member of the quadrangular set

$$Q\begin{pmatrix} O & A & U \\ I & B & C \end{pmatrix}.$$

It is a worthwhile exercise to demonstrate that the existence of C is independent of the choice of lines through O, U, I.

§ 13.31 Hilbert's Theorem.

Theorem 9 (*Hilbert's Theorem*)*:* If the commutative law of multiplication $ab = ba$ holds, then Pappus' Theorem holds (and conversely).

Proof: In the manner prescribed in Fig. 16, construct

$$\mathbf{C} = \mathbf{O} + ab\mathbf{I}$$
$$\mathbf{C}' = \mathbf{O} + ba\mathbf{I}.$$

Let $KC \cap K'C' = C''$.
Now, if $ab = ba$ then $C = C' = C''$,
i.e. B, C'', A are collinear.
But $BC''A$ is the Pappus line of the Pappus triads H, K, K'; P, Q', Q.
$\therefore$ the multiplicative commutativity of the numbers implies the truth of Pappus' Theorem.

The Converse is left as an exercise (Exercise 3.15).

Hilbert's Theorem reveals the enormously powerful undercurrents of Pappus' Theorem. It signifies that a plane is Pappian if and only if the co-ordinates are elements of a field. Geometrically, the

algebra of a field is realised in the structure of a Pappian plane.

Without the validity of the commutative law of multiplication (i.e. in a division ring), the plane is non-Pappian but Desarguesian which, though structurally incomplete, is a rich mine for exploration.

§ 13.4 Field of Points. Introduction of Co-ordinates in the Line and Plane. Briefly, what we did in the immediately preceding theory was to give a geometrical description of the laws of an algebraic field.

To achieve this, we exhibited the construction of points, which, combined in certain defined ways, possessed all the formal appearances of a field. Consequently, the points so constructed could be regarded as constituting a *field of points*. Just as the symbol ∞ is excluded from the algebraic field, so the point I ($= P_\infty$), with which it is associated, is excluded from the geometric field.

Now, it is possible to construct the field of points without appealing to algebraic methods and, using only the synthetic background of Chapters 1 and 2, with Desargues' and Pappus' Theorems added as axioms, if necessary. Let us examine this approach briefly.

No algebraic field is now given. The problem is to construct a field by purely geometrical means. Constructions are the same as those given previously, i.e. multiplication and addition of points are defined as before. Proofs of the field properties are now synthetic, however. As before, the existence of a division ring of points depends on the validity of Desargues' Theorem in the plane, while the division ring takes on the more rigid structure of a field if Pappus' Theorem holds.

Summarising the situation, we say that the geometry of the Pappian plane implies the existence of a field of points, whose laws of combination are the laws governing an algebraic field. Otherwise expressed *the concept of number, and of the additive and multiplicative combinations of number, is inherent in the synthetic axioms.*

Fields associated with different lines are isomorphic.

Fuller details of the geometric constructions and their algebraic relationships may be found in, say, Baker [7], Coxeter [27], Fishback [40], Hodge and Pedoe [51] (which contains diagrams of the special cases), O'Hara and Ward [79], Robinson [92], Seidenberg [98], or Veblen and Young [113]. Methods of approach to the geometric-algebraic relationships vary greatly in these books.

The field of points, which has been constructed by purely geometric means, can be exploited to introduce co-ordinates for the line. Of course, there is no need to do this from our point of view, as we have defined homogeneous co-ordinates in terms of vector spaces and thus endowed them with a formidable generality. Nevertheless, it is worth having a brief look at the other approach as it affords elegant evidence of the power of purely geometric methods.

Our goal in the next section is to examine briefly how real co-ordinates can be associated with points on a projective line.

§ 13.5 Real Projective Line. Let us now go back to the harmonic sequence which was constructed a little earlier. Take the field F to be the field R of real numbers. Then the existence of the infinite harmonic sequence establishes a 1–1 correspondence between the points of the harmonic sequence (excluding P_∞) and the positive and negative integers of R.

If the projective line (on which the harmonic sequence exists) is to resemble the real Euclidean line with which we are familiar, we need to introduce points on the line which can be made to correspond to rational numbers and, eventually, to irrational numbers. First let us see how the rational numbers enter the picture.

Using our previous notation and constructions, we construct points $P_{\frac{1}{2}}, P_{\frac{1}{3}}, P_{\frac{1}{4}}, P_{\frac{1}{5}}, \ldots$ which arise as harmonic conjugates of $P_2, P_3, P_4, P_5, \ldots$ w.r.t. P_1 and P_{-1}. Following Robinson [92], we observe that the three points $P_0, P_\infty, P_{\frac{1}{n}}$ determine the harmonic sequence

$$\ldots P_{-\frac{3}{n}}, P_{-\frac{2}{n}}, P_{-\frac{1}{n}}, P_0, P_{\frac{1}{n}}, P_{\frac{2}{n}}, P_{\frac{3}{n}}, \ldots$$

Ultimately, we arrive at the point $P_{\frac{m}{n}}$ where m, n are positive or negative integers,

$$\text{i.e. } r = \frac{m}{n} \text{ is a rational number.}$$

The set of these rational points is called a *harmonic net* or Möbius' *net of rationality* determined by P_0, P_1, P_∞. (Dually there is a net of rationality determined by three concurrent lines.) (See Fig. 15(c).)

We have therefore established a 1–1 correspondence between the field of rational numbers and the field of points based on P_0, P_1, P_∞ but excluding P_∞.

A basic feature of the real Euclidean line is that points on it can be ordered, i.e. placed in order, so that if any three points A, B, C are given we can say that one of these points is 'between' the other two. No such concept exists in the geometry derived from our Axioms *PP*1–*PP*3. How, then, may we introduce the analogue of this concept into the projective line?

As we shall see in Chapter 5, *any* three distinct collinear points may be made to correspond to *any* three distinct collinear points. Consequently, a relationship involving only three points cannot be invariant and therefore is meaningless in Projective Geometry. On the other hand, the harmonic relation involving a set of *four* distinct collinear points is a projective invariant (as will be shown in Chapter 5—or see Part 1).

In effect, a harmonic set can be exploited in connection with an undefined relation called *separation* (a projective invariant), and the axioms relating to separation are called *Axioms of Order*. These axioms automatically imply an infinite number of points on the line and hence exclude finite geometries from further consideration. A line on which the separation relation has been defined is called an *ordered projective line*.

To complete the real (ordered) projective line, an *Axiom of Continuity* is added to ensure the existence of a point on the line (excepting P_∞), corresponding to a given irrational number. The line (an ordered, continuous projective line) can now, except for P_∞, be co-ordinatised by the real numbers. Indeed, it has all the attributes of the real Euclidean line, except the notion of distance. (The exceptional character of P_∞ is removed by the introduction of homogeneous co-ordinates.)

Co-ordinates may also be introduced into the plane and into higher space on the basis of the constructions for addition and multiplication. This is a story which is told elsewhere. (Consult the texts mentioned a little earlier.) In establishing co-ordinates in the plane, the objective is to demonstrate that collinear points satisfy a homogeneous linear equation.

The co-ordinates we have thus introduced are non-homogeneous though they may easily be made homogeneous in the usual way.

The method by which we defined homogeneous co-ordinates took no account of notions of order and continuity. Therefore, a real projective line (§ 11.1), remains for us a line whose co-ordinates are given by real numbers.

Our Chapter is concluded with a brief statement of the significance of Pappus' Theorem, and with an indication of possible co-ordinate systems in the projective plane other than the classical system of homogeneous co-ordinates.

§ 13.6 Importance of Pappus' Theorem.

Geometrically: It implies the validity of Desargues' Theorem,
Algebraically: It implies the commutativity of multiplication (i.e. the existence of a field).

§ 13.7 Other Systems of Co-ordinates in the Projective Plane. Homogeneous co-ordinates whose basic theory has been expounded above are not the only analytic tools for describing π. Two other useful systems are

(i) the ternary ring system of Marshall Hall, and
(ii) the homomorphism system of Emil Artin.

Of these, the former will be examined in Chapter 11.

Suggested topics for further study

(1) Analytical treatment of Desargues' and Pappus' Theorems by Hilbert [50].
(2) Line Geometry in three dimensions (Václav Hlavatý: 'Differential Line Geometry', Noordhoff (1953)).
(3) Analytical Projective Geometry in n dimensions.
(4) Fuller study of harmonic sequences and the Möbius net of rationality (Robinson [92]).
(5) Proofs of the Algebra and Geometry results stated in § 13.
(6) Life and achievements of Galois (Bell [10]).

EXERCISES 3

1. Are the lines $x + y + z = 0$, $x + y + 2z = 0$, $2x + 2y + z = 0$ concurrent? If so, where?

2. An arbitrary line $lx + my + nz = 0$ cuts the side YZ of the triangle XYZ in L. What are the co-ordinates of L? What is the equation of LX? Determine the co-ordinates of similarly defined points M,N on ZX, XY respectively. Write down the equations of MY, NZ.

3. Show that the equations of the six sides of a complete quadrangle may be written as $x_i \pm x_j = 0$ $(i < j;\ i, j = 1, 2, 3)$, where the homogeneous co-ordinates are taken as x_1, x_2, x_3. State the dual of this result.

Points $D(0, y_1, z_1)$, $E(x_1, 0, z_1)$, $F(x_1, y_1, 0)$ are taken on the sides YZ, ZX, XY respectively of the triangle of reference such that DX, EY, FZ are concurrent at P. Also XU, YU, ZU (U being the unit point) cut the sides YZ, ZX, XY in A, B, C respectively. Answer the following Exercises 3.4–3.8:

4. Determine the co-ordinates of A, B, C.

5. Deduce that $\mathbf{U} = \mathbf{X} + \mathbf{Y} + \mathbf{Z} = \mathbf{A} + \mathbf{X} = \mathbf{B} + \mathbf{Y} = \mathbf{C} + \mathbf{Z}$.

6. Obtain the co-ordinates of P.

7. What are the equations and line co-ordinates of the lines XA, YB, ZC?

8. Where, in the diagram, are the points:
(i) $(1, 1, 2)$ (ii) $(0, 1, 2)$ (iii) $(0, 1, -1)$?

9. What is the equation of the line joining the point of intersection of the lines

$$3x - y - 4z = 0,\ 2x + y - z = 0$$

to the point of intersection of the lines

$x + y + 2z = 0$, $3x - y + 2z = 0$?

10. Find the equation of the line joining the intersection of $x + z = 0$ and $y + z = 0$ to the point $l + m + n = 0$.

11. Calculate the co-ordinates of the point of intersection of $x + y + z = 0$ and the line through $l + n = 0$, $m + n = 0$.

12. Analytically, prove that the diagonal points of a complete quadrangle are non-collinear. (For the proof, look at Fishback [40] pp. 107–109.)

13. Duplicate the construction of Fig. 15 by means of two other arbitrary lines, a', b' through A, B and an arbitrary line i' through I, obtaining points P', Q', H', K', C'. Using Desargues' Theorem, prove that $C' = C$, i.e. C is independent of the lines a, b, i.

14. Triangles ABC, DEF are in perspective from P (not on the sides of either triangle). If AE, BF, CD are also concurrent, show *analytically* that AF, BD, CE are likewise concurrent. (Refer to Exercise 2.9 of which this is the analytical treatment.)

15. Establish the Converse of Hilbert's Theorem, namely: If Pappus' Theorem holds, then the commutativity of the law of multiplication holds.

16. Choosing $A(1, 0, 0)$, $B(1, 0, 1)$, $C(m, 0, 1)$, $A'(0, 1, 0)$, $B'(0, n, 1)$, $C'(0, 1, 1)$, $X(0, 0, 1)$, in Fig. 7a (i.e. $AA'X$ is triangle of reference), verify that

$$N = AB' \cap A'B \equiv (1, n, 1)$$
$$M = AC' \cap A'C \equiv (m, 1, 1)$$
$$L = BC' \cap B'C \equiv (k, 1 - k, 1)$$

where $k = (1 - n)(1 - m^{-1}n)^{-1}$ and the numbers belong to a division ring. Hence show that L, M, N are collinear, i.e. that $nm = mn$ (Pedoe [85]). Note that in this exercise XA' has equation $x = 0$, not $z = 0$ as in Fig. 7a for Example 6.

17. Co-ordinates of A, B, C are (a, b', c'), (a', b, c'), (a', b', c) respectively, and XYZ is the triangle of reference. Determine the equation of the line through the three points $BC \cap YZ$, $CA \cap ZX$, $AB \cap XY$. Further, show that $abc = a'b'c'$ implies that

(i) XB, YC, ZA are concurrent;
(ii) XC, YA, ZB are concurrent.

18. Work carefully through the given analytical proof of Pappus' Theorem in the case where the co-ordinate field is non-commutative (e.g. it might be the ring of quaternions). You should then see that Pappus' Theorem is violated in this division ring.

19. Suppose, in the Desargues configuration, we are given that $V \equiv (\alpha, \beta, \gamma)$ and $\mathbf{A}' = \mathbf{V} + \lambda\mathbf{A}$, $\mathbf{B}' = \mathbf{V} + \mu\mathbf{B}$, $\mathbf{C}' = \mathbf{V} + \nu\mathbf{C}$ (none of α, β, γ, λ, μ, ν being zero) and that ABC is the triangle of reference. Then prove that the equation of the Desargues line is

$$\frac{x}{\lambda} + \frac{y}{\mu} + \frac{z}{\nu} = 0.$$

[Note that, since the equation of BC is $x = 0$, the x co-ordinate of $BC \cap B'C'$ is 0, so that $BC \cap B'C' = \mathbf{B}' - \mathbf{C}' \equiv (0, \mu, -\nu)$.]

20. Two points X, Y in [3] have co-ordinates (x_1, x_2, x_3, x_4), (y_1, y_2, y_3, y_4). Define the numbers $p_{ij} = x_i y_j - x_j y_i (= -p_{ji})$ where $i, j = 1, 2, 3, 4$. (Clearly $p_{ii} = 0$.) These are the *line* (*ray*) *co-ordinates* of the line p determined by X, Y. Prove

(i) $p_{12}p_{34} + p_{13}p_{42} + p_{14}p_{23} = 0$;
(ii) $p_{12}q_{34} + p_{13}q_{42} + p_{14}q_{23} + p_{34}q_{12} + p_{42}q_{13} + p_{23}q_{14} = 0$ implies that the lines p, q are coplanar,
where q_{ij} are the line-co-ordinates of the line q determined by the points $X'(x_1', x_2', x_3', x_4')$, $Y'(y_1', y_2', y_3', y_4')$.

Given (Maxwell [10] p. 78) the points $A(1, 3, 5, 7)$, $B(3, 1, -1, -3)$, $C(8, 3, 2, 1)$ and $D(2, 1, 0, -1)$, do the lines AB, CD intersect?

Numbers p_{ij} are the co-ordinates of a line in Line Geometry. Among them are five independent ratios. This fact, taken with result (i) shows that only four of them are actually independent. Interpreting this situation, we assert that Line Geometry is essentially four dimensional. Furthermore, result (i) asserts that not every set of six numbers are line co-ordinates of a line in [3].

CHAPTER 4

COLLINEATIONS

[Socrates] '... in all departments of knowledge, as experience proves, anyone who has studied geometry is infinitely quicker of apprehension than one who has not.'

(Plato)

AMONG the most important aspects of a study of Geometry is the part played by transformations. In our treatment, which will feature algebraic methods, we witness a simple and elegant, yet powerful, application of the theory of matrices. Ultimately, the theory of transformations is recognised as being intimately connected with group theory.

§ 14 TRANSFORMATIONS IN THE PLANE

§ 14.1 Transformation of Co-ordinates. With respect to two distinct triangles of reference PQR, $P'Q'R'$ let a general point X have co-ordinates (x, y, z), (x', y', z') respectively. In terms of P', Q', R', suppose P, Q, R have co-ordinates (a_1, b_1, c_1), (a_2, b_2, c_2), (a_3, b_3, c_3) respectively. That is,

$$\begin{aligned} \mathbf{P} &= a_1\mathbf{P}' + b_1\mathbf{Q}' + c_1\mathbf{R}' \\ \mathbf{Q} &= a_2\mathbf{P}' + b_2\mathbf{Q}' + c_2\mathbf{R}' \\ \mathbf{R} &= a_3\mathbf{P}' + b_3\mathbf{Q}' + c_3\mathbf{R}' \end{aligned}$$

Then
$$\begin{aligned} \mathbf{X} &= x\mathbf{P} + y\mathbf{Q} + z\mathbf{R} \; (= x'\mathbf{P}' + y'\mathbf{Q}' + z'\mathbf{R}' \text{ also}) \\ &= x(a_1\mathbf{P}' + b_1\mathbf{Q}' + c_1\mathbf{R}') + y(a_2\mathbf{P}' + b_2\mathbf{Q}' + c_2\mathbf{R}') \\ &\quad + z(a_3\mathbf{P}' + b_3\mathbf{Q}' + c_3\mathbf{R}') \end{aligned}$$

i.e.
$$\begin{aligned} x'\mathbf{P}' + y'\mathbf{Q}' + z'\mathbf{R}' &= (a_1x + a_2y + a_3z)\mathbf{P}' + (b_1x + b_2y + b_3z)\mathbf{Q}' \\ &\quad + (c_1x + c_2y + c_3z)\mathbf{R}' \end{aligned}$$

whence
$$x' = a_1x + a_2y + a_3z,\; y' = b_1x + b_2y + b_3z,$$
$$z' = c_1x + c_2y + c_3z$$

i.e.
$$\boxed{\begin{bmatrix} x' \\ y' \\ z' \end{bmatrix} = \begin{bmatrix} a_1 & a_2 & a_3 \\ b_1 & b_2 & b_3 \\ c_1 & c_2 & c_3 \end{bmatrix} \begin{bmatrix} x \\ y \\ z \end{bmatrix}} \tag{2}$$

or
$$\boxed{\mathbf{X}' = \mathbf{AX}} \tag{2'}$$

with inverse
$$\boxed{\mathbf{X} = \mathbf{A}^{-1}\mathbf{X}'} \tag{2''}$$

($A = |\mathbf{A}| \neq 0$, i.e. $\mathbf{A}$ is non-singular),

i.e.
$$\begin{cases} Ax = A_1x' + B_1y' + C_1z' \\ Ay = A_2x' + B_2y' + C_2z' \\ Az = A_3x' + B_3y' + C_3z' \end{cases}$$

where $A_1, A_2, \ldots, C_3$ are the cofactors of $a_1, a_2, \ldots, c_3$ in $\mathbf{A}$. (Remember that $\mathbf{A}^{-1} = \dfrac{\mathbf{A}^*}{A}$ where $\mathbf{A}^*$ is the adjoint of $\mathbf{A}$, i.e. the transpose of the matrix of the cofactors of $\mathbf{A}$.)

Alternatively, writing $\mathbf{A}^{-1} = \mathbf{B}$ we may express the non-singular linear transformations (2′), (2″) as

$$\begin{cases} \mathbf{X} = \mathbf{BX}' , \\ \mathbf{X}' = \mathbf{B}^{-1}\mathbf{X}, \end{cases}$$

respectively.

Consequently,

a change of co-ordinate system in the plane is represented by a non-singular linear transformation of co-ordinates, and conversely.

Clearly, this result readily holds in $[n]$, and, in particular, in $[1]$.

When equation (2′) connects the co-ordinates of the *same* point referred to two different co-ordinate systems (triangles of reference), the transformation merely renames the point in another system. For this reason, it is sometimes called an *alias transformation* (and represents the passive aspect of the linear transformation (2′)).

On the other hand, equation (2′) also tells us how to pass from the point P to the point P' whose co-ordinates are given by $\mathbf{AX}$ in the original system of co-ordinates, i.e. points are transformed (= made to correspond) according to the law $\mathbf{X}' \rightarrow \mathbf{AX}$. Points are thus assigned new positions. Interpreted thus, equation (2′) describes an *alibi transformation* (and represents the active aspect of the transformation). There is generally a change of triangle of reference in this interpretation. Obviously, one can move freely from one interpretation to the other, as the same algebraic apparatus applies in both cases.

An alibi transformation is also called a *collineation* or a *projective transformation.*

Schematically, the situation is summarised thus:

Type of transformation	Elements transformed	Effect on triangle of reference
(Alias) transformation of co-ordinates	co-ordinates	changed
(Alibi) transformation of points (= collineation)	points	changed

Our main concern will be with collineations.

First we examine two special transformations. Consider

$\mathbf{A} = \text{diag.}\left(\frac{1}{a}, \frac{1}{b}, \frac{1}{c}\right) \qquad (a \neq 0, b \neq 0, c \neq 0)$

which clearly transforms (1, 0, 0) into $\left(\frac{1}{a}, 0, 0\right) \equiv (1, 0, 0)$ and similarly for the other vertices of the triangle of reference. That is, it leaves the triangle of reference unaltered.

Suppose that the co-ordinates of any point in a given co-ordinate system are (a, b, c). Then, clearly, under the above transformation

$$(a, b, c) \rightarrow (1, 1, 1).$$

Thus

given a triangle of reference, it is always possible to find a co-ordinate system in which the co-ordinates of the unit point are (1, 1, 1),

that is, any point not on the sides of the triangle of reference can be taken as the unit point since a transformation can be found which casts it into this form.

Likewise, the transformation

$\mathbf{B} = \text{diag.}\ (l, m, n) \qquad (l \neq 0, m \neq 0, n \neq 0)$

will keep the triangle of reference fixed while at the same time it will transform a general line $lx + my + nz = 0$ into the unit line $x + y + z = 0$. Otherwise expressed

given a triangle of reference, it is always possible to find a co-ordinate system in which the equation of the unit line is $x + y + z = 0$.

However, since $\mathbf{A} \neq \mathbf{B}$ in general, the two simplifying processes for unit point and unit line cannot both be effected simultaneously when the point and line are both arbitrary. Once the triangle of reference and unit point are given, a geometrical construction determines the unit line.

Note: In the above presentation, $\mathbf{X}$ does *not* represent the vertex (1, 0, 0) of the triangle of reference, which will be vectorially expressed as a 3×1 matrix when necessary. Accordingly, there should be no confusion if we now take XYZ as triangle of reference.

§ **14.2 Similarity Transformations.** Suppose the two aspects of the transformation (2′) are carried out simultaneously (Fig. 17).

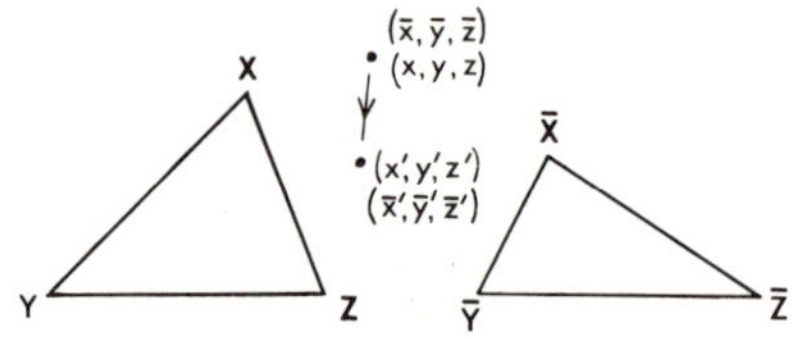

FIG. 17. Similarity transformation

Make the transformation of co-ordinates

$$\overline{\mathbf{X}} = \mathbf{AX} \qquad (|\mathbf{A}| \neq 0) \tag{i}$$

from the triangle of reference XYZ to the triangle of reference $\overline{X}\overline{Y}\overline{Z}$.
The collineation

$$\mathbf{X}' = \mathbf{TX} \qquad (|\mathbf{T}| \neq 0) \tag{ii}$$

w.r.t. $\overline{X}\overline{Y}\overline{Z}$ becomes

$$\begin{aligned}\overline{\mathbf{X}}' &= \mathbf{AX}' && \text{from (i)}\\ &= \mathbf{ATX} && \text{(ii)}\\ &= \mathbf{ATA}^{-1}\overline{\mathbf{X}} && \text{(i)}\\ &= \mathbf{B}^{-1}\mathbf{TB}\overline{\mathbf{X}} \text{ if } \mathbf{B} = \mathbf{A}^{-1}\end{aligned}$$

i.e. $\mathbf{ATA}^{-1}$ ($=\mathbf{B}^{-1}\mathbf{TB}$) *specifies the collineation given by* **T** *referred to the new triangle of reference.*

Collineation matrices **A**, $\mathbf{ATA}^{-1}$ ($= \mathbf{B}^{-1}\mathbf{TB}$) are *similar* and $\mathbf{ATA}^{-1}$ is said to be a *similarity transformation* of **T** by **A**. It needs to be shown (Exercise 4.2) that the similarity transformation is non-singular, i.e. that $|\mathbf{ATA}^{-1}| \neq 0$.

Note: Next, let π, π' be two distinct planes and suppose $\mathbf{X}' = \mathbf{PX}$ denotes a non-singular correspondence between points $X \in \pi$ and $X' \in \pi'$. Make a change of co-ordinate system $\mathbf{X} = \mathbf{AY}$ in π. Then the equation of the non-singular correspondence is

$$\begin{aligned}\mathbf{X}' &= \mathbf{PAY}\\ &= \mathbf{Y}\end{aligned}$$

provided $\mathbf{A} = \mathbf{P}^{-1}$. This means that suitable choice of co-ordinate system in π implies that the correspondence can be so expressed that corresponding points of π, π' have the same co-ordinates. Since a co-ordinate system in [2] is uniquely determined once the triangle of reference and unit point are chosen, then the above *correspondence between planes is uniquely given when four points of one plane, no three of them collinear, are made to correspond to four points of the other plane, no three of them collinear.*

[Similarly, it could be shown that this type of correspondence between two coplanar lines is uniquely defined when three pairs of corresponding points are known. See the next Chapter, Theorem 14 (the Fundamental Theorem).]

When $\pi' = \pi$, we have the special case discussed above.

Properties of projective correspondences when $\pi' \neq \pi$ carry over to collineations ($\pi' = \pi$) generally, with the exception that it is not usually possible to reduce the equation of a collineation to the form $\mathbf{X}' = \mathbf{Y}$ as a change of co-ordinate system in π affects both $\mathbf{X}'$ and $\mathbf{X}$.

§ 15 COLLINEATIONS. GROUPS

Concentrate now on the alibi aspects of (2′). Strictly speaking, if we enlarge our concept of a collineation, we could think of points in one plane π being transformed into points of another plane π' by the collineation but for our purposes we will take $\pi' = \pi$. That is, we are dealing with self-collineations. That is, we are talking about the collineation as a mapping of π onto itself.

Collineations are always assumed to be non-singular.

Because x', y', z' in (2) are homogeneous (projective) co-ordinates, we could multiply each co-ordinate by a scalar multiple $k \neq 0$. For simplicity, we choose $k = 1$. This is equivalent to ignoring the factor, since two transformations with proportional coefficients are regarded as identical. That is, $\mathbf{A}$, $\mathbf{B}$ define the same collineation only if $\mathbf{B} = \mu\,\mathbf{A}$ ($\mu \neq 0$).

A collineation determines a matrix to within a scalar factor, while a matrix determines a collineation uniquely.

What effect has the transformation (2″) on XYZ?

Clearly

$$\begin{bmatrix} a_1 & a_2 & a_3 \\ b_1 & b_2 & b_3 \\ c_1 & c_2 & c_3 \end{bmatrix} \begin{bmatrix} 1 \\ 0 \\ 0 \end{bmatrix} = \begin{bmatrix} a_1 \\ b_1 \\ c_1 \end{bmatrix}$$

that is, X is transformed into the point whose homogeneous co-ordinates are given by the first column of $\mathbf{A}$. Similarly for Y, Z. Columns of $\mathbf{A}$ are thus independent and $\mathbf{A}$ is non-singular.

Evidently then, a collineation carries points into points, i.e. makes points correspond to points. But what effect has it on lines?

Under the transformation (2″) the line (1″)

$$\mathbf{LX} = \mathbf{O}$$

is transformed into the line

$$\mathbf{LA}^{-1}\mathbf{X}' = \mathbf{O}$$

that is

$$\mathbf{MX}' = \mathbf{O}$$

where $\mathbf{M} = \mathbf{LA}^{-1}$.

Under the collineation with matrix $\mathbf{A}$, *a line with line-co-ordinates given by* $\mathbf{L} = [l, m, n]$ *is transformed into a line whose line co-ordinates are given by* $\mathbf{LA}^{-1}$.

This property of transforming lines into lines gives the collineation its name which, incidentally, is due to Möbius.

Gathering these two features of a collineation into a definition we have

Definition: A collineation is a 1–1 linear transformation given by the equations

$$\begin{cases} \mathbf{X} \to \mathbf{X}' = \mathbf{AX} & \text{(points)} \\ \mathbf{L} \to \mathbf{L}' = \mathbf{LA}^{-1} & \text{(lines) (Fig. 18).} \end{cases}$$

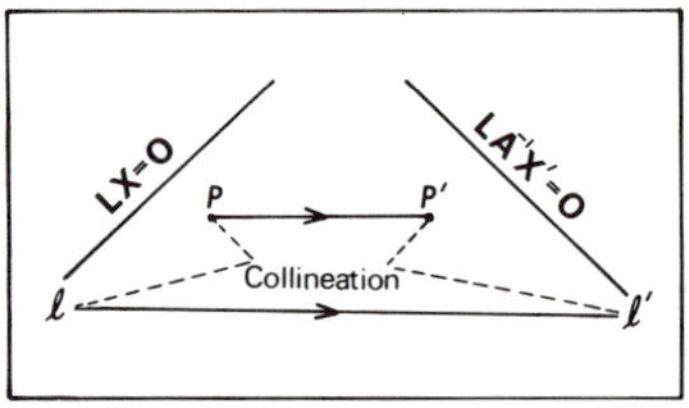

FIG. 18. Collineation

The property of carrying lines into lines distinguishes collineations from other types of 1–1 transformations, e.g. *correlations* which transform point into line and line into point, and *standard quadratic transformations* $x' : y' : z' = yz : zx : xy$ which are non-linear (quadratic); correlations are linear.

§ 15.1 Projective Linear Group (Collineation Group). Group theory, which plays a dominant and crucial role in a deeper understanding of the nature of Projective Geometry, is now related to collineations. Apart from a knowledge of elementary group-theoretic ideas, the only other group concepts required are those of coset, invariant (or normal) subgroup, factor group, and homomorphism (including isomorphism).

First, we recall that a collineation determines a matrix up to a scalar factor, while a matrix determines a collineation uniquely. Thus, if a collineation is given by $\mathbf{X}' = k\mathbf{AX}$ (scalar $k \neq 0$), the matrix $\mathbf{A}$ determines the collineation uniquely, whereas the collineation is equally well given by the matrices $k\mathbf{A}$. Clearly, there is a close relationship between the set of all matrices and the set of all collineations. We assume that the elements of all the matrices under consideration belong to some field F.

Among the set of all 3×3 non-singular matrices there are the diagonal matrices

$$\text{diag.}\ (k, k, k) = \begin{bmatrix} k & . & . \\ . & k & . \\ . & . & k \end{bmatrix} = k\mathbf{I} = \mathbf{I}_k \text{ (say)}$$

where $\mathbf{I}(= \mathbf{I}_1$, i.e. $k = 1)$ is the 3×3 unit matrix, and $k \neq 0$. These diagonal matrices correspond to the collineation determined by $\mathbf{X}' = k\mathbf{X}$.

In discussing the group theory aspects of collineations, we require the following results:

(i) 3×3 non-singular matrices form a group, the *full linear group* $L(3, F)$, under the operation of matrix multiplication;

(ii) (non-singular) collineations form a group, the *collineation group* (= *projective linear group*) $PGL(2, F)$, under the operation of one collineation followed by another;

(iii) diagonal matrices $k\mathbf{I}$ form an invariant subgroup H of $L(3, F)$;

(iv) $PGL(2, F)$ is isomorphic ($\cong$) to the factor group of $L(3, F)$ w.r.t. H, written as $L(3, F)/H$.

The symbol $L(3, F)$ signifies that the matrices are 3×3 matrices with elements in the field F. The symbol $PGL(2, \text{F})$ means that the Projective Geometry group (linear) relates to two dimensions and that the coefficients occurring in the equations of the collineation belong to the field F.

Result (ii) and, by implication, result (i) are established in Theorem 10. (Indeed, the reader ought to be familiar with the basic relationship (i) between matrices and groups embodied in the symbol $L(3, F)$.) Result (iii) is established as a Lemma, while Theorem 11 takes care of (iv). All the groups under discussion are infinite.

Perhaps it is worthwhile recalling the notion of a factor group for later reference. Given an invariant subgroup N of a group G, the cosets of N, together with a law of combination of cosets, form a group called the factor group of G w.r.t. N, written G/N.

Preliminaries over, we are now in a position to develop the sequence of results.

Theorem 10: Collineations form a (non-Abelian) group.

Proof: Consider the totality of collineations represented by matrices $\mathbf{A}_1$, $\mathbf{A}_2$, $\mathbf{A}_3, \ldots$

Let $\mathbf{X}' = \mathbf{A}_1\mathbf{X}$, $\mathbf{X}'' = \mathbf{A}_2\mathbf{X}'$ be two collineations. Define the operation of combining two collineations (i.e. the product of two collineations) as one collineation followed by another. This operation is equivalent to matrix multiplication.

$\therefore$ $\mathbf{X}'' = \mathbf{A}_2\mathbf{X}' = \mathbf{A}_2\mathbf{A}_1\mathbf{X} = \mathbf{AX}$ where $\mathbf{A} = \mathbf{A}_2\mathbf{A}_1$
$\therefore$ collineations obey the law of closure.

Clearly, collineations obey the associative law since $\mathbf{A}_1(\mathbf{A}_2\mathbf{A}_3) = (\mathbf{A}_1\mathbf{A}_2)\mathbf{A}_3$.

An identity collineation exists with matrix $\mathbf{I}$ (3×3 unit matrix) which transforms nothing at all, i.e. leaves things as they are.

With every (non-singular) collineation, matrix $\mathbf{A}$, there is associated an inverse collineation, with matrix $\mathbf{A}^{-1}$, which reverses the geometry of the transformation.

Since $\mathbf{A}_1\mathbf{A}_2 \neq \mathbf{A}_2\mathbf{A}_1$ in general, the commutative law is not valid
$\therefore$ collineations form an infinite non-Abelian group, $PGL(2, F)$.

Corollary: All non-singular 3×3 *matrices form a (non-Abelian) group* $L(3, F)$.

The truth of this statement is established by using the argument of Theorem 10 applied to all 3×3 non-singular matrices. Each representative 3×3 matrix used in Theorem 10 (e.g. $\mathbf{A}_1$) is only one of an infinite number of matrices (e.g. $k\mathbf{A}_1, k \neq 0$) all determining the same collineation.

Next, let us consider the diagonal matrices $k\mathbf{I}$.

Lemma: Diagonal matrices form an invariant subgroup of $L(3, F)$.

Proof: It is easy to verify that the diagonal matrices $k\mathbf{I}$ do form a subgroup of $L(3, F)$.

Let $\mathbf{A}$ be any matrix of $L(3, F)$.

$$\begin{aligned} \therefore\ \mathbf{A}^{-1}(k\mathbf{I})\mathbf{A} &= k\mathbf{A}^{-1}\mathbf{IA} \\ &= k\mathbf{I} \qquad \because\ \mathbf{A}^{-1}\mathbf{IA} = \mathbf{A}^{-1}\mathbf{A} = \mathbf{I} \end{aligned}$$

$\therefore$ diagonal matrices $k\mathbf{I}$ form an invariant subgroup H of $L(3, F)$.

A combination of the above results leads to the main theorem relating groups of matrices and collineations.

Theorem 11: $PGL(2, F) \cong L(3, F)/H$.

Proof: Since H is an invariant subgroup (Lemma) we may form cosets from it and so produce a factor group. Consider a matrix $\mathbf{A} \in L(3, F), \notin H$.

Multiply every element of H by $\mathbf{A}$ on the right (say) to form the (right) coset $H\mathbf{A}$ of which a typical member is $(k\mathbf{I})\mathbf{A} = k\mathbf{A}$. That is, the coset $H\mathbf{A}$ corresponds to just one collineation (call it A) with representative matrix $\mathbf{A}$.

Picking out a matrix $\mathbf{B} \in L(3, F), \notin H, \notin H\mathbf{A}$, we form another coset $H\mathbf{B}$ corresponding to the collineation B with representative matrix $\mathbf{B}$.

Proceeding in this way we eventually use up all the matrices of $L(3, F)$ and all the collineations of $PGL(2, F)$.

We note that the product of cosets, $H\mathbf{A}.H\mathbf{B} = H\mathbf{AB}$, i.e. a (unique) coset, corresponds to the (unique) collineation AB (product of the collineations A, B).

This is so because

$$\begin{aligned} HA.HB &= HHAB \qquad (AH = HA \because H \text{ is an invariant subgroup}) \\ &= HAB \qquad \because HH = H. \end{aligned}$$

Using Theorem 10, its Corollary and the Lemma together with the idea of isomorphism, we have that

$$PGL(2, F) \cong L(3, F)/H.$$

Note: (i) *Each coset of matrices of* $L(3, F)$ *w.r.t.* H *corresponds to just one collineation.* The invariant subgroup H itself (i.e. $H\mathbf{I}$) corresponds to the identity collineation I.

(ii) In the technical language of group theory, the correspondence between elements of $L(3, F)$, namely matrices, and elements of $PGL(2, F)$, namely collineations, is a *homomorphism of* $L(3, F)$ *onto* $PGL(2, F)$. *Kernel of this homomorphism is* H.

Figs. 19, 19(a) give pictorial interpretations of some of these group-theoretic ideas. Extension of Theorem 11 to $[n]$ is a simple matter.

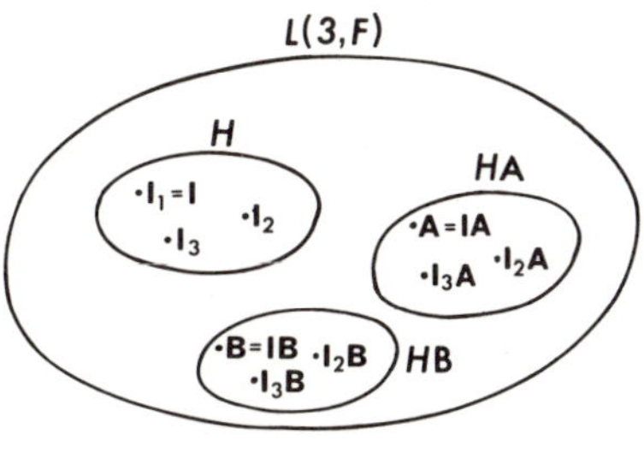

FIG. 19. Structure of the projective linear group

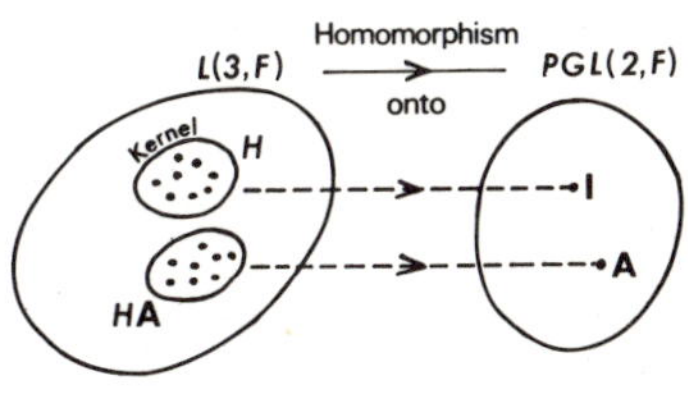

FIG. 19. (a) Homomorphism of $L(3, F)$ onto $PGL(2, F)$

We are now in a position to give a definition of plane Projective Geometry in terms of group theory, a definition which is readily extendable to $[n]$.

Projective Geometry in [2], with co-ordinates over a given field F, i.e. $P_2(F)$, is the theory of these plane projective properties which remain invariant under the transformations (2′) of the group $PGL(2, F)$. In short

Plane Projective Geometry is the invariant theory of $PGL(2, F)$.

It was this far-reaching connection between geometry and groups which was one of Klein's greatest contributions to knowledge. Klein, in delivering his inaugural dissertation on accepting a Chair of Mathematics at Erlangen in 1872, enunciated his famous *Erlanger Programm* namely

a geometry is a set of properties invariant under a certain group of transformations.

Seen in this light, the various sub-geometries of Projective Geometry are related in the same way to that geometry as certain sub-groups are related to the parent group. But more about this at a later, more appropriate time. A precise statement of Klein's *Erlanger Programm* is given in § 38.8.

Let it be recorded in passing that correlations and collineations together do form a group, of which the collineations form a subgroup of index 2. Correlations themselves form only a coset; the reason they lack the group structure is that the identity in the correlation-collineation group is a collineation (not a correlation, since it leaves points as points, and lines as lines). Also, it may be proved that the product of two correlations is a collineation, i.e. the multiplicative closure law is invalid for correlations.

As remarked in Veblen and Young [113] 1, p. 268, the Principle of Duality in [2] may be regarded as a consequence of the presence of correlations in this group.

§ 15.2 Invariant Points of a Collineation. Eigenvalues. Invariant (= self-corresponding) points $\mathbf{X} = \begin{bmatrix} x \\ y \\ z \end{bmatrix}$ of the plane collineation given by (2′) must be expressed by the matrix equation (λ a scalar)

$$\mathbf{AX} = \lambda\mathbf{X} \qquad (\alpha)$$

i.e.

$$(\mathbf{A} - \lambda\mathbf{I})\mathbf{X} = \mathbf{O}.$$

To determine $x : y : z$, we need to solve the three linear equations in x, y, z contained in this matrix equation. This requires the *characteristic equation* of $\mathbf{A}$, namely

$$|\mathbf{A} - \lambda\mathbf{I}| = 0.$$

Hence, λ must be an *eigenvalue* of $\mathbf{A}$ (i.e. a root of the characteristic equation) and the co-ordinate vector $\mathbf{X}$ must be an *eigenvector* corresponding to this root. The characteristic polynomial of $\mathbf{A}$ is $|\mathbf{A} - \lambda\mathbf{I}|$

Regarding invariant elements, let us note that

(i) two invariant points lie on an invariant line,

(ii) two invariant lines intersect in an invariant point.

While a line of invariant points is certainly an invariant line, an invariant line does not necessarily consist of nothing but invariant points. That is, the term 'invariant line' will mean either a line of invariant points or a line containing (at least) two invariant points.

A specific example, analysed in detail, will reveal some of the mechanism of collineations.

EXAMPLE 10: Find the invariant points of the collineation

$$x' : y' : z' = 2y : -x + 3y + z : 3z.$$

[Convention: Zeros in matrices are represented by dots, for simplicity.]

Solution: Matrix of the transformation is $\mathbf{A} = \begin{bmatrix} . & 2 & . \\ -1 & 3 & 1 \\ . & . & 3 \end{bmatrix}$ with eigen-matrix $\begin{bmatrix} -\lambda & 2 & . \\ -1 & 3-\lambda & 1 \\ . & . & 3-\lambda \end{bmatrix}$ leading to characteristic equation

$$(3-\lambda)(\lambda^2 - 3\lambda + 2) = 0$$

$\therefore$ eigenvalues are $\lambda = 1, 2, 3$.

Now substitute the values 1, 2, 3 in turn for λ in the equation (α) with the given matrix **A**.

$$\lambda = 1 \text{ in } (\alpha) \Rightarrow \begin{bmatrix} . & 2 & . \\ -1 & 3 & 1 \\ . & . & 3 \end{bmatrix} \begin{bmatrix} x \\ y \\ z \end{bmatrix} = 1 \begin{bmatrix} x \\ y \\ z \end{bmatrix}$$

i.e. $2y = x,\ -x + 3y + z = y,\ 3z = z$

i.e. $2y = x,\ z = 0$

$\therefore\ \lambda = 1 \Rightarrow (2, 1, 0)$ is an invariant point.

$$\lambda = 2 \text{ in } (\alpha) \Rightarrow \begin{bmatrix} . & 2 & . \\ -1 & 3 & 1 \\ . & . & 3 \end{bmatrix} \begin{bmatrix} x \\ y \\ z \end{bmatrix} = 2 \begin{bmatrix} x \\ y \\ z \end{bmatrix}$$

i.e. $2y = 2x,\ -x + 3y + z = 2y,\ 3z = 2z$

i.e. $y = x,\ z = 0$

$\therefore\ \lambda = 2 \Rightarrow (1, 1, 0)$ is an invariant point.

$$\lambda = 3 \text{ in } (\alpha) \Rightarrow \begin{bmatrix} . & 2 & . \\ -1 & 3 & 1 \\ . & . & 3 \end{bmatrix} \begin{bmatrix} x \\ y \\ z \end{bmatrix} = 3 \begin{bmatrix} x \\ y \\ z \end{bmatrix}$$

i.e. $2y = 3x,\ -x + 3y + z = 3y,\ 3z = 3z$

$\therefore\ \lambda = 3 \Rightarrow (2, 3, 2)$ is an invariant point.

Thus the invariant points of the collineation are (2, 1, 0), (1, 1, 0), (2, 3, 2).

Note: (i) From the eigenvectors $[2, 1, 0]^T$, $[1, 1, 0]^T$, $[2, 3, 2]^T$ form the matrix

$$\boldsymbol{\Delta} = \begin{bmatrix} 2 & 1 & 2 \\ 1 & 1 & 3 \\ . & . & 2 \end{bmatrix} \text{ with inverse } \boldsymbol{\Delta}^{-1} = \tfrac{1}{2} \begin{bmatrix} 2 & -2 & 1 \\ -2 & 4 & -4 \\ . & . & 1 \end{bmatrix}.$$

$$\therefore\ \boldsymbol{\Delta}^{-1}\mathbf{A}\boldsymbol{\Delta} = \tfrac{1}{2} \begin{bmatrix} 2 & -2 & 1 \\ -2 & 4 & -4 \\ . & . & 1 \end{bmatrix} \begin{bmatrix} 2 & 2 & 6 \\ 1 & 2 & 9 \\ . & . & 6 \end{bmatrix} = \begin{bmatrix} 1 & . & . \\ . & 2 & . \\ . & . & 3 \end{bmatrix}$$

i.e. $\boldsymbol{\Delta}^{-1}\mathbf{A}\boldsymbol{\Delta}$ reduces $\mathbf{A}$ to a diagonal form whose elements are the eigenvalues of $\mathbf{A}$.

(ii) Trace $\mathbf{A}$ = trace $\boldsymbol{\Delta}^{-1}\mathbf{A}\,\boldsymbol{\Delta} = \Sigma$ (eigenvalues) $= 6\ (= 1 + 2 + 3 = 0 + 3 + 3)$. (Trace $\mathbf{A}$ = sum of the diagonal elements.)

Of course, (i) and (ii) are only particular illustrations of well-known theorems of matrix theory.

Let us analyse just what this particular collineation and its inverse with matrix $\mathbf{A}^{-1} = \frac{1}{6}\begin{bmatrix} 9 & -6 & 2 \\ 3 & . & . \\ . & . & 2 \end{bmatrix}$ do in the plane. It is easy to verify that the following transformations occur for points

$$\begin{array}{rcl} & \mathbf{A} & \\ X & \longrightarrow & Y \\ Y & \longrightarrow & (2, 3, 0) \\ Z & \longrightarrow & (0, 1, 3) \\ (1, 1, 1) & \longrightarrow & (2, 3, 3) \\ (3, 1, 0) & \longrightarrow & X \\ (1, 0, 1) & \longrightarrow & Z \end{array}$$

and for lines

$$\begin{array}{rcl} x = 0 & \longrightarrow & 9x - 6y + 2z = 0 \\ y = 0 & \longrightarrow & x = 0 \\ z = 0 & \longrightarrow & z = 0 \\ x + y + z = 0 & \longrightarrow & 6x - 3y + 2z = 0 \\ x - 3y + z = 0 & \longrightarrow & y = 0. \end{array}$$

Thus the line $z = 0$ is invariant. (Other invariant lines are $x - 2y + 2z = 0$ and $2x - 2y + z = 0$.) Notice that not all points on $z = 0$ are invariant. In fact, only two are, namely $(2, 1, 0)$, $(1, 1, 0)$, as proved in Example 10. This means (as we remarked earlier) that

two invariant points determine an invariant line

the remaining non-invariant points on the invariant line being permuted among themselves by the collineation. Expressed otherwise, the existence of just two invariant points is enough to determine an invariant line. However, it may happen that every point on the line is invariant, in which case it is convenient to speak of the line as being *pointwise invariant* or point-by-point invariant.

Diagrammatically, the situation is as shown in Fig. 20, where starred points are invariant.

Evidently

(i) the columns of $\mathbf{A}$ give the transforms of the points X, Y, Z respectively (as was previously remarked), and

(ii) the rows of $\mathbf{A}^{-1}$ give the transforms of the lines $x = 0$, $y = 0$, $z = 0$ respectively.

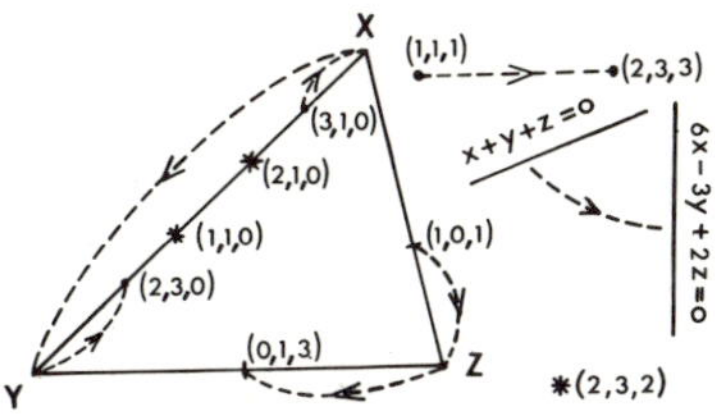

FIG. 20. Analysis of invariant elements of a given collineation

Sometimes we may wish to determine the point (or line) which is transformed by the collineation into a given point (or line). This problem is covered in the procedures which we now outline.

Representing the co-ordinates of a point (or line) before and after the collineation in matrix form, as previously,

	before collineation	after collineation
point	$\mathbf{X}$	$\mathbf{X}'$
line	$\mathbf{L}$	$\mathbf{L}'$

we may solve the various types of problem which arise according to the following scheme:

Given $\mathbf{X}$,	to find $\mathbf{X}'$,	use $\mathbf{X}' = \mathbf{AX}$
$\mathbf{X}'$	$\mathbf{X}$	$\mathbf{X} = \mathbf{A}^{-1}\mathbf{X}'$
$\mathbf{L}$	$\mathbf{L}'$	$\mathbf{L}' = \mathbf{LA}^{-1}$
$\mathbf{L}'$	$\mathbf{L}$	$\mathbf{L} = \mathbf{L}'\mathbf{A}$.

Choosing the unit point and unit line, for simplicity of illustration, we readily verify that, under the collineation given by $\mathbf{A}$,

$$\begin{cases} \text{the unit point} & \rightarrow (2, 3, 3) \text{ (noted earlier)} \\ (5, 3, 2) & \rightarrow \text{the unit point} \\ \text{the unit line} & \rightarrow 6x - 3y + 2z = 0 \text{ (noted earlier)} \\ -x + 5y + 4z = 0 & \rightarrow \text{the unit line.} \end{cases}$$

Evidently

(i) the rows of $\mathbf{A}$ give the lines which are transformed into $x = 0$, $y = 0$, $z = 0$ respectively,

and

(ii) the columns of $\mathbf{A}^{-1}$ give the points which are transformed into X, Y, Z respectively.

§ 15.3 Canonical Equations for Collineations. Types of collineations in the plane may be distinguished from each other according to the inter-relationship of the points which they leave invariant. These, in turn, may be examined by means of the canonical forms to which the transformation matrices can be reduced. Altogether, in $P_2(C)$, *there are just six different types of plane collineations.*

The six collineations, including the identity, are distinguished from each other by two main facts:
(i) the multiplicity of the eigenvalue λ_i, and
(ii) the rank of the matrix $\mathbf{A} - \lambda_i\mathbf{I}$.
Details are to be found in, say, Levy [64], Semple and Kneebone [99], Springer [147], Todd [111], Veblen and Young [113] 1, and Winger [120]. (See Fig. 21, which excludes the identity collineation. What are the duals of the invariant figures of the various types of collineation?)

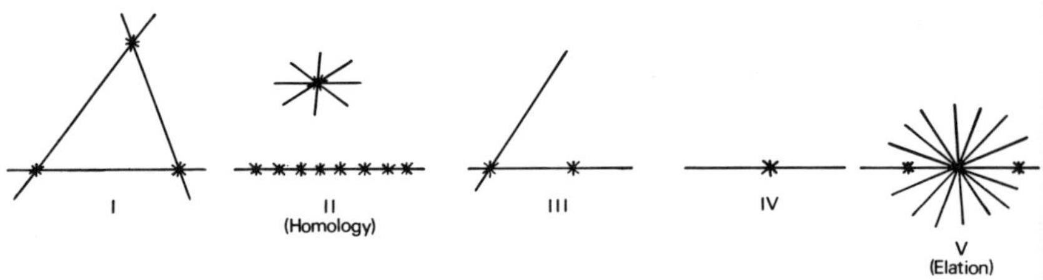

FIG. 21. Types of (non-identity) collineations in $P_2(C)$

Firstly, consider the *general collineation* (2′) in which it is assumed that the characteristic roots of $\mathbf{A}$ (= eigenvalues of $|\mathbf{A} - \lambda\mathbf{I}| = 0$) are $\lambda_0, \lambda_1, \lambda_2$ which are all distinct, i.e. there are no repeated roots of the characteristic equation. Consequently, the collineation has three distinct, independent invariant points (why?) (and dually the lines joining them in pairs are invariant). [See Example 10.] Taking these points as vertices of the triangle of reference, we may reduce the general collineation to its canonical matrix form

$$\text{diag.}\ (\lambda_0, \lambda_1, \lambda_2).$$

Observe that in $P_2(R)$, it is possible to have all three roots real, or only one root real.

When $\lambda_0 = \lambda_1 = \lambda_2$ we have the *identity collineation* which necessarily leaves invariant all the points, and all the lines, of the planes, i.e. the entire plane.

When $\lambda_0 = \lambda_1 \neq \lambda_2$, we obtain a *homology*, a type of collineation which is discussed more fully in § 16. (See Figs. 21, 21(a).)

Of the three remaining types of collineation, one of them, the *elation* (Figs. 21, 21(b)), will also be discussed in some detail in § 16.

The names 'homology' and 'elation' are due to Poncelet and Lie (1842–1899, Norwegian) respectively.

While the six types of collineation exhaust all possibilities in $P_2(C)$, the situation is different in the projective plane over other fields. For example, in $P_2(R)$, the collineation with matrix

$$\begin{bmatrix} \cdot & -1 & \cdot \\ 1 & \cdot & \cdot \\ \cdot & \cdot & 1 \end{bmatrix}$$

leaves only one point fixed, namely (0, 0, 1), whereas in $P_2(C)$ it preserves the invariance of three fixed points (since, in the field C, a cubic equation must have three roots), the three invariant points being (0, 0, 1), (1, i, 0), (1, $-i$, 0). Again, in the case of $P_2(GF(3))$, say, the collineation with matrix

$$\begin{bmatrix} \cdot & \cdot & 2 \\ 1 & \cdot & \cdot \\ \cdot & 1 & \cdot \end{bmatrix}$$

has no fixed points at all, since the characteristic equation is $\lambda^3 = 2$ which has no roots in the given finite field, i.e. the characteristic polynomial $\lambda^3 - 2$ is irreducible in this field.

Sorting out these ideas we can assert:

In $P_2(R)$, a collineation fixes at least one point and at least one line;

In $P_2(GF(p^k))$, a collineation may have no invariant points or lines.

§ 16 CENTRAL COLLINEATIONS. HOMOLOGIES AND ELATIONS

§ **16.1** *Definitions:* A *central* (or *perspective*) *collineation* is a collineation leaving invariant a point C (the *centre*) and a line of points (the *axis* l) pointwise invariant. For brevity, it may be referred to as a (C, l) *collineation.*

When $C \in l$, the central collineation is an *elation.*

When $C \notin l$, the central collineation is a *homology.*

Unlike other collineations, the central (perspective) collineations are expressible by means of two planar perspectivities in [3].

(See Hall [45] or Pedoe [85], and Part 1 of this book.)

Let us now obtain the canonical equation of a homology in π.

Take the centre of the homology (call it α) to be at Z (0, 0, 1) and the axis l (line of invariant points) to be $z = 0$.

Conditions imposed on the elements of the homology matrix **A** are

$$\begin{cases} Z \text{ invariant} : a_{13} = a_{23} = 0 \\ X\ (\in z = 0) \text{ invariant} : a_{21} = a_{31} = 0 \\ Y\ (\in z = 0) \text{ invariant} : a_{12} = a_{32} = 0 \\ Q\ (x_1, x_2, 0)\ (\in z = 0) \text{ invariant} : a_{11} = a_{22}. \end{cases}$$

All these statements should be verified by the reader. Canonical form for a homology is thus

$$\text{diag. } (1, 1, k) \qquad (k \neq 0, \neq 1)$$

where $k\ (= \frac{a_{33}}{a_{11}})$ is the *modulus* of the homology. Clearly, the homology matrix has a repeated eigenvalue.

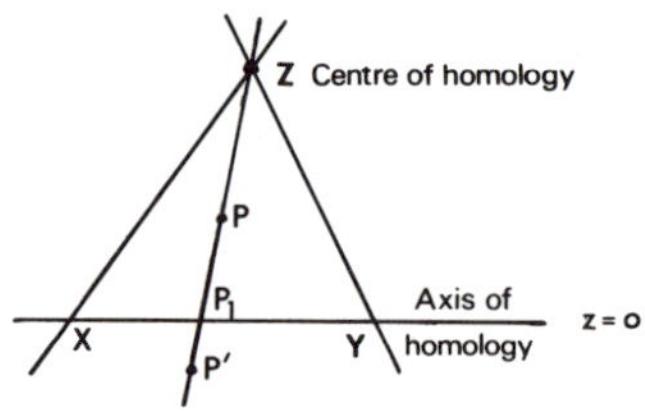

FIG. 21. (a) Homology

Let $P\ (p, q, r)$ be any point in π. Then $P' = P\alpha$—see comment on notation below—has co-ordinates (p, q, kr). Now, by (1), PP' has equation

$$\begin{vmatrix} x & y & z \\ p & q & r \\ p & q & kr \end{vmatrix} = 0,$$

i.e. $qx - py = 0$
which passes through $Z\ (0, 0, 1)$.

$\therefore$ Z, P, $P\alpha$ are collinear for all $P\ (\notin l, \neq Z)$.

Suppose $ZP \cap l = P_1$. Then P_1 has co-ordinates $(p, q, 0)$ and it is an invariant point. Consequently, ZP, joining the invariant points Z and P_1, is an invariant line.

Hence,

a homology is characterised by its leaving invariant

- *centre C*
- *all points on the axis l*
- *all lines of the pencil through C.*

(See Fig. 21(a).) Among others, Coxeter [28] p. 53 gives a synthetic derivation of a homology.

When $k = -1$, the homology is called a *harmonic homology*. If a collineation (matrix **A**) is such that $\mathbf{A}^2 = \mathbf{I}$, the collineation is said to be an *involutory collineation*. Eigenvalues of an involutory collineation are ± 1. Obviously, a harmonic homology is an involutory collineation (and it is the only type of collineation which is involutory).

Notation: While employing the symbol $P\alpha$ to represent the transformation of the point P by the homology α, the author suggests as an alternative representation P^{α}, as used by Pedoe [85]. For consistency with the symbolism employed in other operations on points (e.g. matrices, projectivities (§ 18)), the symbol αP could be used.

Consider next the case of elations (for which the centre lies on the axis of the elation).

Choose the centre of the elation (call it β) to be C $(a, b, 0)$ and the axis l of the elation to be $z = 0$.

Now the invariance of any point $(p, q, 0)$ on l (in particular, the invariance of C $(a, b, 0)$ on l) implies, in **A**,

$$a_{12} = a_{21} = a_{31} = a_{32} = 0, a_{11} = a_{22}.$$

Also, if P (p, q, r) is any point of π, then the co-ordinates of $P' = P\beta$ are $(a_{11}p + a_{13}r, a_{11}q + a_{23}r, a_{33}r)$. Requiring $C, P, P\beta$ to be collinear as in (1) imposes the conditions $a_{11} = a_{33}$, $a_{13} : a_{23} = a : b$. Dividing by a_{11} ($\neq 0$), we derive the canonical elation matrix

$$\begin{bmatrix} 1 & . & a \\ . & 1 & b \\ . & . & 1 \end{bmatrix}.$$

Hence,

an elation is characterised by the invariance of

$\left\{\begin{array}{l}\textit{all points (including the centre } C\ (a, b, 0) \text{ in this case) } \textit{on the axis } l\ (z = 0) \\ \textit{all lines of the pencil through } C.\end{array}\right.$

(See Fig. 21(b).)

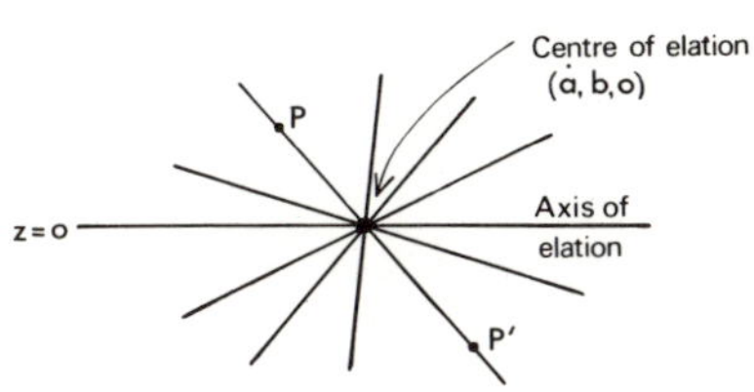

FIG. 21. (b) Elation

Summarising the basic features of homologies and elations, we say *a (central) collineation which fixes every point on a line also fixes a point C and every line of the pencil through C.*

Stated otherwise: *A central collineation γ is completely determined by its centre C, its axis l, and the transformation $P \to P\gamma$ of any point $P \notin l$, $P \neq C$, such that C, P, $P\gamma$ are collinear.*

What is the dual of the first statement?

A purely geometrical approach to collineation theory may be found in Hall [45] or Pedoe [85]. Recent researches by the contemporary German mathematician, Reinhold Baer ('Homogeneity of projective planes', *Amer. J. Math.* 64 (1942) 137–152) and others have revealed the deep meaning of central collineations for an understanding of the structure of the plane.

Elations may be determined synthetically, as in Coxeter [28] p. 54.

From the matrix expressions for homologies and elations it is simple to deduce the result that

Theorem 12: (C, l) collineations form a group (written $G(C, l)$).

Evidently, $G(C, l)$ is a subgroup of $PGL(2, F)$.

EXAMPLE 11: Determine the nature of the collineation

$$x' : y' : z' = y : -x + 2y : -3x + 3y + z.$$

Find its centre and axis.

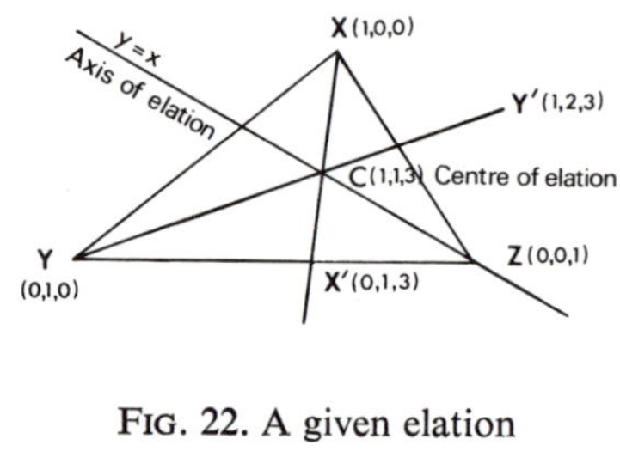

FIG. 22. A given elation

Solution: Write $\mathbf{A} = \begin{bmatrix} \cdot & 1 & \cdot \\ -1 & 2 & \cdot \\ -3 & 3 & 1 \end{bmatrix}$

whence $|\mathbf{A} - \lambda\mathbf{I}| = 0$ leads to $\lambda = 1, 1, 1$ and the rank of the matrix $\mathbf{A} - \mathbf{I}$ is 1.

Now $\mathbf{AX} = 1.\,\mathbf{X}$ gives $y = x$,
i.e. the invariant line is $y = x$.

Any point (x, x, z) on this line $\xrightarrow{\mathbf{A}}$ (x, x, z), i.e. remains invariant.

Any point $P(x_1,\ x_2,\ x_3) \xrightarrow{\mathbf{A}} P'(x_2,\ -x_1 + 2x_2,\ -3x_1 + 3x_2 + x_3)$

whence by (1) PP' has equation $x(x_3 - 3x_2) + y(3x_1 - x_3) + z(x_2 - x_1) = 0$ (after division throughout by $x_1 - x_2 \neq 0$) which is satisfied by the point (1, 1, 3), i.e. (1, 1, 3) is the centre and as it lies on the axis $y = x$, the collineation is an elation.

$$\text{Under the elation, } X \xrightarrow{\mathbf{A}} X'\ (0, 1, 3)$$

$$Y \xrightarrow{\mathbf{A}} Y'\ (1, 2, 3)$$

and, of course, $XX' \cap YY' = C$ (Fig. 22). Being on the axis $y = x$, Z is necessarily invariant.

Observe that $\mathbf{A}^{-1} = \begin{bmatrix} 2 & -1 & . \\ 1 & . & . \\ 3 & -3 & 1 \end{bmatrix}$ ($|\mathbf{A}| = 1$) from which we verify that $y = x$ is invariant, i.e. $[1, -1, .]\ \mathbf{A}^{-1} = [1, -1, .]$.

§ **16.2 Central collineations and Desargues' Theorem.** By now, it will have become apparent that central collineations bear a close relationship to Desargues' Theorem in which the centre of perspective and the Desargues line are the centre and axis of the collineation respectively.

It can be proved (Exercise 4.11) that

(1) *the existence of a central collineation with a given centre and axis implies the validity of Desargues' Theorem;*

(2) (conversely) *the validity of Desargues' Theorem for all pairs of perspective triangles implies that all possible* (C, l) *collineations exist.*

Details of these proofs may be found in Hall [45] and Pedoe [85]. When the central collineation is an elation, the general Desargues' Theorem is called the *minor Desargues' Theorem.* In modern researches into the structure of the projective plane (and of the accompanying algebraic apparatus) elations and the minor Desargues' Theorem are of vital significance.

§ 17 TRANSITIVITY. TRANSLATION PLANES

Definition: Given a centre C and an axis l in π, if all possible central collineations with centre C and axis l exist, then π is (C, l) *transitive.*

We prove the result:

Theorem 13: If π is both (C, l) transitive and (C', l) transitive ($C, C' \in l$, $C \neq C'$) then π is (C'', l) transitive for every choice of $C'' \in l$.

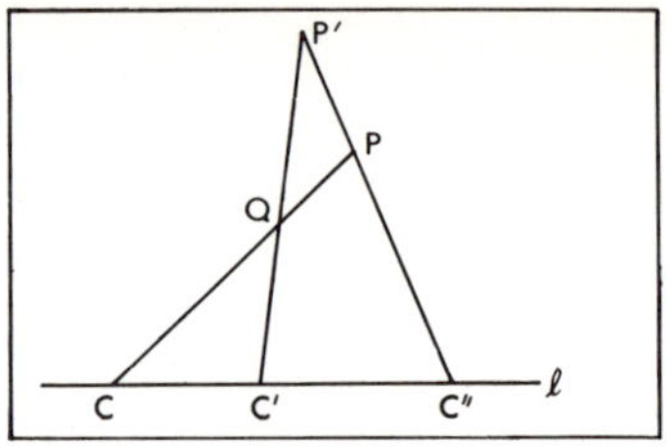

FIG. 23. Translation plane

Proof: Let $\alpha \in G(C, l)$, $\alpha' \in G(C', l)$ where α, α' are elations. Write $P\alpha = Q$, $Q\alpha' = P'$.

$\because$ by Exercise 4.17, $\alpha'' = \alpha\alpha'$ is an elation.

Define $Q = PC \cap P'C'$ and C'', P, P' collinear ($C'' \in l$).

By hypothesis, $P\,\alpha = Q$, $Q\,\alpha' = P'$

$\because$ $P\,\alpha'' = (P\,\alpha)\alpha' = Q\,\alpha' = P'$

$\because$ the centre of α'' is C''

$\because$ π is (C'', l) transitive (Fig. 23).

Definition: π is a *translation plane* w.r.t. l if it is (C, l) transitive for every $C \in l$.

Translation planes play an important part in the modern theory of the foundations of Projective Geometry (the structure of π). Later, in § 38.22, we will see that it is reasonable to call these planes 'translation planes'. See also Exercise 4.18.

Suggested topics for further study.

(1) Canonical forms of collineations (Semple and Kneebone [99], Todd [111], Veblen and Young [113] 1, Winger [120]).
(2) Correlations. Polarities (= correlations of period 2) (Coxeter [27], Semple and Kneebone [99]).
(3) Standard quadratic transformations (Primrose [89], Semple and Kneebone [99]).
(4) Anti-collineations.
(5) Transformation of co-ordinates.

EXERCISES 4

1. Similar matrices have the same characteristic polynomial. Prove.
2. A similarity transformation matrix is non-singular. Show.

3. Let $S(\lambda)$ be the set of eigenvectors of the linear transformation σ (of the vector space V into itself) whose matrix is $\mathbf{A}$, corresponding to the eigenvalue λ. Prove that $S(\lambda)$ is a vector subspace of V. (It is called the *eigenspace* of σ w.r.t. λ.) For further details, consult Evar D. Nering: 'Linear Algebra and Matrix Theory' (Wiley, 1963) p. 94. The set of eigenvalues is called the *spectrum* of σ.

4. Determine the collineation which transforms the points

$(1, 0, 0), (0, 1, 0), (0, 0, 1), (1, 1, 1),$

into the points

$(0, 1, 0), (0, 0, 1), (1, 0, 0), (1, 1, 1)$

respectively.

5. In Example 10

(i) Which point is transformed into $(1, 1, 1)$?

(ii) Which line is transformed into the unit line $x + y + z = 0$?

(iii) What is the transform of the point $(1, -1, 2)$?

(iv) What is the transform of the line $x = y + 2z$?

6. Again, in Example 10

(i) Verify the Cayley-Hamilton Theorem: If $\mathbf{A}$ is a square matrix and $f(x)$ is its characteristic polynomial (a cubic in this case), then $f(\mathbf{A}) = \mathbf{O}$;

(ii) Verify that the eigenvalues of $\mathbf{A}^{-1}$ are the reciprocals of the eigenvalues of $\mathbf{A}$, namely, $1, \frac{1}{3}, \frac{1}{2}$.

More generally, it may be proved (Levy [64] p. 174) that if l is an invariant line of a collineation with matrix $\mathbf{A}$ and if $\mu\mathbf{L} = \mathbf{L}\mathbf{A}^{-1}$ (λ being an eigenvalue of $\mathbf{A}$) then $\mu\lambda = 1$.

7. Establish the theorem: A collineation in π which leaves invariant every point on each of two distinct lines is the identity.

8. Exhibit the truth of the theorem: A collineation in π which leaves invariant every point on one line and two points not on the line is the identity.

Exercises 4.7, 4.8 guarantee that the maximum number of points left invariant by a collineation, which keeps invariant all points on a given line, is at most one which is not on the line.

9. Cosets form a group. Prove.

10. Generalise the theory of collineations to $[n]$. Especially consider the case $n = 1$.

11. Given a collineation with centre O and axis MN mapping A on A', let the distinct triangles ABC, $A'B'C'$ be in perspective from O and $M = CA \cap C'A'$, $N = AB \cap A'B'$. Define $L = BC \cap B'C'$. Show that the existence of the collineation involves Desargues' Theorem, i.e. that L, M, N are collinear.

Discover the invariant points of the following Exercises 4.12–4.16 (if a collineation is central, obtain its centre and axis):

12. $x' : y' : z' = y + z : z + x : x + y$

13. $x' : y' : z' = x + 2z : y - 3z : z$

14. $x' : y' : z' = y : x : z$

15. $x' : y' : z' = x : y : x - y + z$

16. $x' : y' : z' = x : x + 2y + 5z : x + y + 4z.$

17. Prove that the product of two elations having the same axis l but different centres C, C' is an elation with axis l and centre C'' ($\neq C, C'$).

18. Elations with a common axis form a commutative group. Show. (Include the identity collineation.) See Hall [45] pp. 358–359.

[Because of the relationship between elations in the projective plane and translations in the ordinary Euclidean plane, this elation group is called the *translation group*.]

19. Homologies with a given centre and axis (plus the identity) form a group. Prove.

20. Prove that the product of two homologies with different centres and a common axis is an homology or an elation (Patterson [83]).

CHAPTER 5

PROJECTIVITIES, CROSS-RATIO AND INVOLUTION

'. . . von Staudt, having adjoined to the visible Universe a more extensive region, proves that the enlarged domain is a coherent and manageable whole; that there is no essential difference between the elements recognised by the bodily senses and those apprehended by pure intellect.'

(Charlotte A. Scott)

BEFORE making the next big development, namely the definition and fundamental properties of a conic derived entirely from synthetic methods, we need to be equipped with some more mathematical techniques. This brings us to a set of related concepts (projectivity, cross-ratio and involution) which are essentially one-dimensional in nature, though occurring in a two-dimensional setting. Generally speaking, we will be concerned with the relationships existing among points of one or more lines.

§ 18 PROJECTIVITIES

A projectivity is an extension of the simpler idea of a central perspectivity which has previously been encountered in connection with Desargues' Theorem and with homologies and elations.

§ 18.1 Perspectivity. First we define a *range* to be a set of collinear points. The line of collinearity is called the *base* of the range. Consider the two distinct coplanar lines a, b (Fig. 24). We say that the 1–1 correspondence between points $A_i \in a$ and points $B_i \in b$ shown in Fig. 24 is a (*central*) *perspectivity* ($i = 1, 2, 3, \ldots$). The ranges of points $A_i \in a$ and $B_i \in b$ are said to be *in perspective* having *centre* (*vertex*) *of perspectivity* V. Alternatively, we may speak of the points of a (or b) being the *projection* of the points of b (or a) from V. The point $X = a \cap b$ is called a *self-corresponding point* of the perspectivity.

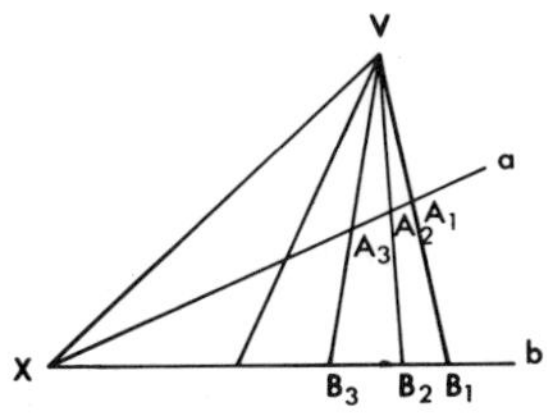

FIG. 24. Perspectivity

Symbolically, the perspectivity may be written

$$(A_1, A_2, \ldots) \overset{V}{\overline{\wedge}} (B_1, B_2, \ldots)$$

or simply

$$(A) \overset{V}{\overline{\wedge}} (B)$$

where (A), (B) designate the sets of points on a, b respectively.

Concurrent lines VA_1B_1, VA_2B_2, ... form the *pencil* of lines $V(A_1, A_2, \ldots)$ through V which, of course, in this case is the same as the pencil $V(B_1, B_2, \ldots)$. (An individual line of the pencil may be called a *ray*).

Dual of a range of points A_1, A_2, ... on the line l is a pencil of lines a_1, a_2, ... through the point L. (What is the space dual of this range of points? Answer: An *axial pencil of planes*, i.e. the set of planes all intersecting in a line.)

What is the plane dual of a perspectivity? (Answer: An *axial perspectivity* whose *axis of perspectivity* is the dual of the centre of the perspectivity. Draw a diagram to illustrate this concept.)

Notice (Example 5, Fig. 13) that, for four collinear points, *the harmonic property is invariant under a perspectivity.*

§ 18.2 Projectivity. Following Poncelet, we say that a chain of perspectivities (finite in number)

$$\begin{aligned}(A_1, B_1, C_1, \ldots) &\overset{V_1}{\overline{\wedge}} (A_2, B_2, C_2, \ldots)\\ &\overset{V_2}{\overline{\wedge}} (A_3, B_3, C_3, \ldots)\\ &\cdots\cdots\cdots\\ &\overset{V_{n-1}}{\overline{\wedge}} (A_n, B_n, C_n, \ldots)\end{aligned}$$

determines a *projectivity* (also called a *projective transformation*). Fig. 25 describes this projectivity. A line like l_2, which connects two perspectivities of the chain on l_1 and l_3, is called an *intermediary line*. The construction need not be restricted to just one plane. All that is required is that two successive lines l_i, l_{i+1} and their centre of perspectivity V_i should be coplanar ($i = 1, \ldots, n-1$).

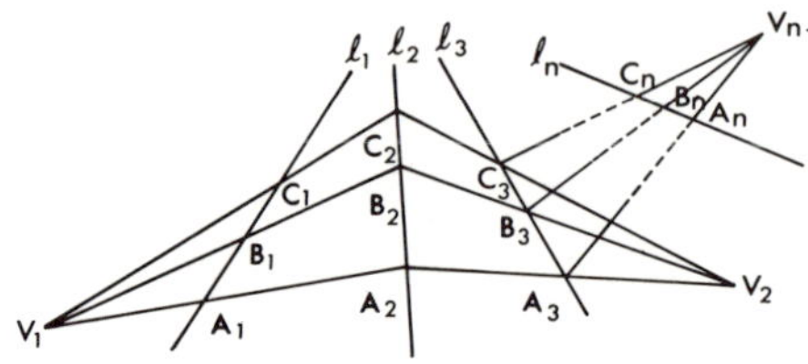

FIG. 25. Projectivity

Ranges of points on any two lines of the chain of perspectivities are said to be *projectively related* or, more simply, just *related.* (Enunciate the dual of this concept as an exercise. Illustrate it by means of a suitable diagram.) Thus related ranges are connected by a projectivity, which is essentially a 1–1 and onto correspondence, i.e. the points of two related ranges are in 1–1 correspondence.

Points $X_1, X_2, \ldots$ and $Y_1, Y_2, \ldots$ belonging to two related ranges may, in von Staudt's symbolism, be expressed succinctly as

$$(X) \barwedge (Y)$$

where $(X) = \{X_1, X_2, \ldots\}$, $(Y) = \{Y_1, Y_2, \ldots\}$. Heed should be taken of the small but important difference between the symbol $\overline{\barwedge}$, for a perspectivity and that, $\barwedge$, for a projectivity. In the definition of a projectivity (above), we may accordingly write $(A_1, B_1, C_1, \ldots) \barwedge (A_n, B_n, C_n, \ldots)$.

Remarks:

(i) Clearly, a perspectivity is a special case of a projectivity.

(ii) Every projectivity Π obviously has an inverse Π^{-1}: $\Pi\Pi^{-1} = \Pi^{-1}\Pi = \mathrm{I}$, where I is the *unit projectivity* which leaves things as they are.

(iii) A projectivity Π_1 followed by a projectivity Π_2 (written $\Pi_2\Pi_1$) yields a third projectivity Π_3 ($= \Pi_2\Pi_1$) called the *product* of Π_1 and Π_2. Can you prove that $(\Pi_1\Pi_2)^{-1} = \Pi_2^{-1}\Pi_1^{-1}$?

(iv) All this suggests that projectivities have the group structure.

(v) *Under a projectivity, the harmonic property is invariant.*

§ 18.3 Fundamental Theorem of Projective Geometry.

EXAMPLE 12: Any three distinct points on a given line may be related to any three distinct points on another line by at most *two* perspectivities.

Solution: Given $A_1, B_1, C_1 \in l_1$, and $A_2, B_2, C_2 \in l_2$. Assume that none of these six points is $l_1 \cap l_2$.

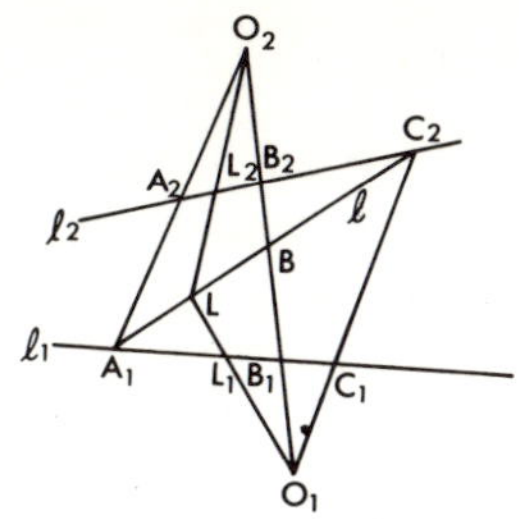

FIG. 26. Projectivity as the product of two perspectivities

Construct the points $O_1 = C_1C_2 \cap B_1B_2$, $O_2 = A_1A_2 \cap B_1B_2$, $B = A_1C_2 \cap O_1O_2$ (Fig. 26). Then

$$(A_1, B_1, C_1) \overset{O_1}{\barwedge} (A_1, B, C_2)$$

$$\overset{O_2}{\barwedge} (A_2, B_2, C_2)$$

$$\text{i.e. } (A_1, B_1, C_1) \barwedge (A_2, B_2, C_2).$$

Similarly, a projectivity exists which takes A_2, B_2, C_2 into any three specified points of l_1.

Observe how the above construction enables us to determine pairs of corresponding points $L_1 \in l_1$, $L_2 \in l_2$.

Corollary: If the lines l_1, l_2 coincide, then the projectivity is expressible as the product of three perspectivities. (How?)

Remarks:

(i) Is it necessary for l_1, l_2 to be coplanar in Example 12? Why?

(ii) Is the intermediary line l unique?

(iii) A group of operations having the property of transforming three given elements into three specified elements is said to be *triply-transitive.*

Does Example 12 establish an upper limit to the number of points on each of two lines necessary to determine a projectivity? Yes, since the invariance of the harmonic relation connecting a set of four collinear points, being a special relation, implies that one cannot always make four given points on one line correspond to four given points in another line in the manner of a projectivity.

So the required number is <4. Furthermore, it must be >2, since it is easy to construct a projectivity connecting points $A_1, B_1 \in l_1$, and $A_2, B_2 \in l_2$. Consequently, our non-rigorous argument suggests that the desired number is 3.

Collecting these ideas together, we assert (and prove algebraically—cf. Todd [111] pp. 22–23):

Theorem 14 (*Fundamental Theorem of Projective Geometry*):
A projectivity between the points on two distinct coplanar lines is uniquely determined by three pairs of corresponding points.

Proof: Paralleling the work in § 14, let $\mathbf{X}' = \mathbf{PX}$ be a projectivity relating points $X \in l$ to points $X' \in l'$ $(l, l' \in \pi)$ where $\mathbf{X} = \begin{bmatrix} x_1 \\ x_2 \end{bmatrix}$, $\mathbf{X}' = \begin{bmatrix} x_1' \\ x_2' \end{bmatrix}$ and $\mathbf{P}$ is a non-singular 2×2 matrix, i.e. $\mathbf{P} = \begin{bmatrix} a_1 & a_2 \\ b_1 & b_2 \end{bmatrix}$ where $a_1b_2 - a_2b_1 \neq 0$.

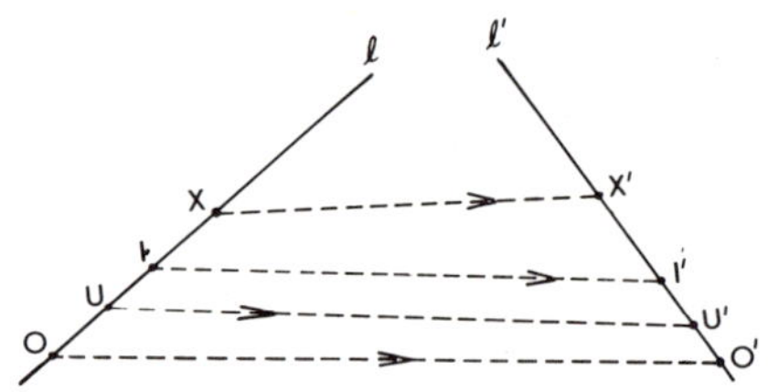

FIG. 27. Fundamental theorem of projective geometry

On l, make the transformation of co-ordinates $\mathbf{X} = \mathbf{AY}$ (Fig. 27) where $\mathbf{Y} = \begin{bmatrix} y_1 \\ y_2 \end{bmatrix}$ and $\mathbf{A}$ is a non-singular 2×2 matrix. Then

$$\begin{aligned} \mathbf{X}' &= \mathbf{PAY} \\ &= \mathbf{Y} \qquad \text{provided } \mathbf{A} = \mathbf{P}^{-1} \end{aligned}$$

Thus, by an appropriate choice of co-ordinate system in l, the equation of the non-singular projectivity can be put in the form $\mathbf{X}' = \mathbf{Y}$, whereby corresponding points of l, l' have the same co-ordinates.

Now, a co-ordinate system on a line is completely determined by two base reference points O, I and the unit point U, i.e. by three points. Making these points correspond on the two lines (O to O', I to I', U to U'), we have proved the Theorem. (Notice that $\mathbf{Y} = \mathbf{PX}$.)

In its geometrical form this Theorem is due to von Staudt. Texts such as Pedoe [85], Robinson [92], and Veblen and Young [113] give detailed proofs of the Fundamental Theorem by purely geometrical means. Our approach has followed the method used by Todd [111]. Subsequently in this section, another treatment will be offered. Seidenberg [98] provides much detailed information about the close relationships existing among Desargues' Theorem, Pappus' Theorem, and the Fundamental Theorem, e.g. the equivalence of Pappus' Theorem and the Fundamental Theorem.

At this stage, the reader ought to consult the statement of the problem in Exercises 5.10, 5.11 in which a certain line (the *cross-axis of the projectivity*) is associated with the lines of two related ranges. This cross-axis coincides with the Pappus line for two related triads. Its existence tells us how to construct pairs of corresponding points on two related ranges when three pairs of corresponding points have been assigned.

A chain of perspectivities, as we indicated, defines a projectivity. Reversely, suppose we are given a projectivity, we ask the question: What is the minimum number of perspectivities necessary to describe it? The answer to this query, clearly a useful one to raise, is contained in the following Theorem.

Theorem 15: A projectivity between two distinct coplanar lines is either a perspectivity or can be expressed as the product of two perspectivities.

Proof: If the projectivity is already a perspectivity, then the first part of the Theorem is true. The second part is an immediate consequence of Example 12 and Theorem 14.

§ 18.4 Group of Projectivities. Next, make the hypothesis that the lines l_1, l_n in Fig. 25 coincide ($l_1 = l_n = l$, say). Thus we have a projectivity connecting corresponding pairs of points of two related ranges on the same line l.

Projectivities, i.e. projective transformations of a line into itself, form a group, the *linear projective group* $PGL(1, F)$ w.r.t. the field F, just as in the case for projective transformations in two dimensions. Without repeating the argument used in Theorem 11, which it parallels (for 2×2 non-singular matrices) and which the reader should revise, we assert:

Theorem 16: Projectivities form an infinite non-Abelian group $PGL(1, F) \cong L(2, F)/H$ where H is the invariant subgroup of $L(2, F)$, the full linear group over F, corresponding to the projectivities $\mathbf{X}' = k\mathbf{X}$ ($k \neq 0$ a factor of proportionality).

Various interesting subgroups of $PGL(1, F)$ occur. Commutative projectivities may exist, constituting an Abelian subgroup. Projectivities on l which (say) leave each of two given points invariant form a subgroup of $PGL(1, F)$. Veblen and Young [113] 1, p. 68, give further information about these special subgroups and, in addition, establish $PGL(1, F)$ geometrically. (Recall § 18.2 (iv).) However, we return now to the algebraic statement of a projectivity.

§ 18.5 Fundamental Equation of a Projectivity. Consider the projectivity (Theorem 14)

$$\mathbf{X}' = \mathbf{A}\mathbf{X}$$

$$\text{i.e.}\boxed{\begin{cases} x'_1 = ax_1 + bx_2 \\ x'_2 = cx_1 + dx_2 \end{cases} \qquad (|\mathbf{A}|\ (= ad - bc) \neq 0)} \tag{2a}$$

Introducing non-homogeneous parameters (co-ordinates) $x = \dfrac{x_1}{x_2}$, $x' = \dfrac{x_1'}{x_2'}$, we can express (2a) as

$$x' = \frac{ax + b}{cx + d} \tag{i}$$

Solving for x in terms of x', we deduce

$$x = \frac{dx' - b}{-cx' + a} \tag{ii}$$

corresponding to the inverse projectivity given by

$$\mathbf{X} = \mathbf{A}^{-1}\mathbf{X}'.$$

Equation (i) or (ii) yields

$$cxx' - ax + dx' - b = 0 \qquad (ad - bc \neq 0)$$

or, in a slightly changed notation,

$$\boxed{pxx' + qx + rx' + s = 0 \qquad (ps - qr \neq 0)} \tag{3}$$

with none of p, q, r, s being zero in the general case. Equation (3) is the *fundamental equation of a projectivity*. It is a *bilinear* equation since x (or x') can be expressed in terms of x' (or x) as a linear fraction. From this point of view, $PGL(1, F)$ may be called the *linear fractional group*, which is associated with the name of Möbius.

Remarks:

(1) Suppose (3) factorises as $(x - \alpha)(x' - \beta) = 0$. Then it is a simple deed to show that $ps = qr$, which is contrary to hypothesis. Consequently, (3) does not factorise. Durell [33] p. 131 proves that the implication of $ps = qr$ is that the relationship between x and x' is no longer 1–1.

(2) Dually, a set of results exists for two projectively related pencils of lines. What are they?

(3) Projectivity (3) is completely determined when the actual values of p, q, r, s are known. Dividing (3) through by any one of these constants, say p, we derive

$$\boxed{xx' + Qx + Rx' + S = 0} \tag{3'}$$

$\left(Q = \dfrac{q}{p}, R = \dfrac{r}{p}, S = \dfrac{s}{p}\right)$. Given three pairs of values of x, x'—say x_i, x_i'—for variable corresponding points X_i, X_i' $(i = 1, 2, 3)$, we may solve (3) completely. (See Example 14.) This proves the Fundamental Theorem algebraically by a variation of the method in Theorem 14.

§ 18.6 Double Points of a Projectivity. When the projectivity relates two ranges on the same line, there will be some points of one range which coincide with their corresponding points on the other range. Such points are called *self-corresponding* or *double* (or *united*) *points* of the projectivity. Just how many double points can exist?

For a double point to exist, we must have $x' = x$. Thus the equation determining the double points is, from (3),

$$\boxed{px^2 + (q + r)x + s = 0} \tag{3a}$$

i.e. a *projectivity has two double points* corresponding to the roots of (3a).

What restrictions are placed on the coefficients in (3a) when

(i) $x = 0$, (ii) $x = \infty$ respectively? (Answers: (i) $s = 0$, (ii) $p = 0$.)

Projectivities may be classified according to the nature of these double roots. This classification may for simplicity be set out schematically as follows:

Nature of roots	Name (type) of projectivity
distinct and real	hyperbolic
distinct and complex	elliptic
coincident	parabolic

One deduction from this table is the fact that in $P_2(R)$ elliptic projectivities cannot exist.

Dually, the double elements of a projectivity in [2] are *double lines*, and in [3] are *double planes*. You should draw diagrams to clarify these ideas.

Every collineation in $P_2(R)$ is a projectivity. Is this true in $P_2(C)$? [No! But why? (Robinson [92] pp. 150–152)]. In $P_2(C)$, the transformation $z' = \bar{z}$ (complex conjugate of z) is called an *anti-projectivity*.

Because, in the author's experience, students sometimes find the concept of projectivity an elusive one to grasp, some pictorial and numerical instances of projectivities will now be provided.

Firstly, the reader's attention is invited to Figs. 28, 29 which show the double points E, F of a hyperbolic projectivity, and the double point E of a parabolic projectivity, respectively. In each case, the projectivity is determined by the three pairs of points $A, A'; B, B'; C, C'$. Another parabolic projectivity is shown in Fig. 15(c) where the double point is P_∞.

Next we turn to solve some representative problems.

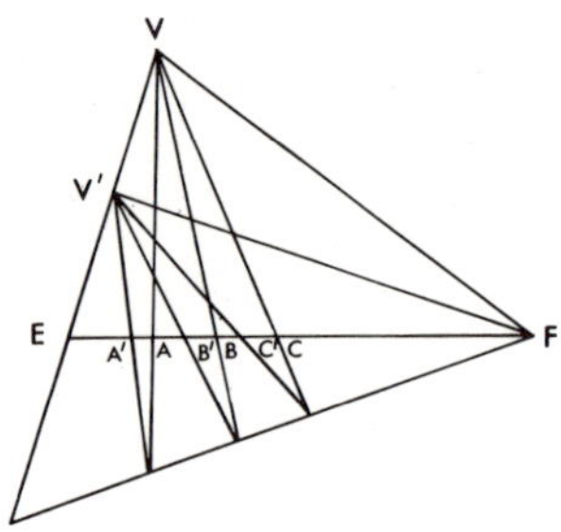

FIG. 28. Hyperbolic projectivity

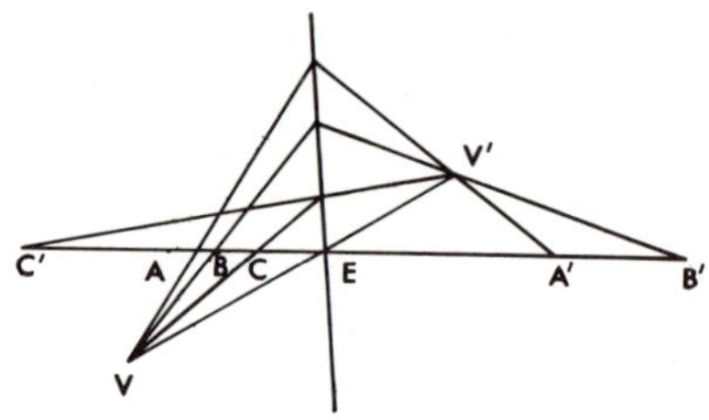

FIG. 29. Parabolic projectivity

EXAMPLE 13: Given the projectivity $xx' - 6x + x' + 6 = 0$, determine the values of x' corresponding to the values $1, 0, \infty$ of x, and the values of x corresponding to the values $1, 0, \infty$ for x'. What are the double points of the projectivity?

Solution: Calculation of corresponding values is relatively simple and is set out in the table hereunder:

x	-1	0	1	$1\frac{2}{5}$	2	∞
x'	∞	-6	0	1	2	6

Perhaps it is worth emphasising that to discover x' when $x = \infty$, we divide the equation of the projectivity by x first to obtain $x' - 6 + \dfrac{x'}{x} + \dfrac{6}{x} = 0$, and then put $x = \infty$.

Double points occur when $x' = x$, i.e. when $x^2 - 5x + 6 = 0 = (x-2)(x-3)$.

Thus the double points are given by the parameters 2, 3. Since these are real and distinct, the projectivity is hyperbolic.

EXAMPLE 14: Three values $-2, 0, 3$ of x with corresponding values $7, -1, 2$ for x' satisfy a 1–1 algebraic correspondence. Determine the equation of this projectivity and its double points. What value of x corresponds to $x' = \infty$ and what value of x' corresponds to $x = \infty$?

Solution: Let the projectivity (which, by Theorem 14, is uniquely fixed by three pairs of corresponding points) have equation

$$xx' + qx + rx' + s = 0.$$

Substitute for the three pairs of corresponding values in turn.

Then
$$-14 - 2q + 7r + s = 0 \quad \text{(i)}$$
$$r - s = 0 \quad \text{(ii)}$$
$$6 + 3q + 2r + s = 0 \quad \text{(iii)}$$

Solving (i), (ii), (iii), we find $q = -3$, $r = s = 1$.

Equation of the projectivity is thus

$$xx' - 3x + x' + 1 = 0$$

with double points given by $x^2 - 2x + 1 = 0 = (x - 1)^2$.

That is, the double points are given by values 1, 1 so the projectivity is parabolic.

Computation gives $\begin{cases} x' = 3 & \text{when } x = \infty \\ x = -1 & \text{when } x' = \infty. \end{cases}$

Suitable choice of double points leads to simplified standard (canonical) forms for the equation of the projectivity. One such canonical form ($x' = kx$) is determined in Example 15. Another ($x' = x + c$) is left as Exercise 5.5.

EXAMPLE 15: If the parameters of the double points of a hyperbolic projectivity are 0, ∞, the canonical equation of the projectivity is $x' = kx$ (k a constant).

Solution: In equation (3a), $\begin{cases} x = 0 \text{ implies } s = 0 \\ x = \infty \text{ implies } p = 0 \end{cases}$

$\therefore$ the canonical equation of the projectivity (3) reduces to

$$qx + rx' = 0$$

i.e. $$x' = kx \qquad \left(k = -\frac{q}{r}\right).$$

§ 18.7 Summary of the Nature of a Projectivity. A projectivity, which is a 1–1 correspondence between points of two lines (distinct or coincident) formed from a finite chain of perspectivities, is uniquely determined by three pairs of corresponding points. Algebraically, a projectivity is expressed in matrix form as $\mathbf{X}' = \mathbf{A}\mathbf{X}$ ($\mathbf{A}$ non-singular) with canonical forms $x' = kx$ or $x' = x + c$. Projectivities form an infinite non-Abelian group.

§ 18.8 Cyclic Projectivities. Given a projectivity Π:

$$\Pi P = P'$$
$$\Pi^2 P = \Pi P' = P''$$
$$\Pi^3 P = \Pi P'' = P'''$$
$$\cdots\cdots\cdots\cdots$$

we may have either
(i) an infinite sequence of distinct points, or
(ii) after a certain finite number of repetitions we return to P, i.e. $P^{(n)} = \Pi^n P = P$.

This second possibility gives rise to *cyclic projectivities.*

Clearly, in (ii), $\Pi^n = \mathrm{I}$ (the unit projectivity which leaves points as they are). We call n the *period* of the projectivity if it is the smallest integer for which $\Pi^n = \mathrm{I}$.

An example of a projectivity of period 3 (i.e. $\Pi^3 = \mathrm{I}$) occurs when, given just three collinear points A, B, C we define the projectivity Π to be such that

$$(A, B, C) \barwedge (B, C, A).$$

Plainly, three applications of this projectivity produce the original order A, B, C.

Most valuable of the periodic projectivities are those with period 2 (i.e. $\Pi^2 = \mathrm{I}$. Such a projectivity is called an *involution.* A good deal more will be said a little later (§ 20) about the properties associated with this concept.

Revert for a moment to Example 12. This showed that, under a projectivity, *any three* collinear points correspond to *any three* collinear points.

Thus, a relationship involving just three points (such as 'betweenness', as in Fig. X(c), which is a projective figure—refer also to § 11, Remark (5)), cannot be invariant and consequently cannot have any meaning in Projective Geometry. Hence, an invariant relation connecting a set of collinear points requires more than three points. Such a relationship is that of cross-ratio, involving a set of four collinear points, to which we now give our attention. (The invariance of cross-ratio under a projectivity means that any four collinear points can only be transformed into four collinear points if the corresponding cross-ratios are the same.)

§ 19 CROSS-RATIO

Having brought in the idea of a projectivity, we are now in a position to explain the related concept of cross-ratio. This is done abstractly.

Suppose we are given four collinear points A, B, C, D and two reference points O, I on the line. Write $\mathbf{A} = \mathbf{O} + a\mathbf{I}$, $\mathbf{B} = \mathbf{O} + b\mathbf{I}$, $\mathbf{C} = \mathbf{O} + c\mathbf{I}$, $\mathbf{D} = \mathbf{O} + d\mathbf{I}$, with the parameters a, b, c, d respectively.

§ **19.1 Definition:** The *cross-ratio* (or *anharmonic ratio*) of the range A, B, C, D (written (AB, CD) in the notation of Möbius, 1827, or, in terms of the parameters $(a, b; c, d)$) is the number defined to be

$$\boxed{(AB, CD) = \frac{(a-c)(b-d)}{(a-d)(b-c)} = (a, b; c, d)} \tag{4}$$

Taking homogeneous co-ordinates of $A, \ldots, D$ to be $(a_1, a_2), \ldots, (d_1, d_2)$ and writing them non-homogeneously as $a = \frac{a_1}{a_2}, \ldots, d = \frac{d_1}{d_2}$, we may express (4) in terms of the homogeneous co-ordinates of the four points, namely

$$\boxed{(AB, CD) = \frac{(a_1c_2 - a_2c_1)(b_1d_2 - b_2d_1)}{(a_1d_2 - a_2d_1)(b_1c_2 - b_2c_1)}} \tag{4a}$$

Notice particularly the arrangement of the co-ordinates a, b, c, d in the factors of (4). They are associated, as differences, in the representation accompanying Fig. 30.

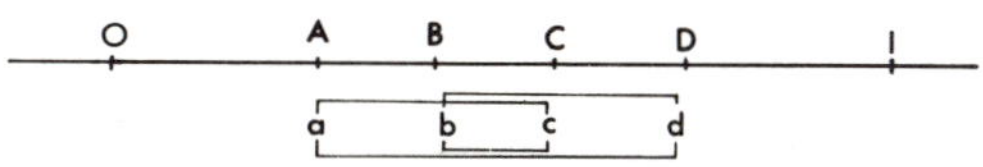

FIG. 30. Cross-ratio

How would you define, dually, the cross-ratio of a pencil of four lines? An interesting special case of (4) is now discussed (Example 16).

EXAMPLE 16: Evaluate the cross-ratio (IO, UX) of the four collinear points $O(0, 1)$, $I(1, 0)$, $X(x_1, x_2)$, $U(1, 1)$.
Solution: Substituting the co-ordinates $(0, 1)$, $(1, 0)$, (x_1, x_2), $(1, 1)$ in (4a) we have

$$\begin{aligned}(IO, UX) &= \frac{(1.1 - 0.1)(0.x_2 - 1.x_1)}{(1.x_2 - 0.x_1)(0.1 - 1.1)} \\ &= \frac{(1)(-x_1)}{(x_2)(-1)} \\ &= \frac{x_1}{x_2} \\ &= x.\end{aligned}$$

The detailed computation shows how use of homogeneous co-ordinates avoids difficulties in manipulating ∞. Less rigorously, we would, in practice, proceed as follows:

Take the parameter in each case to be the ratio $\dfrac{\text{first co-ordinate}}{\text{second co-ordinate}}$. Then the four points have parameters $0\left(=\frac{0}{1}\right)$, $\infty\left(=\frac{1}{0}\right)$, $x\left(=\frac{x_1}{x_2}\right)$, $1\left(=\frac{1}{1}\right)$ respectively.

$$\therefore\ (IO, UX) = (\infty, 0; 1, x)$$

$$= \frac{(\infty - 1)(0 - x)}{(\infty - x)(0 - 1)} = 1.\frac{-x}{-1} \quad \text{using a property of } \infty$$

$$= x.$$

Projective (homogeneous) co-ordinates of a point X on a line are thus related to the base reference points O, I and the unit point U. A similar situation exists in the plane in connection with the triangle of reference; see Semple and Kneebone [99] p. 24.

Example 16 shows that if we are given three distinct collinear points A, B, C there is a unique distinct point D for which $(AB, CD) = k$ where k is a prescribed number. We have taken A, B, C as I, O, U respectively and $D = X$, $k = x$.

§ 19.2 Fundamental Properties. Simple properties of cross-ratio flow readily from (4) and the following results are left for verification:

I $(AB, CD) > 0$ if the range is A, B, C, D

< 0 if the range is A, C, B, D (i.e. roles of B, C are interchanged).

Knowledge of Property III below enables us to write down three other ranges of which the cross-ratio is negative, namely: B, D, A, C; C, A, D, B; D, B, C, A.

II $(AB, CD) = 0$ if $A = C$ or $B = D$

$= 1$ if $A = B$ or $C = D$

$= \infty$ if $A = D$ or $B = C$.

Usually, A, B, C, D are distinct points, so that 0, 1, ∞ are excluded as cross-ratio values. Occasionally, however, it is convenient to consider coincidences (e.g. in § 35.3).

III Associated with four collinear points, there exist only six different values of the $4! = 24$ possible cross-ratios, viz.

$$\lambda, \frac{1}{\lambda}, 1 - \lambda, \frac{1}{1 - \lambda}, \frac{\lambda - 1}{\lambda}, \frac{\lambda}{\lambda - 1}$$

(a result discovered by Möbius). Observe that

$$(AB, CD) = (BA, DC) = (CD, AB) = (DC, BA)$$

i.e. *the value of the cross-ratio is unaltered if the letters are interchanged in pairs.* See Example 17.

IV *Cross-ratio is independent of the co-ordinate system*, i.e. independent of the choice of O and I and depends only on A, B, C, D. [For if O', I' are two distinct points on OI, let $\mathbf{O}' = \mathbf{O} + p\mathbf{I}$, $\mathbf{I}' = \mathbf{O} + q\mathbf{I}$, $\mathbf{A} = \mathbf{O}' + a'\mathbf{I}'$, etc. whence $a = \dfrac{p + a'q}{1 + a'}$ etc. Eventually, we have

$$\frac{(a-c)(b-d)}{(a-d)(b-c)} = \frac{(a'-c')(b'-d')}{(a'-d')(b'-c')}.]$$

V Sets of four corresponding points on two related ranges are *equi-cross*, i.e. have the same cross-ratio (and conversely). [Let $\mathbf{A} = \mathbf{O} + a\mathbf{I}$ and $\mathbf{A}' = \mathbf{O}' + a'\mathbf{I}'$, etc. and substitute in (3) whence $a = -\dfrac{ra' + s}{pa' + q}$, etc. After calculation, we have

$$(AB, CD) = (A'B', C'D').]$$

Expressed otherwise (cf. § 18.2(v))

cross-ratio is invariant under a projective transformation (projectivity).

VI Suppose $a = 0$, $b = \infty$, i.e. $A = O$, $B = I$. [Note that $\mathbf{B} = \mathbf{O} + b\mathbf{I}$ may be written as $\frac{1}{b}\mathbf{B} = \frac{1}{b}\mathbf{O} + \mathbf{I}$. But (by § 10) $\frac{1}{b}\mathbf{B} = \mathbf{B}$ so $\mathbf{B} = \frac{1}{b}\mathbf{O} + \mathbf{I} = \mathbf{I}$ when $b = \infty$.]

$$\therefore \text{ from (4a), } (AB, CD) = (0, \infty; c, d) = \frac{(0-c)(\infty-d)}{(0-d)(\infty-c)},$$

$$\text{i.e.} \quad \boxed{(AB, CD) = \frac{c}{d}} \qquad (4')$$

VI(a) Next, suppose further that A, B, C, D is a *harmonic range*, i.e. D is the harmonic conjugate of C w.r.t. A, B

$\therefore$ $\mathbf{C} = \mathbf{O} + c\mathbf{I} = \mathbf{A} + c\mathbf{B}$ by the above

and $\mathbf{D} = \mathbf{O} + d\mathbf{I} = \mathbf{A} - c\mathbf{B}$ by Example 5,

i.e. $d = -c$ or $c + d = 0$ (sum of parameters is zero)

$\therefore$ for a harmonic range, we must have

$$\boxed{(AB, CD) = -1} \qquad \text{from } (4')$$

$$= (0, \infty; d, -d)$$

$$\text{i.e. } \lambda = -1.$$

Or we might say that $H(AB, CD)$ is valid.

Möbius' six cross-ratio values now reduce to only three, viz.

$$-1, \tfrac{1}{2}, 2 \; (\because \lambda = -1).$$

Note that, for a harmonic range,

$$(AB, CD) = (BA, CD) = (AB, DC)$$

i.e. *the value of a harmonic cross-ratio is unaltered if only one pair of letters has its order changed.*

VII *Any transversal cuts a pencil of lines in four points whose cross-ratio is equal to that of the pencil* (Fig. 31). (What is the dual of this statement?) Obviously, there is an isomorphic, i.e. 1–1 correspondence, between the two sets of parameters describing points and lines, so the statement is sensible. See Faulkner [38] pp. 24–27.

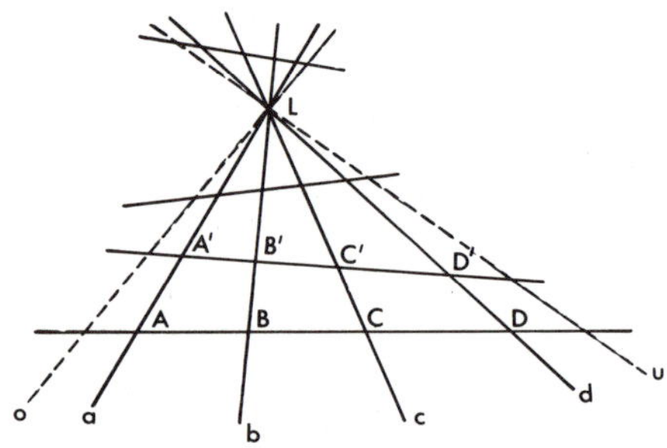

FIG. 31. Equi-cross ranges. Cross-ratio of a pencil

If the vertex of the pencil of lines a, b, c, d is L and if transversals cut these lines of the pencil in points A, B, C, D; A', B', C', D'; . . ., we may express the idea of the cross-ratio of the pencil as

$$L(ab, cd) = (AB, CD) \; (= (A'B', C'D') = \ldots).$$

Alternative ways of writing the cross-ratio of the pencil are $L(AB, CD)$ or just (ab, cd) remembering that in this symbolism a, b, c, d represent lines, not parameters of the lines.

Consequently: two transversals cut the pencil in equi-cross ranges and their point of intersection is self-corresponding (a result known to Pappus). Dually, two pencils in (axial) perspective are related, i.e. two vertices joined to a range of points form equi-cross pencils and their join is self-corresponding. (Consult Figs. 31(a), (b).)

Finally, it might be mentioned that there exists a theory of the cross-ratio of four planes of an axial pencil of planes (§ 18.1) which is similar to the theory for a range of points and a pencil of lines.

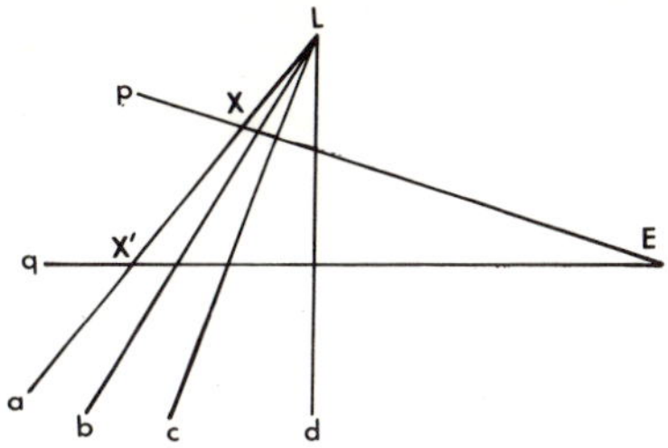

FIG. 31. (a) Related ranges in central perspective

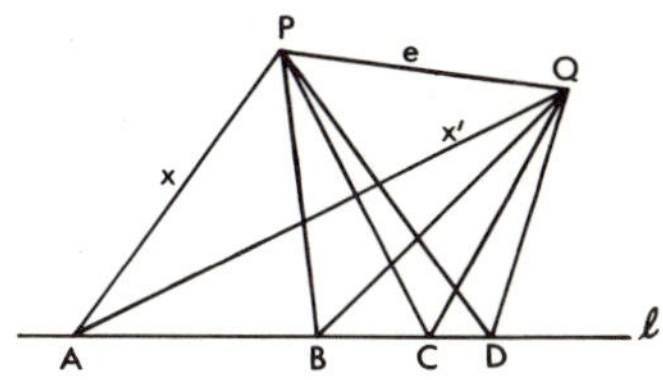

FIG. 31. (b) Related pencils in axial perspective

§ 19.21 Möbius Cross-Ratios. The six cross-ratios listed in III are worthy of some further comment. Labelling them $f_i(\lambda)$ $(i = 1, \ldots, 6)$, since they are indeed all functions of λ, we may derive the following basic facts about them:

(i) $f_1(\lambda)f_2(\lambda)f_3(\lambda)f_4(\lambda)f_5(\lambda)f_6(\lambda) = 1$;

(ii) $\sum_{i=1}^{6} f_i(\lambda) = 3$;

(iii) Writing $(AB, CD) = \sec^2\theta$, we may express the remaining Möbius cross-ratios in terms of the other trigonometrical functions of θ, a result due to the Irish mathematician, John Casey (1820–1891), in about 1871. Of course, they are also expressible in terms of the hyperbolic functions by virtue of *the Gudermannian relation* $\tan \phi = \sinh x$ (Christof Gudermann, 1798–1851, German). Interrelationships among the Möbius cross-ratios are represented by two useful schematic diagrams in O'Hara and Ward [79] p. 191;

(iv) the *cross-ratio group** is isomorphic to the (non-Abelian) permutation group on three letters (and therefore also to the group of rotational symmetries of the equilateral triangle). How would you explain the group law of combination (= product) of two cross-ratio functions?

Earlier, in III, the reader was invited to observe that $(AB, CD) = (BA, DC)$. It seems worthwhile looking at this from a purely geometrical point of view. Notice that the following proof requires three perspectivities to produce the required result.

* The group of the six cross-ratio/functions $f_i(\lambda)$ $(i = 1, \ldots 6)$.

EXAMPLE 17: $(AB, CD) = (BA, DC)$.

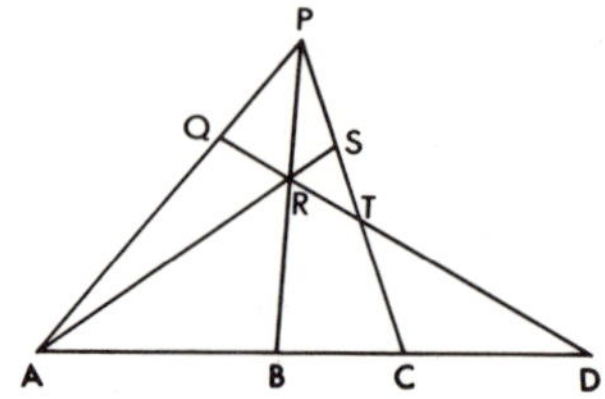

FIG. 32.

Solution (*geometrical*)*:* Through A, C, D draw any lines to establish the points P, Q, T (Fig. 32). Let $PB \cap QD = R$, $AR \cap PC = S$. Then

$$(A, B, C, D) \overset{P}{\overline{\wedge}} (Q, R, T, D)$$
$$\overset{A}{\overline{\wedge}} (P, S, T, C)$$
$$\overset{R}{\overline{\wedge}} (B, A, D, C)$$

i.e. $(A, B, C, D) \overline{\wedge} (B, A, D, C)$
$\therefore$ by § 19 *V*, it follows that
$(AB, CD) = (BA, DC)$.

§ 19.3 Applications of Cross-Ratio. Because of the usefulness of cross-ratio as a technique in a variety of circumstances, we pause for a while to illustrate its usage. The first example selected is a simple, yet important, result.

EXAMPLE 18: If there are more than two self-corresponding points of the projectivity, then every point is a self-corresponding point, i.e. the projectivity is the identity.

Solution: Let E, F, G be three self-corresponding points and let X, X' be a pair of corresponding points.
$\therefore (EF, GX) = (EF, GX')$ by § 19 V.
This can only mean that X, X' coincide. (Refer to Comment (i), Example 19.)

Coming now to Desargues' Theorem (Example 19) we preface our proof by a word of warning. Invariably, the average student experiences difficulty in handling the technique involved, so, with this in mind, a detailed analysis of the problem is offered. Readers ought to examine the proof step by step, following the argument used in the analysis.

EXAMPLE 19 (*Desargues' Theorem in the plane*):
Solution: In Fig. 5(a), let $A'C'$ meet BC, BB', AB in X, Y, Z respectively. (Complete the diagram yourself.)

$$\begin{aligned} \therefore\ M(AA', BL) &= (CX, BL) && \text{(i)}\\ &= C'(CX, BL) && \text{(ii)}\\ &= C'(VY, BB') && \text{(iii)}\\ &= A'(VY, BB') && \text{(iv)}\\ &= (AZ, BN) && \text{(v)}\\ &= M(AA', BN) && \text{(vi)} \end{aligned}$$

$\therefore$ ML, MN, are the same line.
$\therefore$ L, M, N are collinear (Desargues line).

Comments on the technique: (i) If $(AB, CD) = (AB, CD')$ for collinear points A, B, C, D, D', then it readily follows from (4) that $D' = D$, i.e. if three elements of two equi-cross ranges are identical, then the fourth elements coincide, a fact already used in Example 18. Dually, if $(ab, cd) = (ab, cd')$ for concurrent lines a, b, c, d, d', then likewise $d' = d$, i.e. if three members of two equi-cross pencils are identical, then the fourth members coincide.

This simple fact forms the basic philosophy of the technique. Given three concurrent lines MA, MA', MB, if we can show that $M(AA', BL) = M(AA', BN)$ for two pencils, then it must follow that ML and MN are the same line, i.e. that L, M, N are collinear.

(ii) Occasionally students ask: How do you know to start with $M(AA', BL)$? Part of the answer follows from the elaboration of the technique explained in (i) above, always with the proviso that our choice of pencil is not unique. Vertex of the pencil could be L or N, with appropriate variations for the lines of the pencil. Experience in manipulating cross-ratio also produces the necessary wisdom (or craftiness!).

Analysis of the steps in the proof: (i) The right-hand side arises as a section of the given pencil by the transversal BL, in conjunction with § 19 VII.

(ii) Here, a pencil has been formed from a given range and suitable vertex. Actually, many equi-cross pencils may be formed from the given collinear range and arbitrary vertices, but our choice of vertex has to be useful.

(iii) A section of this pencil has been taken by the transversal BV, in conjunction with § 19 VII.

(iv) Switch to another suitable vertex, keeping the collinear range fixed. (Refer to the dual in § 19 VII.)

(v) Take a section of this pencil by the transversal AB.

(vi) The stage is set. Build this range up into the necessary pencil.

Passing now to Pappus' Theorem (Example 20), we employ exactly the same technique as was used in Desargues' Theorem. Detailed elaboration of the reasoning involved is left as an exercise.

EXAMPLE 20 (*Pappus' Theorem*):
Solution: In Fig. 7(a), let $AC' \cap BA' = U$, $BC' \cap CA' = V$. (Again, complete the diagram yourself.)

$$\begin{aligned} \therefore\ M(AB, CL) &= (C'B, VL) \\ &= C(C'X, A'B') \\ &= A(C'X, A'B') \\ &= (UB, A'N) \\ &= M(AB, CN) \end{aligned}$$

$\therefore$ L, M, N are collinear (Pappus line).

EXAMPLE 21 (*Harmonic Theory of the Complete Quadrangle*):
Solution: In Fig. 13, let $RQ \cap PS = T$.

$$\begin{aligned} \therefore\ (AB, CD) &= Q(AB, CD) \\ &= (SP, TD) \\ &= R(SP, TD) \\ &= (BA, CD) \\ \therefore\ (AB, CD) &= -1 \text{ by § 19 VI (a) above.} \end{aligned}$$

Clearly $(SP, TD) = -1$ also.
In pencil form, $R(AB, CD) = R(SP, TD) = -1$
i.e. AD, SD are transversals of a harmonic pencil.
The harmonic property of the complete quadrangle (see § 11, Remark (3)) may be stated thus: *Each païr of opposite sides of the complete quadrangle forms a harmonic pencil with two sides of the diagonal triangle.*

You are advised to make a list of all the harmonic ranges associated with the complete quadrangle.

§20 INVOLUTION

Probably the idea of an involution originated with Desargues, who defined it metrically in 1639. Since then, it has become an indispensable tool in Geometry. Previously, in § 18, we were introduced to an involution which was defined there as a cyclic projectivity Π of period 2. Symbolically, we have, for the involution $\mathscr{I}(=\Pi)$,

$$\begin{aligned} &\mathscr{I}X = X' \\ &\mathscr{I}^2X = \mathscr{I}X' = X = \mathrm{I}X \text{ (where I is the unit projectivity).} \\ &\therefore \mathscr{I}^2 = \mathrm{I} \end{aligned}$$

The basic feature of an involution is that if X' is determined by X, then X is determined by X'.

We proceed to expand this idea in more detail.

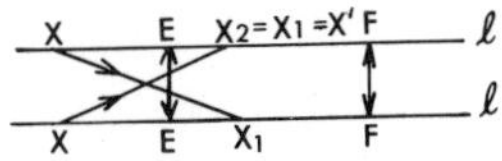

FIG. 33. (a) Involution

Suppose E, F are the double points of two cobasal ranges on the line l. A point X belongs to both ranges (Fig. 33(a)), where, for convenience, the two ranges on l are drawn on separate lines.

Now, to X on the first range corresponds X_1 on the second range, and to X on the second range corresponds X_2 on the first range.

Generally, $X_1 \neq X_2$.

But, if $X_1 = X_2 = X'$, we have an *involution* on l (and the two ranges coincide), i.e. *an involution is a special projectivity between two coincident sets of points* (*i.e. pairs of points of the same set*) *on the same line.*

Otherwise expressed:

Whereas a projectivity is a relation between elements of two sets of objects (possibly coincident), *an involution is a relation between pairs of elements of the* same set.

X, X' are a *point-pair* (*also called* '*mates*') *in the involution* and E, F are the *double points of the involution*, i.e. $E \leftrightarrow E$, $F \leftrightarrow F$, $X \leftrightarrow X'$.

The involution is *hyperbolic* or *elliptic* according as the double points are real or complex. Why cannot the double points of an involution coincide? For the answer, see Involution Property II below.

Dually, we have an *involution pencil*, viz. a special projectivity between two coincident sets of lines (i.e. pairs of lines of the same set) through the same point.

Six concurrent lines are cut by a transversal in an involution. Can you prove that any other transversal cuts these six lines in an involution? Use Involution Property III below.

Properties of an Involution

I Now $(EF, XX') = (EF, X'X)$ since X, X' are mutually corresponding and E, F are self-corresponding

$\therefore$ $(EF, XX') = -1$ by Example 21

$\therefore$ *any pair in involution is harmonically separated by the double points.*

II Since an involution is a particular case of a projectivity, (3) gives

$$pxx' + qx + rx' + s = 0.$$

Also, $px'x + qx' + rx + s = 0$ since the points X, X' are interchangeable.

Subtract $\therefore$ $(q - r)(x - x') = 0$.

$\therefore$ $q = r$ since X, X' are distinct points (i.e. $x \neq x'$)

$\therefore$ the *fundamental equation* of an involution is

$$\boxed{pxx' + q(x + x') + s = 0} \tag{5}$$

or, on dividing by $p \neq 0$, $xx' + Q(x + x') + S = 0$ in an obvious notation,
i.e. $(x - a)(x' - a) = c^2$ where $Q = -a$, $S = a^2 - c^2$,
i.e. $\xi\xi' = c^2$ where $\xi = x - a$, $\xi' = x' - a$.
The ratios of the three constants p, q, s in (5) are known if two pairs of corresponding values of x, x' are known
$\therefore$ *an involution is uniquely determined by two point-pairs.*
Equally well, it is determined by (i) its double points, or (ii) one double point and one pair of corresponding pairs.
For double points, we have $x' = x$, so that

$$\boxed{px^2 + 2qx + s = 0} \tag{5'}$$

Particular notice should be taken of the fact that the double points of an involution cannot coincide (i.e. we cannot have a 'parabolic' involution). If they did coincide, then the condition for equal roots in (5′) is $ps - q^2 = 0$ and this is specifically excluded from a projectivity ($r = q$ in the case of an involution).

Can you think of any construction by which the double points are determined?

III *Cross-ratio property:* The necessary and sufficient condition for the three point-pairs A, A'; B, B'; C, C' to be in involution is that

$$(AB, CC') = (A'B', C'C).$$

To prove that point-pairs belong to an involution, it is often useful to use this property.

IV *Complete quadrangle property:* Any line meets the three pairs of opposite sides of a complete quadrangle in three pairs of points in involution. (A proof of this is given as a Corollary to Theorem 23, Desargues' Involution Theorem in § 23.) Property IV enables us to construct the sixth point of an involution, given five points. Carry out this construction as an exercise.

Pretty clearly the six points in an involution arising from the Complete Quadrangle Property are precisely the six points of a quadrangular set (§ 5). Reverting momentarily to Fig. 13, we see that the line AB meets the opposite sides of the complete quadrangle formed from P, Q, R, S in the point-pairs A, A; B, B; C, D. That is, A, B are double points of an involution of which C, D are a pair, i.e. $H(AB, CD)$ is true.

Note: In Property I above, let the parameters of X, X' be x, x'. We may ake the parameter for E to be 1; and then the parameter for F is -1,

∴ algebraically, I becomes $(1, -1; x, x') = -1$
i.e. $(1 - x)(1 + x') = (1 - x')(1 + x)$
i.e. $x' = \frac{1}{x}$ on simplifying.

This is the standard involutory relation, i.e. if repeated it yields the original. [The term 'involution', of obscure origin, was coined by Desargues in 1639, though not very aptly. O'Hara and Ward [79] suggest *reciprocal projectivity* as a more appropriate term.]

Verify III above taking parameters for A, A', B, B', C, C' to be $a, \frac{1}{a}, b, \frac{1}{b}, c, \frac{1}{c}$ respectively.

Ceva's Theorem (Exercise 5.30), from the projective viewpoint, is a fruitful illustration of the interaction of cross-ratio and involution.

Theorem 17: Four collinear points A, B, C, D determine three distinct involutions, and the double points of these involutions are ordinary points in each of the other two involutions.

Proof: Since an involution is uniquely determined by two pairs of corresponding points, the four given points define the following three possible involutions with double points as specified:

Symbol	Two pairs of points	Double points
$\mathscr{I}_1$	$A, B; C, D$	E_1, F_1
$\mathscr{I}_2$	$A, C; B, D$	E_2, F_2
$\mathscr{I}_3$	$A, D; B, C$	E_3, F_3

Now $(AB, E_1F_1) = -1 = (CD, E_1F_1)$ by Involution Property I
∴ $(AB, E_1F_1) = (CD, F_1E_1)$ (i) by Cross-Ratio Property VI
$= (DC, F_1E_1)$ (ii) by Cross-Ratio Property VI

∴ (i) implies that $A, C; B, D; E_1, F_1$ are in involution (i.e. $\mathscr{I}_2$)
(ii) implies that $A, D; B, C; E_1, F_1$ are in involution (i.e. $\mathscr{I}_3$)

where Involution Property III has been invoked.
∴ the double points E_1, F_1 of $\mathscr{I}_1$ are ordinary points of both $\mathscr{I}_2$, $\mathscr{I}_3$. Similarly, double points E_2, F_2 of $\mathscr{I}_2$ are ordinary points of both $\mathscr{I}_3$, $\mathscr{I}_1$, and double points E_3, F_3 of $\mathscr{I}_3$ are ordinary points of both $\mathscr{I}_1$, $\mathscr{I}_2$.

Comments:(i) $(E_1F_1, E_2F_2) = -1$ and two other similar results.

(ii) $\mathscr{I}_1$, $\mathscr{I}_2$ (say) have real double points whereas $\mathscr{I}_3$ has imaginary double points (Hatton [47]).

(iii) Another statement of the Theorem is: *Two involutions on the same line have one common pair.*

Theorem 18: A projectivity interchanging two points (on a line) must be an involution.

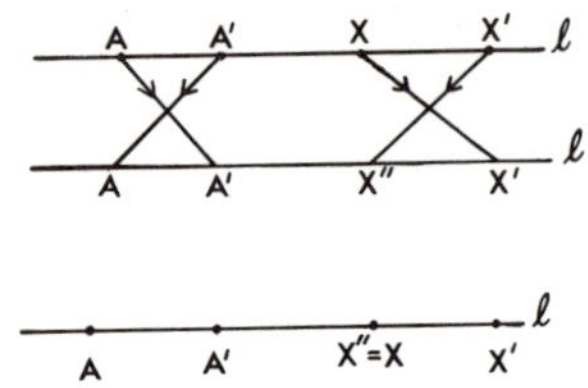

FIG. 33. (b)

Proof: Call the projectivity Π on the line l.

$$\text{Let}\ \begin{cases} \Pi A \;= A' \text{ signifying } A \xrightarrow{\Pi} \mathbf{A}' \text{ (Fig. 33(b))} \\ \Pi A' = A \\ \Pi X \;= X' \\ \Pi X' = X'' \end{cases}$$

i.e. A and A' are interchanged. We wish to show that a second pair X, X' are also interchanged, i.e. that $X'' = X$.

Now $(A, A', X, X') \barwedge (A', A, X', X'')$ given

But $(A, A', X, X') \barwedge (A', A, X', X)$ by Example 17

$\therefore (A', A, X', X) \barwedge (A', A, X', X'')$

$\therefore X'' = X$

$\therefore \Pi$ is an involution

EXAMPLE 22: Any projectivity between points on a line may be expressed as the product of two involutions.

Solution: Set $\Pi A = A'$, $\Pi A' = A''$, where Π is the given projectivity.

Consider the involution $\mathscr{I} : \mathscr{I} A = A''$, $\mathscr{I} A' = A'$ (so that $\mathscr{I} A'' = A$)

$\therefore \mathscr{I}\Pi A = A'$, $\mathscr{I}\Pi A' = A$, i.e. $\mathscr{I}\Pi$ interchanges A, A'

$\therefore \mathscr{I}\Pi$ is an involution, $\mathscr{J}$, by Theorem 18, i.e. $\mathscr{I}\Pi = \mathscr{J}$.

$\therefore \mathscr{I}^2\Pi = \mathscr{I}\mathscr{J}$

$\therefore \quad \Pi = \mathscr{I}\mathscr{J} \because \mathscr{I}^2 = \mathrm{I}$ (the unit projectivity)

Many interesting properties of products of involutions are to be found in, say, Coxeter [28] and the pleasure to be derived from their investigation should be sufficient motivation for further investigation.

EXAMPLE 23: Two projectivities α, β commute if $\alpha\beta P = \beta\alpha P$, for all points P. Granted that α, β are involutions, prove that $\alpha\beta = \beta\alpha$ if and only if $\alpha\beta$ is also an involution.

Solution: Now $\alpha^2 = \beta^2 = \mathrm{I}$ since α, β are involutions (I is the unit projectivity). If $\alpha\,\beta = \gamma$ is an involution, then $(\alpha\,\beta)^2 = \mathrm{I}$

i.e. $\alpha\beta\alpha\beta = \mathrm{I}$

i.e. $\beta\alpha\alpha\beta\alpha\beta = \beta\alpha\mathrm{I}$ pre-multiplying both sides by $\beta\alpha$

$\alpha\beta = \beta\alpha \quad \because \alpha^2 = \beta^2 = \mathrm{I}$ (given)

i.e. α, β commute.

Conversely, $\alpha\beta = \beta\alpha$ implies $(\alpha\beta)^2 = \mathrm{I}$, i.e. $\alpha\beta$ is an involution.

Note: $\beta\gamma = \beta\alpha\beta = \beta\beta\alpha = \beta^2\alpha = \alpha$ and $\gamma\alpha = \beta$ similarly. Under the conditions of the problem, the involutions α, β, γ and the unit I are elements of the (Abelian) Klein group (*Vierergruppe*).

EXAMPLE 24: If $(A, B, C, D, E, F) \barwedge (B, C, D, A, E, F)$, prove that A, C and B, D are two pairs of an involution of which the double points are E and F.

Solution: $(A, B, C, D, E, F) \barwedge (B, C, D, A, E, F)$ under a projectivity Π

$\barwedge (C, D, A, B, E, F)$ under Π^2 (i.e. Π repeated)

But under Π^2, $A \leftrightarrow C$, $B \leftrightarrow D$, $E \leftrightarrow E$, $F \leftrightarrow F$.

$\therefore$ by Theorem 18, the projectivity must be an involution.

$\therefore$ A, C; B, D are point-pairs and E, F are double points of this involution.

EXAMPLE 25: Five points A, B, C, I, J lie on a line l, and A is a double point of the involution $\mathscr{I}_1$ determined by I, J and B, C. Prove

$$(IJ, AB) = (IJ, CA).$$

If, further, B is a double point of the involution $\mathscr{I}_2$ determined by I, J and C, A, prove that C is a double point of the involution $\mathscr{I}_3$ determined by I, J and A, B.

Solution: $(IJ, AB) = (JI, AC)$ from $\mathscr{I}_1$

$= (IJ, CA)$ by § 19 III (i)

Also $(IJ, AB) = (JI, CB)$ from $\mathscr{I}_2$ (ii)

$\therefore$ (i), (ii) $\Rightarrow (IJ, CA) = (JI, CB)$

$\therefore$ C is a double point of the involution $\mathscr{I}_3$ determined by I, J and A, B.

In the next Chapter, we will use a projectivity to define a conic (which, from our previous experience, we know to be a curve represented by a quadratic equation). At first sight, this seems an astonishing expectation. However, the equation (3) of a projectivity provides a connecting link.

Suggested Topics for Further Study

(1) Interrelationships among Desargues' Theorem, Pappus' Theorem and the Fundamental Theorem (Seidenberg [98]).

(2) Von Staudt's Algebra of 'Throws' (*Würfe*) (Coxeter [27], Veblen and Young [113], von Staudt [115] p. 166, Scott, Charlotte A.: 'On von Staudt's *Geometrie der Lage*', *Math. Gaz.*, 1 (1896–1900) 307–314, 323–331, 363–370).

EXERCISES 5

1. Is the product of two perspectivities another perspectivity? Under what circumstances may this happen?

2. Discover the equation of the projectivity given by the three pairs of corresponding values 0, -1; -1, $-1\frac{2}{3}$; ∞, -3. Determine the nature of this projectivity, e.g. is it hyperbolic?

3. If k_1, k_2 are the eigenvalues of the matrix $\mathbf{A}$ of the transformation

$$\rho x_1' = ax_1 + bx_2$$
$$\rho x_2' = cx_1 + dx_2$$

where $\rho \neq 0$ is a factor of proportionality, show that k_1, k_2 give the double points and that, corresponding to k_1, the double root is given by

$$(x_1, x_2) = (-b, a - k_1) = (d - k_1, c).$$

4. What value of x corresponds to $x = \infty$ in the fundamental equation of the projectivity?

5. Coincident double points of a projectivity are taken to be ∞, ∞. Establish that the canonical equation of the projectivity is then $x' = x + c$ (c a constant).

6. Related ranges (on distinct lines) with a self-corresponding point are in perspective. Prove.

7. Double points of the projectivity $xy - 8x + y + 12 = 0$ have parameters α, β. Evaluate the cross-ratio $(xy, \alpha\beta)$.
[This interesting result tells us that, when the double points (with parameters α, β) of a projectivity are distinct, the equation of the projectivity relating elements of the same set may be expressed as $(xy, \alpha\beta) = k$ where x, y are the parameters of variable pairs of corresponding points of the projectivity, and k is a constant.]

8. Given two pairs of corresponding points and one double point of two related ranges on the same line, can you construct the other double point?

9. Set up two perspectivities to obtain the projectivity $(A, B, C) \barwedge (B, A, C)$ for three collinear points A, B, C.

10. If $(A, B, C, \ldots, P) \barwedge (A', B', C', \ldots, P')$ and the corresponding points are on two distinct lines l, l', then the intersections of cross-joins such as $AX' \cap A'X$ are collinear. Prove. (This line of collinearity is the *cross-axis* of the projectivity.)

11. In Exercise 5.10, let $Z, O (\in l)$ correspond respectively to $O', Z (\in l')$. Verify that OO' is the cross-axis. What happens if $O' = O$, i.e. if Z is self-corresponding?

12. Four skew lines in [3] have, in general, just two transversals. True or false? (Hint: Use Theorem 4.) What happens if there exist more than two transversals?

13. Evaluate the following cross-ratios of distinct points

(i) $(\infty, b; c, d)$
(ii) $(\infty, 0; x, kx)$
(iii) $(a, b; \frac{1}{2}(a + b), \infty)$
(iv) $(a, -a; 1, a^2)$
(v) $(a, b; \frac{1}{2}(a + b), a - b)$
(vi) $\left(1, -1; x, \dfrac{1}{x}\right)$.

14. Given that $(a, b; c, d) = -1$, prove that

(i) if $d = 0$, then $\dfrac{1}{a} + \dfrac{1}{b} = \dfrac{2}{c}$
(ii) if $d = \infty$, then $a + b = 2c$.

15. Allowing $(AB, CD) = (AC, DB) = \lambda$, determine λ. (This is the case of an *equianharmonic range*. What are the values now of the six Möbius cross-ratios?)

16. Solve the cross-ratio equation $(x, 4; 2, 3) = (3, 2x; 1, 4)$ $(x \neq 2)$.

17. If the parameters of A, B and A', B' are given respectively by the equations $ax^2 + 2hx + b^2 = 0$ and $a'x^2 + 2h'x + b'^2 = 0$, then the condition for $(AB, A'B') = -1$ is $2hh' - ab' - a'b = 0$. Prove.

18. Vertex of a plane homology (§ 16) is Z and the axis of homology is XY. Let $P(a, b, c)$, $P'(a, b, kc)$ be corresponding points of the homology. If $ZPP' \cap XY = Q$, prove that

$(Z, Q; P, P') = k$. [When k (the modulus of the homology) $= -1$, the homology is harmonic.] Observe the connection between the exercise and Exercise 5.13(ii).

19. Given that $(AB, CD) = k$, $(AB, CD') = k'$, show that $(BC, DD') = (0, 1; k, k')$.

20. Let PQR be a triangle and O a general point. Suppose $OP \cap QR = A$, $OQ \cap RP = B$, $OR \cap PQ = C$. Further define A', B', C' by $H(QR, AA')$, $H(RP, BB')$ and $H(PQ, CC')$ respectively. Establish the collinearity of A', B', C'.

21. Given a set of five collinear points O, I, A, B, C show that $(OI, AB).(OI, BC).(OI, CA) = 1$.

22. Determine the value of the number k for which the pairs of points with parameters 2, 1; 1, 3; 4, k are in involution.

23. If (1, 0), (0, 1) are corresponding points of an involution, show that its equation becomes $xx' = k$(const.).

24. Reduce to the canonical form $x' = -x$ the equation of an involution in which the double points are given parameters $0, \infty$.

25. One pair of points of an involution always harmonically separates a given pair of points. Prove.

26. In Fig. 15, establish that O, C; A, B are pairs of points in an involution of which I is a double point.

27. Establish, in Fig. 16, that O, I; A, B; U, C are pairs of points in an involution.

28. Lines joining any point O to four collinear points A, B, C, D cut any transversal through D in A', B', C', D respectively, and $BC' \cap OA = O'$, $O'B' \cap AB = K$. If $(AB, CD) = -1$, prove that $(AB, KC) = -2$ (Hatton [47] p. 87).

29. Two pairs of line-pairs determining an involution are given by

$$\phi_1 \equiv a_1x^2 + 2h_1xy + b_1y^2 = 0, \qquad \phi_2 \equiv a_2x^2 + 2h_2xy + b_2y^2 = 0.$$

Double lines of this involution are then given by

$$\frac{\partial\phi_1}{\partial x} \cdot \frac{\partial\phi_2}{\partial y} = \frac{\partial\phi_2}{\partial x} \cdot \frac{\partial\phi_1}{\partial y}.$$

Prove. (Durell [33] p. 167.)

30. (*Ceva's Theorem.*) If ABC is a triangle, O is any point, and $AO \cap BC = D$, $BO \cap CA = E$, $CO \cap AB = F$ and if, further, BC, CA, AB meet an arbitrary line in L, M, N respectively, then $(BC, DL).(CA, EM).(AB, FN) = -1$ (Heading [48]).

[This Theorem was first proved metrically by Giovanni Ceva (1647–1734, Italian). Lines such as AO, BO, CO are known as *Cevians*.]

Later (Chapter 8), we will come across the familiar form of Ceva's Theorem.

CHAPTER 6

THE CONIC : GEOMETRIC THEORY

> 'There exist a small number of very simple fundamental relations which contain the scheme, according to which the remaining mass of theorems [in projective geometry] permit of orderly and easy development.
>
> By a proper appropriation of a few fundamental relations one becomes master of the whole subject; order takes the place of chaos, one beholds how all parts fit naturally into each other, and arrange themselves serially in the most beautiful order, and how related parts combine into well-defined groups. In this manner, one arrives, as it were, at the elements, which nature herself employs in order to endow figures with numberless properties with the utmost economy and simplicity.'
>
> (Steiner)

EVERYONE familiar with ordinary plane geometry knows that only a small, though important, portion of the subject is covered by the incidence properties of point and line. Curves, such as conics, exist and these are obviously of a higher degree of complexity than lines. How may they be introduced into Projective Geometry?

Various equivalent definitions of a conic are at our disposal from previous knowledge. Reviewing these briefly, we have

(a) *the conic as a section of a cone*, a concept of ancient origin dating to Apollonius (third century B.C., Greek);

(b) *the conic defined in terms of the focus and directrix property*, a discovery due to Pappus;

(c) *the conic as a curve represented by an equation of the second degree*, a description made possible only after the advent of Descartes' analytical methods; and

(d) *the conic as a curve cut by any line in two points*, an analytical justification of which is straightforward in the field of complex numbers.

With a little effort, one can show that these equivalent statements are interdependent among themselves.

In Projective Geometry a conic must be so defined as to be free from any metrical notions and to cover all the properties of the special conics (hyperbolas, ellipses, circles, parabolas). Surprisingly, related ranges and pencils can be employed in this definition. This novel and powerful idea is due to Steiner and Michel Chasles (1793–1880, French). Initially the theory may seem strange but as it develops we will find ourselves dealing with familiar objects. In fact, we shall begin by defining two new (dual) objects, point-conic and line-conic, from which we build up the concept of conic.

There is some beautiful mathematics in the reasoning which now follows and it is our hope that the reader will derive from it the pleasure which the study of it merits.

§ 21 PROJECTIVE GENERATION AND ELEMENTARY PROPERTIES OF THE CONIC

§ **21.1 Point-conic. Definition** (due to Steiner [107] in 1832): A *point-conic* is the locus of the points of intersection of corresponding lines (rays) of two non-perspective related pencils with distinct vertices (Fig. 34(a)).

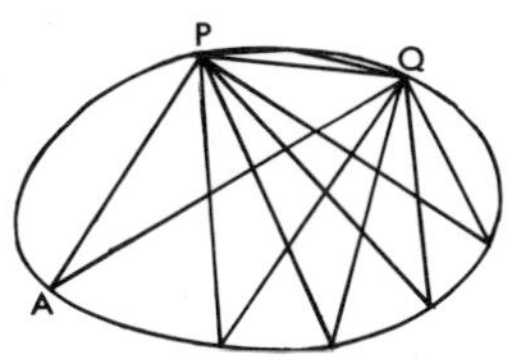

FIG. 34. (a) Steiner's definition of a point-conic

Thus, in Fig. 34(a), the locus of A, the point of intersection of corresponding lines PA, QA of two related pencils with vertices P, Q, is a point-conic. *Note:* A point is either on, or not on, a point-conic. 'Insideness' and 'outsideness' are not projective concepts. A valid projective result is not invalidated merely because a satisfactory diagram cannot be drawn. The projective plane is wider in scope than the real Euclidean plane in which the reader is accustomed to drawing diagrams.

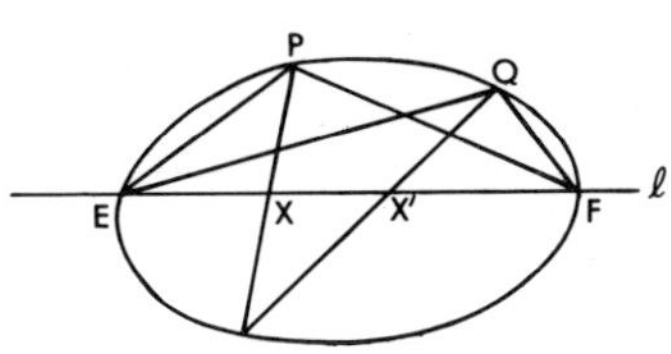

FIG. 35. Intersection of a line and a point-conic

Let a line l meet the point-conic of Fig. 34(a). It meets the two related pencils in two related ranges, of which X, X' are a representative corresponding pair.

But these related ranges will have two self-corresponding points, say E, F, by equation (3a), and they are the intersections (Fig. 35) of corresponding lines of the two pencils (and thus lie on the point-conic). It follows that two points of any line are points of the point-conic, i.e.

I *Any line meets the point-conic in two points.*

When it is desired to draw attention to the fact that a line l meets the point-conic in two points, say E, F, the line may be referred to as the *secant* EF. Further, the pencil from Q has one line through P which must meet the corresponding line of the pencil from P in P,

i.e. P lies on the point-conic.

Similarly, Q lies on the point-conic.

Hence

II *The vertices of the defining pencils lie on the point-conic.*

§ **21.11 Tangent.** From the definition of a point-conic (Fig. 34(a)), a line (secant) through P meets the point-conic in P, A so that PA, QA are corresponding lines. When $A = P$, the line PA meets the point-conic in two coincident points and is then called the *tangent* at P. That is, the tangent at P is the line of the pencil from P corresponding to the line QP of the pencil from Q. See Fig. 35(a), called a *degenerate* (= *singular*) *point-conic.*

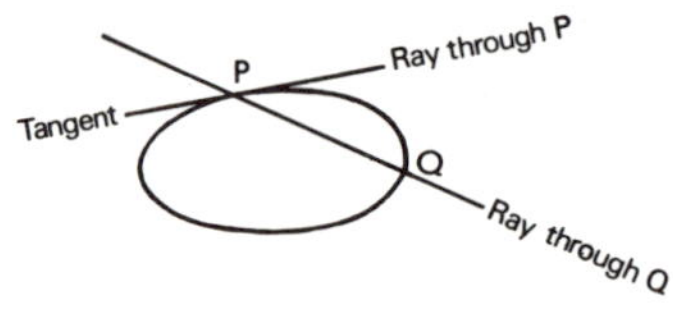

FIG. 35. (a) Tangent

Sometimes we write the tangent at P as PP.

§ **21.12 Degenerate Point-Conic.** Suppose the pencils are so related that the line PQ corresponds to the line QP (i.e. the line joining the vertices of the pencils is self-corresponding). We show that the point-conic *degenerates* into a line-pair (i.e. two lines).

With regard to Fig. 36 let PA, QA and PB, QB be pairs of corresponding lines. Now, AB meets the pencils from P, Q in related ranges.

∴ there are three self-corresponding points A, B, C

∴ every point of AB is self-corresponding (Example 18)

∴ every point of AB is a point of intersection of corresponding lines of related pencils

∴ the point-conic contains AB.

Let X be *any* point of PQ

∴ PX, QX are a pair of corresponding lines, i.e. X lies on the point-conic (by definition)

∴ the point-conic contains PQ.

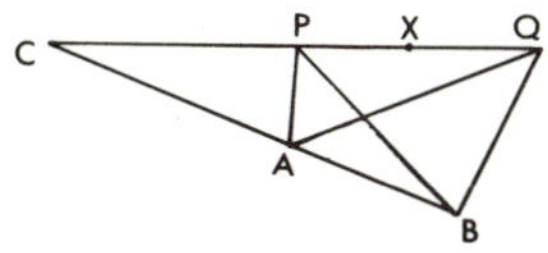

FIG. 36. Degenerate point-conic

There cannot be more than two lines in the degenerate point-conic since an arbitrary line meets the point-conic in two points. Thus (Fig. 36)

III *The point-conic may degenerate into a line-pair (which may coincide).*

Observe that the degenerate point-conic is the points of intersection of corresponding lines of two related pencils in perspective where the line ABC is the axis of perspective and PQ is the line joining the vertices of the perspective pencils.

Note: Unless it is specifically stated to be degenerate, a point-conic will always be assumed to be non-degenerate (= *non-singular*).

§ 21.13 Uniqueness of a Conic. By the dual of Theorem 14 in § 18, two related pencils are completely defined if three pairs of corresponding lines are given. Accordingly, the two vertices of the pencils and the three points of intersection of corresponding lines suffice to determine the general point-conic (provided always that no three of the five points are collinear for then the point-conic would degenerate). That is, we have the basic result (known to Pappus)

Theorem 19: Five points determine a point-conic uniquely.

Theorem 19 means, for instance, that an infinite number of point-conics may be made to pass through four given points. A particular point-conic of this infinite system is uniquely specified if it is made to pass through a (fifth) assigned point, distinct from the given four.

In the construction of the point-conic, we took P, Q as fixed vertices of the two related pencils, and A, B, C, . . . as the points of intersection of corresponding lines of the related pencils. Would we have obtained the *same* conic if we had taken, as the vertices of the two defining pencils, any two points on the conic other than P, Q (say A, B)? The answer is: Yes. However, this requires proof and the reader is referred to O'Hara and Ward [79] p. 99 or Tuller [148] pp. 58–59 for details. Thus,

Any two distinct points on a point-conic can be taken as vertices of the generating (= defining) pencils.

§ **21.14 Line-conic.** By now it has probably occurred to the reader that it might be interesting to dualise the notion of a point-conic, i.e. a set of points satisfying a given condition—this is what we mean by the term 'locus'. Duality leads to the notion of a line-conic, i.e. a set of lines satisfying a given condition—this is what we mean by the term 'envelope'. The concept of a line-conic originated with Chasles in 1828.

Definition: A *line-conic* (*conic-envelope*) is the envelope of the join of corresponding points of two non-perspective related ranges of points on distinct bases (Fig. 34(b)).

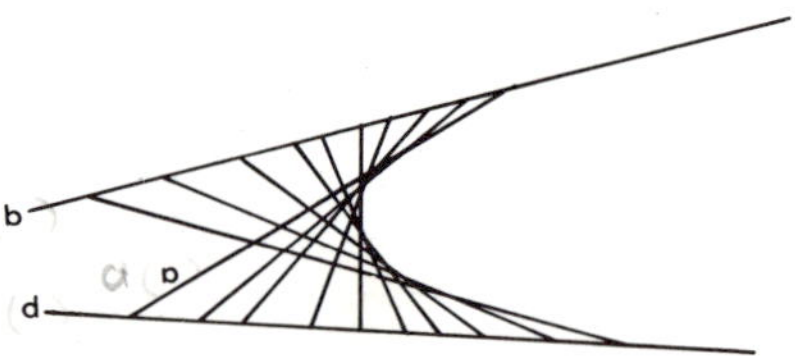

FIG. 34. (b) Chasles' definition of a line-conic

Observe particularly that

a line-conic is a set of lines,

not the illusory 'curve' which these lines appear to indicate.

Dually to the methods used for the point-conic, the following features of a line-conic can be established:

I′ *Two lines through any point are lines of the line-conic.*

II′ *The bases of the defining ranges belong to (lie on) the line-conic.*

III′ *The line-conic may degenerate into a pair of points* (regarded as pencils of lines through the points). *The two points may coincide.*

A degenerate line-conic (III′) would occur if, in the definition, we did not impose the restriction of non-perspectivity on the defining ranges.

Corresponding to Theorem 19, we have

Five lines determine a line-conic uniquely.

Furthermore

Any two distinct lines on a line-conic can be taken as bases of the defining ranges.

Dual of a tangent to a point-conic is a *point of contact* of a line-conic, namely if p, q are the bases of the defining ranges, the point of contact on the line p is the point of the range on p corresponding to the point $q \cap p$ of the range on q. That is, a point of contact is a point which is on two coincident lines of a non-degenerate line-conic.

The following results may be proved (Veblen and Young [113] 1, pp. 116–118):

(i) *Tangents to a point-conic form a line-conic.*

(ii) *Points of contact of a line-conic form a point-conic.*

Clearly, the point common to a tangent and a point-conic is the point of contact of the tangent.

It is now possible for us to amalgamate the dual concepts of point-conic and line-conic into one concept.

Definition: A *conic* is a point-conic and its tangents.

Clearly then

A conic is a self dual concept.

From the earlier discussion it follows, for example, that

(a) a unique conic passes through five points;

(b) a unique conic touches five lines.

Hereafter we will not generally distinguish between point-conic and line-conic unless it is necessary to do so.

Query: How would you describe the space dual of (a) a point-conic, and (b) a conic-envelope? These space-duals are called a *cone of planes* and a *cone of lines* respectively.

§ 21.2 Chasles' Theorem. Taking six points A, B, C, D, P, Q on a point-conic, we have that $P(AB, CD) = Q(AB, CD)$ (by definition, § 19 V dual, and II above). Keeping A, B, C, D fixed and letting P, Q be positions of a variable point on the conic so that $P(AB, CD) =$ constant, we have the simple but valuable result (Fig. 37(a)):

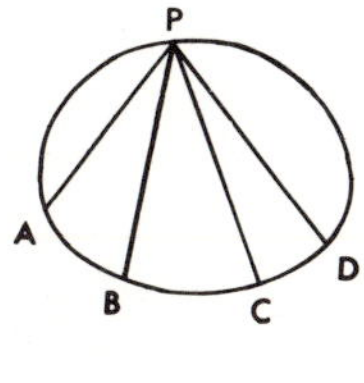

FIG. 37. (a) Chasles' theorem

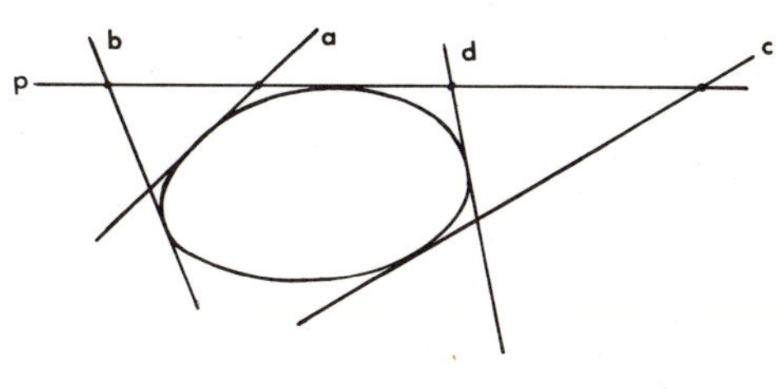

FIG. 37. (b) Dual of Chasles' theorem

Theorem 20 (*Chasles' Theorem*): The cross-ratio of the pencil of four lines joining four fixed points on a conic to a variable point on the conic is constant (and conversely).

On p. 3 of [21], Chasles states his result in these words: '*Si de quatre points d'une conique on mène des droites à un cinquième point de la courbe: le rapport anharmonique de ces droites a une valeur constante, quel que soit le cinquième point.*'

(A similar statement occurs in Chasles [22] for a circle.)

State the Converse of Chasles' Theorem.

§ 21.21 Applications of Chasles' Theorem and its Dual.

EXAMPLE 26: Dualise Chasles' Theorem.

Solution: Dual of Chasles' Theorem: The cross-ratio of the range of the four points of intersection of four fixed tangents to a conic with a variable tangent to the conic is constant.

Fig. 37(b) describes this dualisation.

EXAMPLE 27: A variable line through a fixed point A meets the sides LM, MN, NL of a given triangle LMN in X, Y, Z respectively so that $(ZY, PX) = k$ (constant), where P is a variable point on the line. Prove that the locus of P is a conic through A, L, M, N (Fig. 38).

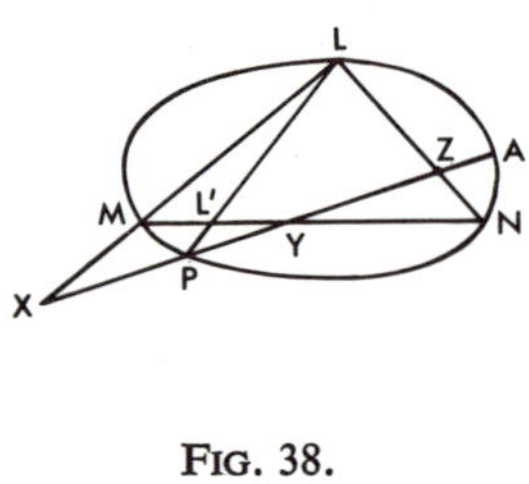

FIG. 38.

Solution:

$$\begin{aligned} &\text{Let} \quad PL \cap MN = L'. \\ &\text{Now} \quad (ZY, PX) = k \\ &\therefore \quad L(ZY, PX) = k \\ &\text{i.e.} \quad (NY, L'M) = k \\ &\therefore \quad P(NY, L'M) = k \\ &\text{i.e.} \quad P(NA, LM) = k \end{aligned}$$

$\therefore$ by the Converse of Chasles' Theorem, the locus of P is a conic through A, L, M, N.

Theorem 21 (*Steiner's Theorem*): If two triangles circumscribe a conic, then their six vertices lie on a conic.

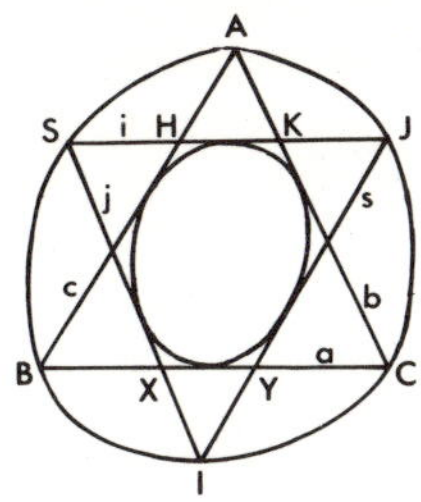

FIG. 39. Steiner's theorem

Proof: Let the two triangles be ABC, IJS (Fig. 39).
Label sides opposite A, B, C in ABC as a, b, c.
Label sides opposite I, J, S in IJS as i, j, s.

$$\therefore A(BC, JS) = (HK, JS)$$
$$= i(cb, sj)$$
$$= a(cb, sj) \text{ by the Dual of Chasles' Theorem}$$
$$= (BC, YX)$$
$$= I(BC, JS)$$

$\therefore A, B, C, I, J, S$ lie on a conic (by the Converse of Chasles' Theorem).

Dually: If two triangles are inscribed in a conic, their six sides touch a conic.

The original proof of this Theorem may be found in Steiner [107] 2, p. 37. Later (Chapter 8) it will be seen that Steiner's Theorem is a generalisation of a well-known result in Euclidean Geometry.

A conic which passes through the vertices of a triangle (i.e. which circumscribes the triangle) may be called a *circum-conic* of the triangle. A conic which touches the sides of a triangle (i.e. which is inscribed in the triangle) is called an *in-conic* of the triangle.

It is a natural thought to extend the concept of a linear involution to an involution on a conic. This we now proceed to do.

§ 22 INVOLUTION ON A CONIC

Very clearly in Fig. 40, which is self-explanatory, there is a 1–1 correspondence between points like P, P' on the conic and the fixed line l respectively. That is, to a range of points on l there corresponds a *range of points on the conic*. Related ranges on l thus give rise to related ranges on the conic, an idea originating in 1838 with Giusto Bellavitis (1803–1880, Italian).

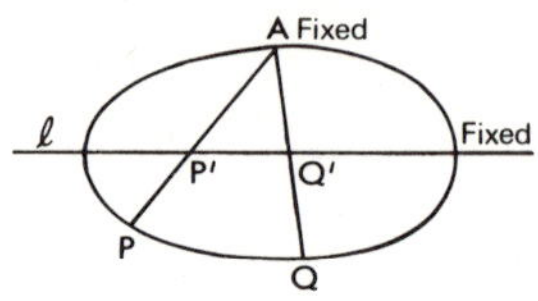

FIG. 40. Involution (pencil) on a conic

If this projectivity on l is an involution (having P', Q' as a pair of corresponding points), then P, Q are a pair in *involution on the conic.* Joining the points of the involution to any given point on the conic yields an ***involution pencil.***

Theorem 22: Secants of a conic drawn from a fixed point meet the conic in pairs of points in involution (and conversely).

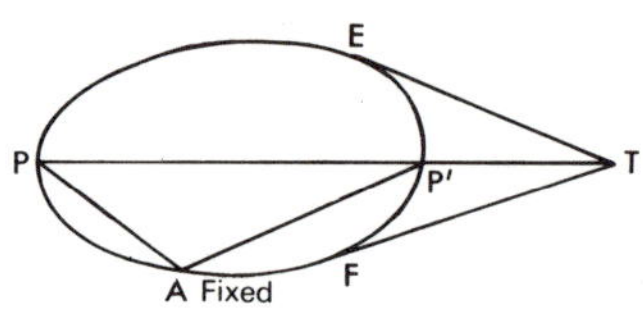

FIG. 41. Involution secants

Proof: Let a secant from a given point T meet the conic in P, P', let E, F be the points of contact of tangents from T to the conic, and let A be a fixed point on the conic (Fig. 41). (Of course, T may be 'inside' the conic, and the diagram is modified accordingly.)

A unique 1–1 correspondence exists between the lines AP, AP'.

∴ AP, AP' are a pair of lines of an involution pencil,

i.e. P,P' are a point-pair of an involution on the conic, of which E, F are the double points.

(Can you prove the converse? It states: Every involution on a conic is given by secants through a fixed point.) W.r.t. Theorem 22, do Exercise 7.5.

EXAMPLE 28: From O, tangents OU, OV are drawn to a conic and a variable secant meets the conic in X, Y. If P is a variable point, let $PU \cap$ secant $= H$, $PV \cap$ secant $= K$ (Fig. 42). Prove $(HK, XY) = -1$. (This result is due to Franz Seydewitz (1807–1852)).

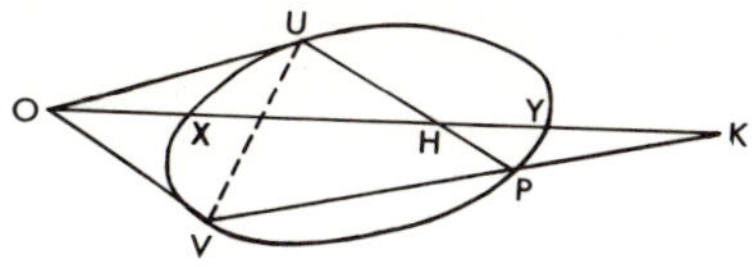

FIG. 42. Seydewitz's theorem

Solution: By Theorem 22, U, U; V, V; X, Y are in involution on the conic. $\therefore$ PU, PV are double lines of the involution pencil of which PX, PY are a pair

$$\therefore P(UV, XY) = -1$$
$$\text{i.e. } (HK, XY) = -1.$$

This Example may be used (e.g. Coxeter [28] p. 77) to establish the validity of Steiner's definition of a conic (see Theorem 26).

From the creative brain of Desargues sprang another of the major theorems of plane Projective Geometry, to which we now devote our attention. (There can be no possible confusion between Desargues' Theorem on Perspective Triangles (Theorem 5), and Desargues' Theorem on Involution (Theorem 23).

§23 DESARGUES' INVOLUTION THEOREM

Theorem 23 (Desargues' Involution Theorem): A system of conics through four fixed points is met by any line in pairs of points in involution (and conversely).

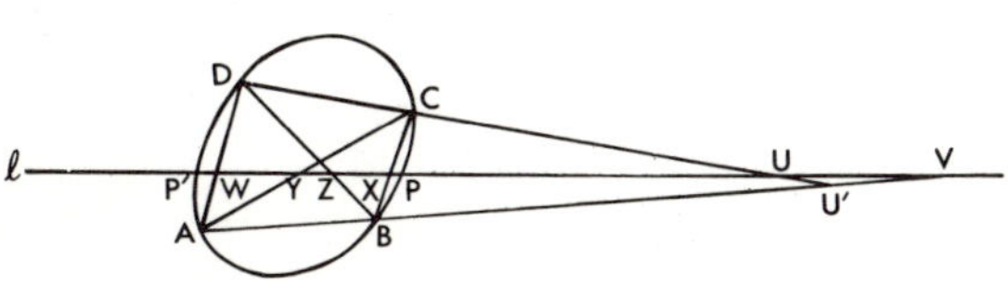

FIG. 43. Desargues' involution theorem

Proof: First recall (Theorem 19) that five points are necessary to determine a conic uniquely. Consequently, an infinite number of conics passes through four fixed points A, B, C, D (or, as we say, such a conic has one degree of freedom). One such conic is met by the arbitrary line l in P, P' (Fig. 43).

Note that l, A, B, C, D are fixed. We vary the conic through A, B, C, D, i.e. P, P' are variable. There are infinitely many possible involutions on l, but we are only concerned with the one determined by the two pairs of points W, X; Y, Z.

Now $(WY, PP') = A(DC, PP')$
$= B(DC, PP')$ by Chasles' Theorem
$= (ZX, PP')$
$= (XZ, P'P)$ by § 19 III

∴ by § 20 III we have that W, X; Y, Z; P, P' are pairs in involution
∴ as the conic varies, points like P, P' belong to this involution.

§ 23.1 Notes on Desargues' Theorem.

(i) '*Cette belle proposition*', as Poudra called Theorem 23, was discovered by Desargues in 1639 and may be found in its original form in his Brouillon project [30] 1, pp. 186 *et seq.*

(ii) The involution on l has two double points
∴ there are just two conics of the system which touch l.

(iii) In the system, there are three degenerate conics, viz. AB, CD; AC, BD; AD, BC (i.e. the pairs of opposite sides of the quadrangle formed from A, B, C, D). These three line-pairs are met by l in three point-pairs belonging to the involution so U, V are a pair in the involution. That is, we have now proved the Complete Quadrangle Property of an involution (§ 20 IV). However, it can be quite simply proved independently of Desargues' Theorem. Omitting the conics and letting $DC \cap AB = U'$, we have

Corollary: $(WY, UV) = A(DC, UU')$
$= B(DC, UU')$
$= (ZX, UV)$
$= (XZ, VU)$ by § 19 III

∴ $W, X;\ Y, Z;\ U, V$ are in involution, by § 20 III (a result known to Pappus [80] Lemma IV, pp. 675–677).

Consequently, Theorem 23 may be extended to state:

If a system of conics circumscribes a complete quadrangle, any line meets the conic and the opposite sides of the quadrangle in pairs of points in involution.

It might be noted in passing that, in Fig. 43, we have $Q\begin{pmatrix} W & Y & U \\ X & Z & V \end{pmatrix}$, i.e. an involution based on a quadrangular set. Moreover, we should record the implicit assumption in Fig. 43 that l does not pass through a vertex of the complete quadrangle.

(iv) The Dual of Desargues' Theorem is Sturm's Theorem (Jacques Sturm, 1803–1855, Swiss).

(v) Theorem 23 deals with an involution of points *on a line* determined by a pencil (= system) of conics and has nothing to do with an involution of points *on a conic.*

§ 23.2 Types of Conic Contact. Ordinarily, two conics intersect in four points A, B, C, D (see § 28 IV). But the conics may be so intimately allied that one or more of these common points coalesce. Various types of contact thus created are shown in Fig. 44. (No more than four points may be in common, for otherwise the conics would coincide entirely by Theorem 19). Osculating contact (= three point contact) derives from the Latin verb *osculare* ≡ 'to kiss'.

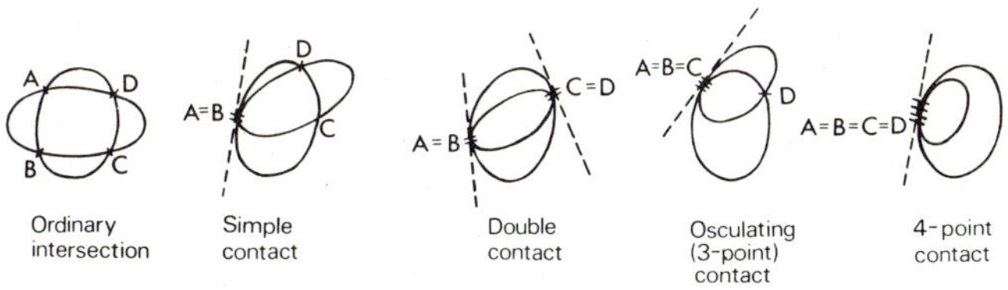

FIG. 44. Types of conic contact

Dually, there exist common coalescing tangents of a pair of conics.

§ 23.3 Poncelet's 'Principle of Continuity'. In 1822, Poncelet [87] 1, p. 66 enunciated his *Principle of Continuity* (*'le principe de continuité'*) which may, roughly, be explained in this manner: any property of a figure which remains a property of the figure as the figure varies in a prescribed way, will be a property of the figure in its limiting position. *Degenerate* cases thus arise. For instance, Figs. 45(a), 45(b) are examples of degenerate quadrangles $AACD$ and $AACC$ through each of which passes an infinite number of conics. Desargues' Involution Theorem is then, by Poncelet's Principle of Continuity, still applicable. Degenerate conics of the systems are AA, CD and AC, AD (Fig. 45(a)) and AA, CC and AC repeated (Fig. 45(b)).

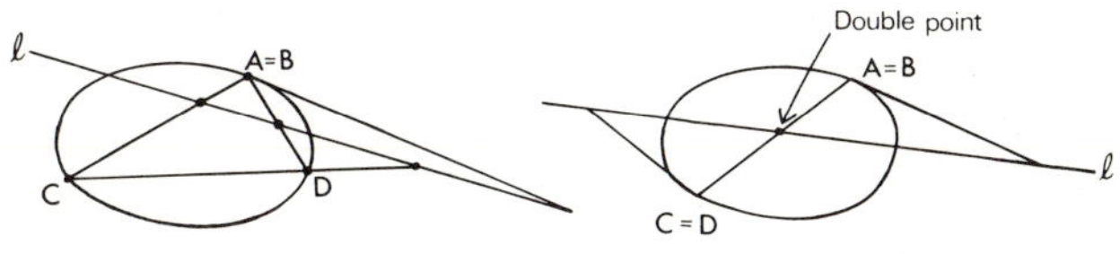

FIG. 45. (a) FIG. 45. (b) Poncelet's 'Continuity Principle' applied to Desargues' involution theorem

A strong word of warning must be sounded here. Nothing in the definition of a projective plane precludes the possibility of a conic consisting of only a finite number of points, so the continuity argument from Analysis or intuition is inadmissible. Accordingly, we interpret Poncelet's idea by thinking not of the limiting processes involved in continuity but rather by thinking of the specialised (= degenerate) case.

Use of such a device is often a cunning method of circumventing an otherwise formidable problem. Herewith is an example—and a very important one—to illustrate the technique of applying Desargues' Theorem to degenerate quadrangles.

EXAMPLE 29: Two conics S_1, S_2 touch at O and intersect at A, B. The tangent at O meets the common tangent at P and Q, at R (Fig. 46). Prove
(i) $(RT, PQ) = -1$;
(ii) $O(AB, PQ) = -1$.

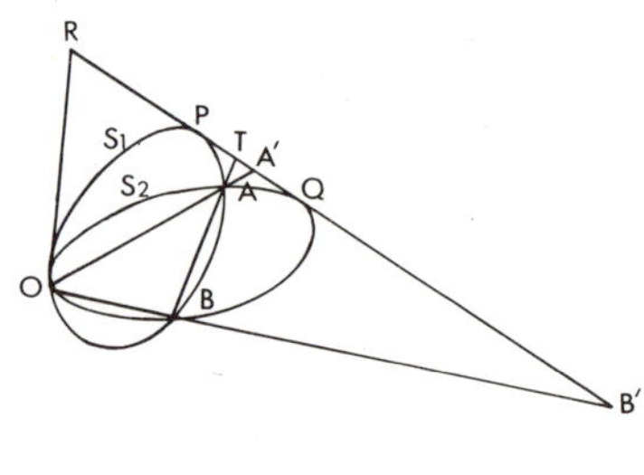

FIG. 46.

Solution: Consider the degenerate quadrangle $OOBA$ inscribed in S_1
$\therefore$ A', B'; R, T; P, P are pairs in an involution α.
Consider the degenerate quadrangle $OOBA$ inscribed in S_2
$\therefore$ A', B'; R, T; Q, Q are pairs in an involution β.
Since two pairs determine an involution completely and since A', B'; R, T are common, the two involutions α, β are the same. Clearly, P, Q are the double points of the involution.
$\therefore$ $(RT, PQ) = -1$ (i) (by § 20 I)
and $(A'B', PQ) = -1$ (by § 20 I)
i.e. $O(AB, PQ) = -1$ (ii).

Proceeding, we encounter next one of the most useful and illustrious theorems in Projective Geometry, due to Pascal. Talented in many regions of human endeavour, all of which he enriched, Pascal well-merited Poncelet's estimation of him as '*un des grands esprit de ce temps*'.

§ 24 PASCAL'S THEOREM

By a *hexagon inscribed in a conic* we mean a figure formed when any six points of the conic are joined up in a given order, beginning and ending with the same point. Keeping one point fixed, we see that there are $\frac{5!}{2} = 60$ different ways of inscribing a hexagon in a conic. Because of the remarkable properties of the inscribed hexagon, Pascal called it the *hexagrammum mysticum* (Latin) ≡ 'mystic hexagon'.

Theorem 24 (*Pascal's Theorem*): If a hexagon is inscribed in a conic, the points of intersection of opposite sides are collinear (*Pascal line*), and conversely.

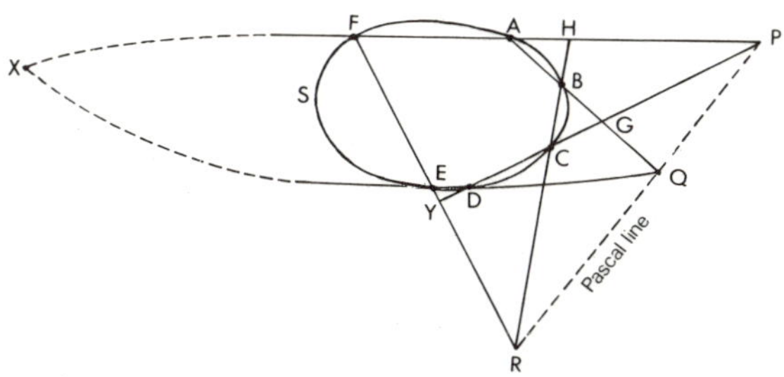

FIG. 47. Pascal's theorem

Proof: In Fig. 47, let $ABCDEF$ be a hexagon inscribed in a conic. Define the points $P = FA \cap CD$, $Q = AB \cap DE$, $R = BC \cap EF$. We are required to establish the existence of the line PQR. Construction points X, Y, G, H are as shown in the diagram.

[In specifying the collinear points P, Q, R, the reader may care to think of them as derived schematically from the hexagon thus:

where the pairs of points with the same linkage give the pairs of opposite sides of the hexagon and hence one of the points P, Q, R. Reproducing the first letter A at the end of the symbolic label for the hexagon (as shown) is possibly a helpful device.]

Method 1: $P(AB, CQ) = (AB, GQ)$
$= D(AB, CE)$
$= F(AB, CE)$
$= (HB, CR)$
$= P(AB, CR)$

$\therefore$ PQ, PR are the same line
i.e. P, Q, R are collinear (*Pascal line*).
(This technique was used in Examples 19, 20.)

Method 2: $A(BD, EF) = C(BD, EF)$ (by Chasles' Theorem)
i.e. $(QD, EX) = (RY, EF)$
i.e. E is a self-corresponding point
$\therefore$ QR, DY, XF must be concurrent. (Why?)
i.e. P, Q, R are collinear.

Method 3: The Theorem follows immediately if we apply Desargues' Theorem (using the transversal PR) first to the conics S; AB, CF; BC, AF, and then to the conics S; CD, EF; CF, DE.

For this purpose, it is desirable to redraw the hexagon so that the Pascal line 'physically' cuts the conic. (Students sometimes experience difficulty with the proof in Method 3.)

§ 24.1 Notes on Pascal's Theorem.

(i) Since there are 60 different Pascal hexagons associated with six given points (e.g. $ACDFEB$ is another), it follows that there are 60 different Pascal lines. Investigating this remarkable figure, we find that the 60 Pascal lines are arranged into 20 sets of three which are concurrent at points called *Steiner points* which lie in sets of four on 15 *Plücker lines*. Initially, research on the mystic hexagon was done by Steiner and Plücker in about 1830 and later in the century Kirkman (1806–1895, English), Cayley (1821–1895, English) and Salmon (1819–1904, Irish) obtained many new results. Thus, the configuration has been a rich and rewarding field of study.

(ii) When the given conic degenerates into a pair of lines, Pascal's Theorem becomes Pappus' Theorem (Theorem 6) and the Pascal line is therefore the Pappus line.

For example, points on the lines ACE, DFB (labelled in these given orders), when cross-joined according to Pappus' Theorem, form the hexagon $ABCDEF$, i.e. the Pappus line is exactly the Pascal line. Other lines lead to other arrangements of the six points of the hexagon. Readers are invited to carry out some of these verifications.

Generalisations of Pascal's Theorem (and Pappus' Theorem) have been made in many ways, e.g. to higher plane curves and higher dimensions.

Just as there is a *cross-axis* (Exercises 5.10, 5.11) for a projectivity relating points on two lines, so there is a cross-axis for a projectivity on a conic and it is found that this cross-axis joins the two distinct double points of the projectivity.

(iii) Degenerate Pascal hexagons occur when some of the six points coincide. Pentagons, quadrangles and triangles inscribed in a conic are thus special cases where Pascal's Theorem is applicable. (By a *pentagon*, we mean a figure formed when five given points are joined up in a given order (producing five lines), beginning and ending with the same point.)

(iv) Dual of Pascal's Theorem (*Brianchon's Theorem*, 1806 (C. J. Brianchon, 1783–1864, French)): If a hexagon circumscribes a conic, the lines joining opposite vertices are concurrent (*Brianchon point*). See Fig. 47(a). [Associated with six tangents to a hexagon, there are, as above, 60 Brianchon points. Degenerate Brianchon hexagons occur when some of the six tangents coincide. When the conic degenerates to two points (regarded as vertices of pencils of lines), we obtain the Dual of Pappus' Theorem.]

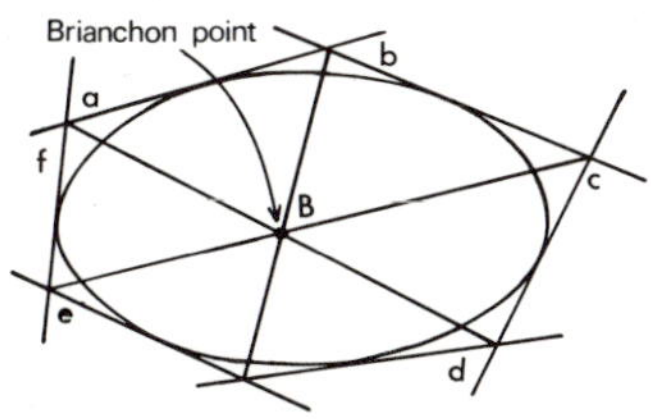

FIG. 47. (a) Brianchon's theorem

(v) Pascal's celebrated Theorem, which he discovered in 1640 at the age of 16 (as a relief from toothache, according to one legend!), was, it is thought, proved by him for the circle in the first instance. Using conical projection, he then obtained it for any conic. In its original form it may be found in Blaise Pascal's '*Essay pour les coniques*', a facsimile of which is to be found in Vera Sanford [96] and a translation of which may be found in Smith [103] pp. 326–330.

(vi) What is the statement of the Converse of Pascal's Theorem? (See Faulkner [38] p. 52, Fishback [40] p. 147.)

Two examples should suffice to illustrate the use of the above results.

EXAMPLE 30 (*Braikenridge-Maclaurin Construction*): Given any five points on a conic, construct the tangent to the conic at any one of them.

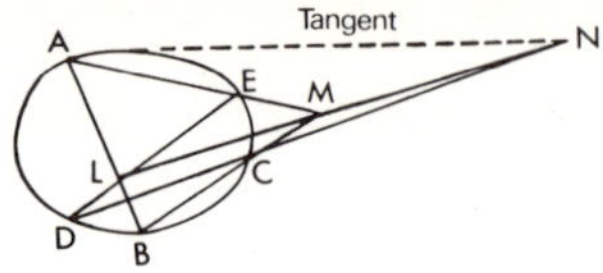

FIG. 48. Braikenridge-Maclaurin construction of a tangent

Solution: Let the five points be A, B, C, D, E. The construction is as shown in Fig. 48 ($AB \cap DE = L$, $BC \cap EA = M$, $CD \cap LM = N$) where AN is the tangent at A. This is so since LMN is the Pascal line of the degenerate Pascal hexagon $AABCDE$ (*Pascal pentagon*).

This construction is an aspect of the construction of a conic given independently in about 1733 by William Braikenridge (*c.* 1700–1759, Scottish) and Colin Maclaurin (1698–1746, Scottish). Claims for priority in discovering this important method led to a disagreeable controversy between Braikenridge and Maclaurin. (See § 26.)

EXAMPLE 31 (*Hessian Line and Hessian Point*): ABC is a triangle inscribed in a conic and $A'B'C'$ is a triangle circumscribing the conic, being tangential at C, A, B. Prove that AA', BB', CC' are concurrent. Further, $BC \cap B'C' = P$, $CA \cap C'A' = Q$, $AB \cap A'B' = R$ (Fig. 49). Prove that P, Q, R are collinear.

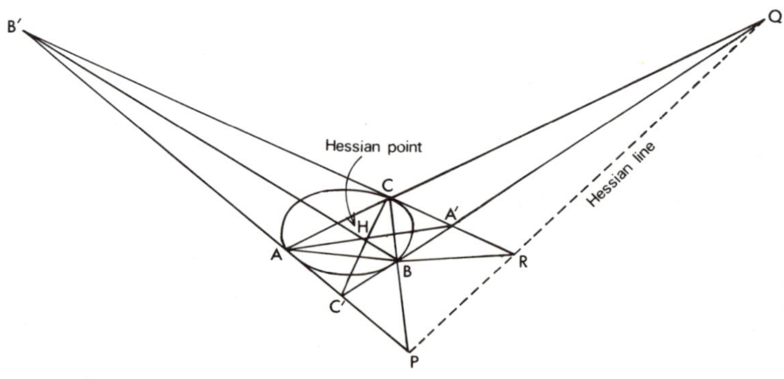

FIG. 49. Hessian point and line

Solution: Apply Pascal's Theorem to the degenerate Pascal hexagon $AABBCC$

$\therefore$ P, Q, R are collinear (*Hessian line*).

Apply Brianchon's Theorem to the degenerate Brianchon hexagon $AB'CA'BC'$

$\therefore$ AA', BB', CC' are concurrent (*Hessian point*).

(*Or*, apply the Converse of Desargues' Theorem to the triangles *ABC*, *A′B′C′*.)

Dorwart [32] p. 118 remarks that the geometrical constructions involving Pascal's and Brianchon's theorems provide approximate solutions for problems in the aircraft industry which are too cumbersome when attempted analytically. One of his reference texts, namely Adams and White [1] p. 282, shows by a diagram how the entire profile of an aeroplane can be described by six conics of the type determined by two tangents (and the points of tangency) and a further point. Cross-sections of the fuselage can be similarly described. 'This method,' say the authors, 'is presently [1961] used very extensively.' A colour photograph of the North American P-51 Mustang Fighter (the first instance of a fuselage developed entirely with conics) is given in Liming [65], frontispiece.

§ 25 POLE AND POLAR

The theory which we are about to propound is basically due to Philippe de la Hire (1640–1718, French) and Gaspard Monge (1746–1818, French), though some of the ideas are of greater antiquity, e.g. Theorem 25 below was known to Apollonius.

§ 25.1 Definitions: The *polar* of a point *P* w.r.t. a conic is the locus of the harmonic conjugate of *P* w.r.t. the pair of points in which a variable line through *P* meets the conic; *P* is called the *pole* (Fig. 50).

As we now show, the polar is a line (and not some other curve).

Theorem 25: The polar of a point w.r.t. a conic intersects the conic in the points of contact of tangents from that point.

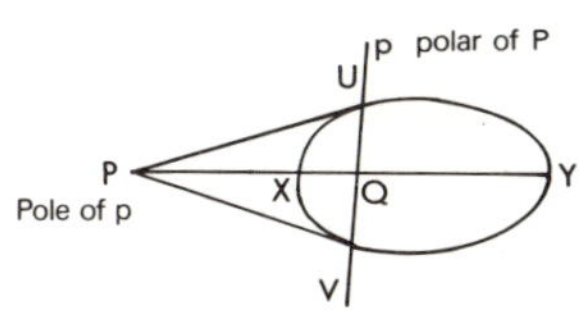

FIG. 50. Pole and polar

Proof: Let *U*, *V* be the points of contact of tangents from *P*, and let any secant through *P* meet the conic in *X*, *Y* and *UV* in *Q* (Fig. 50). The pole may be 'inside' the conic. What then happens to the diagram? See Figs 51, 52.)

$$\begin{aligned}\therefore (PQ, XY) &= U(PQ, XY)\\ &= U(UV, XY)\\ &= V(UV, XY)\\ &= (QP, XY)\\ \therefore (PQ, XY) &= -1 \text{ by § 19 VI(a)}\end{aligned}$$

$\therefore$ Q is the harmonic conjugate of P w.r.t. the conic.

Because the secant through P was arbitrary, the locus of Q is clearly UV, i.e. UV is the polar of P.

Note: When P is on the conic, the polar is the tangent at P.

The property by which pole and polar are related is called a *polarity* w.r.t. the conic. (See § 26.1.)

§ 25.2 Self-Polar Triangle. From the harmonic theory of the complete quadrangle (Example 5), we see that each side of the diagonal triangle XYZ is the polar of the opposite vertex w.r.t. any conic through A, B, C, D (Fig. 51). Then we shall say that XYZ is the *self-polar triangle* w.r.t. all conics through A, B, C, D.

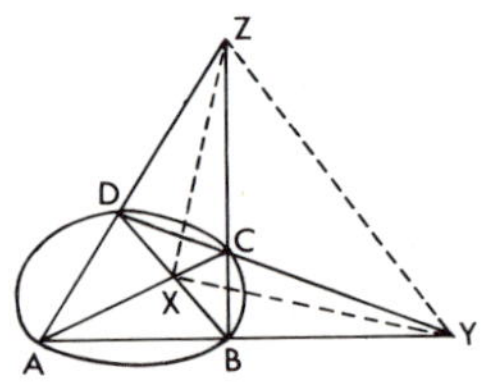

FIG. 51. Self-polar triangle

Hence, we have a method for constructing the polar of a point, say Y.

Obviously, an infinite number of conics passes through A, B, C, D so they all have a common self-polar triangle.

The significance of the concept of a self-polar triangle should not be underestimated. Subsequently, a good deal of use will be made of Fig. 51.

In Fig. 51, draw the tangents at A, B, C, D. What do you know about their points of intersection? Dualise the statements in § 25.2.

§ 25.3 Properties of Pole and Polar.

I *Mutuality property: If the polar p of P passes through Q, then the polar q of Q passes through P*, for

$(PQ, XY) = -1$ since p passes through Q (Fig. 50)

i.e. $(QP, XY) = -1$ (§ 19 VI(a)), which means that q passes through P. Points such that the polar of each passes through the other are called *conjugate points*. Lines such that the pole of each lies on the other are called *conjugate lines*.

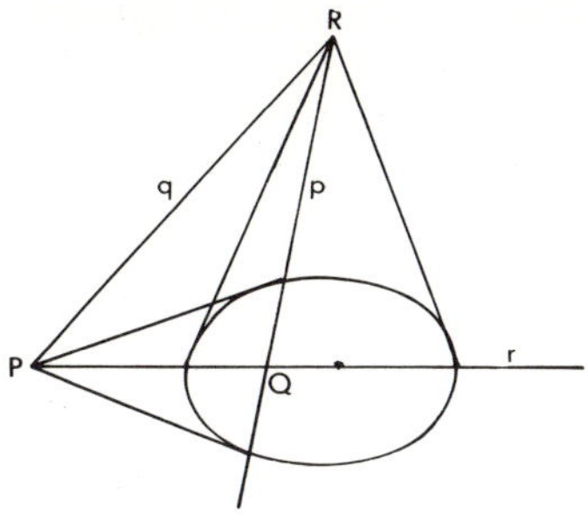

FIG. 52. Conjugate points and lines

Thus, in Fig. 52, P and Q are conjugate points, P and R are conjugate points and R and Q are conjugate points. Examples of conjugate lines are the pairs p, q; p, r and r, q. Why is q the polar of Q?

II *Duality property of pole and polar:* To every result concerning points and their polars w.r.t. a conic, there is a corresponding result concerning lines and their poles.

For example, the Dual of I above is: If the pole P of p lies on q, then the pole Q of q lies on p.

Moreover, the dual of $(PQ, XY) = -1$ in Theorem 25 is: $(pq, xy) = -1$, where p, q, x, y are concurrent lines of which x, y are tangents to a conic. (Illustrate by means of a diagram.)

Question: How is the theory of pole and polar modified in the case of a degenerate conic? Refer to Faulkner [38] pp. 55–56.

Much as we are tempted to develop the theory of pole and polar, we are guided by the greater desire to use only the basic theory as a technique when required. However, a few exercises are left at the end of this Chapter to test the ingenuity and understanding of the interested reader who wishes to proceed further.

§ 26 VON STAUDT AND BRAIKENRIDGE-MACLAURIN DEFINITIONS OF A CONIC

Two further definitions of a conic are worthy of brief mention. For quick reference, we shall call the conic defined jointly by Chasles and Steiner the Chasles-Steiner conic.

§ 26.1 Von Staudt's Definition of a Conic. In 1847, using his concept of a polarity (= an involútory correlation), von Staudt [114] defined a conic as

the locus of self-corresponding points in a hyperbolic polarity. An account in English of this attractive idea is given in, say, Coxeter [27] where it is shown that
(i) every von Staudt conic is a Chasles-Steiner conic, and
(ii) every Chasles-Steiner conic is a von Staudt conic.
An essential step in establishing (i) is to prove:

Theorem 26 (*Seydewitz's Theorem*)*:* If a triangle is inscribed in a conic, any line conjugate to one side meets the other sides in conjugate points.

Proof: Refer to Example 28 and Fig. 42. Suppose UVP is the triangle inscribed in a conic. Then OK passes through O, the pole of UV, i.e. OK and UV are conjugate lines. The Example asks us to show that H (on UP) and K (on VP) are conjugate points, i.e. $(HK, XY) = -1$.
Hence, Example 28 is the same as Seydewitz's Theorem.

Note: Veblen and Young [113] 1, p. 125 Ex. 5, attribute this Theorem and its Converse to von Staudt. What is the statement of the Converse of Theorem 26?

§ **26.2 Braikenridge-Maclaurin Definition of a Conic.** At about the same time (1733) Braikenridge and Maclaurin enunciated the following construction for a conic:

If the sides of a variable triangle pass through three given non-collinear points, while two other vertices move along given lines, the locus of the third vertex is a conic through two of the given points.

Based on some earlier ideas of Newton in his so-called 'organic construction' of a conic, the Braikenridge-Maclaurin method for generating conics led to angry disputation between Braikenridge and Maclaurin. Rather than attempt to award the priority in discovery to one of them, we prefer, according to custom, to give credit jointly to them. Notice that the construction is dependent only on the use of a straight-edge.

Naturally, one wants to be assured that the Braikenridge-Maclaurin conic is equivalent to the Chasles-Steiner conic. This assurance is given in Pedoe [85] pp. 186–187. A glance at Poncelet [87] 1, p. 321 is possibly worthwhile at this stage.

Shortly, we leave the synthetic treatment of conics and examine algebraic aspects of the theory. Mindful of the essential unity of much of Mathematics we will use geometric or algebraic techniques as the context suggests. Juxtaposition of these methods in Chapters 6 and 7 serves to emphasise their mutual relationship, their interchangeability.

Unfortunately, geometers and algebraists have in the past often becom unreasonably dogmatic and pig-headed about such matters. Only a century ago, *Crelle's Journal* (now known also as '*Journal für die reine und angewandte Mathematik*') would not print algebraic proofs of geometric theorems, while *Liouville's Journal* , equally reputable as a scholarly publication, would print nothing but algebraic proofs of geometric theorems.

Suggested topics for further study

((1) Poncelet's 'Principle of Continuity' (Bell [10] pp. 210–211; Poncelet [87] 1, p. 94).
(2) Desargues' own proof of the Involution Theorem (Desargues [30] p. 186 *et seq*).
(3) The von Staudt conic.
(4) The Braikenridge-Maclaurin conic.
(5) Pascal's 'Mystic Hexagon'.
(6) Deeper study of pole and polar.
(7) Pascal: Life and Works (Bell [10]).
(8) Apollonius of Perga's 'Treatise on Conic Sections' ed. T. L. Heath, Heffer (1961).

EXERCISES 6

1. Dualise Seydewitz's Theorem.

2. Two secants PP', QQ' of a conic meet in T; AS is the tangent at a point A of the conic. Prove that AP, AP'; AQ, AQ'; AS, AT are pairs in involution.

3. State Sturm's Theorem, the dual of Desargues' Theorem.

4. Prove that there are two conics which pass through a fixed point and touch four fixed lines.

5. Prove that there are two conics which pass through a fixed point and touch three lines, one at a fixed point.

6. *Castelnuovo's Theorem* [Guido Castelnuovo, 1865–1952, Italian] : If three conics all pass through the same two fixed points then the three lines joining the remaining intersections of the conics taken in pairs are concurrent. Prove.

7. Two conics have double contact at I, J. A tangent to the inner conic at V meets IJ in T and the outer conic in P, Q. Prove that P, Q are harmonic conjugates w.r.t. V, T.

8. If a pentagon $ABCEF$ circumscribes a conic and if CE touches the conic at D, show that AD, BE, CF are concurrent.

(Observe that this gives a method of constructing the point of contact of any of five given tangents to a conic, i.e. the dual of the Braikenridge-Maclaurin Construction, Example 30.)

9. If a quadrangle is inscribed in a conic, prove that the tangents at its vertices meet in pairs on the sides of the diagonal triangle.

10. Show that the Hessian point is the pole of the Hessian line.

11. Do Exercise 6.7 by pole and polar theory.

12. Prove *Lamé's Theorem*: Polars of a point w.r.t. a pencil of conics are concurrent. (Gabriel Lamé, 1795–1870, French.)

13. State the dual of Lamé's Theorem.

14. If two conics have double contact, the polars of any point w.r.t. the two conics meet on the line joining their common points of contact. Can you derive it?

15. (Use of the Converse of Pascal's Theorem.) Let A, B, C, D, E be five points on a conic of which no three are collinear; EF is any line through E but not through any of the other four points. Determine the other point of EF which is on the conic through A, B, C, D, E.

16. Polars, w.r.t. a conic, of four collinear points are concurrent and the cross-ratio of the range is equi-cross with that of the pencil. Prove.

17. Six vertices of two self-polar triangles w.r.t. a conic lie on another conic (Steiner). Prove. Quote the dual. Enunciate the converse.

18. Locus of poles of a line w.r.t. conics of a system is a conic (Steiner). Prove.

[This is the famous *11-point conic* whose existence was established by Steiner in 1846. Consult Faulkner [38] pp. 66–68.]

19. Joins of the intersections of the opposite sides of a complete quadrangle with a conic passing through two vertices of the quadrangle are concurrent. Prove.

[Hint: Use Castelnuovo's Theorem (Exercise 6.6)]. See R. A. Rosenbaum and J. Rosenbaum: 'Some consequences of a well known theorem on conics', *Bulletin of the American Mathematical Society* 55 (10) (1949) 933.

20. Deduce Pascal's Theorem from Castelnuovo's Theorem.

21. *Hesse's Theorem:* If ABC is any triangle, and A', B', C' are the poles of BC, CA, AB respectively w.r.t. any non-singular conic S, then the triangles ABC, $A'B'C'$ are in perspective. (The axis of perspective is the polar of the centre of perspective.)

22. If two pairs of opposite sides of a complete quadrangle are conjugate lines w.r.t. a conic, then so is the third pair (von Staudt).

[This follows from Hesse's Theorem as a Corollary. Occasionally, it is referred to *as* Hesse's Theorem.]

23. Construct the double points of an involution of points on a line. (See Cremona [29].)

24. You are given the following facts about a conic S: A, B are two given fixed points on S, P is a variable point on S, and H is the Hessian Point of the triangle ABP w.r.t. S. Prove:

(i) the locus of H is a conic S';

(ii) S, S' have double contact at A and B.

25. Two conics have double contact. Show that the polars of the points on the first conic, w.r.t. the second conic, envelope a conic having double contact with the first two.

CHAPTER 7

ALGEBRAIC TREATMENT OF THE CONIC

'As long as algebra and geometry proceeded along separate paths, their advance was slow and their applications limited. But when these sciences joined company, they drew from each other fresh vitality and thenceforward marched on at a rapid pace towards perfection.'

(J. Lagrange (1736–1813) Italian-French)

MANY properties of the general conic yield more readily to analytical, rather than synthetic, techniques. Using the Chasles-Steiner definition of a conic, we proceed to derive the standard equation of the second degree for the conic—this is the definition (c) in the list of equivalent definitions of the conic, given in the introductory comments in Chapter 6.

After deriving a few fundamental properties of the conic algebraically, we proceed to obtain some simplified equations of the conic. Finally, the study of a pair of conics considered as a geometrical object is undertaken briefly.

§ 27 ELEMENTARY THEORY

§ 27.1 Equation of the Point-conic. Using homogeneous co-ordinates, we may prove directly from the Chasles-Steiner definition of a conic (§ 21) that its equation is of the second degree in these co-ordinates. Take $A(1, 0, 0)$, $B(0, 1, 0)$ as the vertices of the two related pencils. Now any line through A has equation $\frac{y}{z} = \lambda$ (on using (1)) and a line through B has equation $\frac{x}{z} = \lambda'$ (Fig. 53). If these two pencils are related, their parameters λ, λ' satisfy the bilinear relation

$$p\lambda\lambda' + q\lambda + r\lambda' + s = 0$$

by virtue of (3). Supposing $P \equiv (x, y, z)$ is the point of intersection of corresponding lines, we therefore have

$$p\left(\frac{y}{z}\right)\left(\frac{x}{z}\right) + q\left(\frac{y}{z}\right) + r\left(\frac{x}{z}\right) + s = 0$$

$$\text{i.e. } pxy + qyz + rzx + sz^2 = 0$$

which is of the second degree.

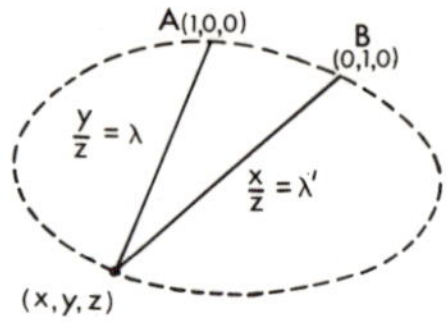

FIG. 53. Genesis of the equation of a conic

Similarly, if we had taken vertices B, C (or C, A) our equation would have had a term in x^2 (or y^2) instead of z^2. Terms in xy, yz, zx would still exist, of course, but with coefficients chosen from among p, q, r. Summarising these results, we see that
a conic which does not pass through the vertices of the triangle of reference is represented by the general homogeneous equation of the second degree in x, y, z:

$$\boxed{S \equiv f(x, y, z) \equiv \quad ax^2 + by^2 + cz^2 + 2fyz + 2gzx + 2hxy = 0} \tag{6}$$

(and conversely). For a proof of the converse, namely
the general homogeneous equation of the second degree in x, y, z *represents a conic which does not pass through the vertices of the triangle of reference*, see Faulkner [38] pp. 98–99. Obviously, a line l' meets the conic in two points corresponding to the two solutions $x : y : z$ of (1′) and (6).

[Generalising this idea, we see that a *curve of order n*, i.e. one whose equation is a polynomial of degree n in x, y, z, is met by any line in n points. Thus, a cubic, a quartic, a quintic, a sextic, . . . curve is met by any line in respectively 3, 4, 5, 6, . . . points.]

The advantage of taking the equation of the conic in the form (6) is that it enables us to express the equation quite simply in matrix form by means of a symmetric matrix.

Expressed in matrix form, equation (6) is

$$\boxed{\mathbf{X}^T\mathbf{A}\mathbf{X} = \mathbf{O}} \tag{6′}$$

where $\mathbf{X}^T$ is the co-ordinate row vector $[x, y, z]$, $\mathbf{O}$ is the 1×1 zero matrix (i.e. the scalar 0) and $\mathbf{A}$ is the matrix of the determinant

$$|\,\mathbf{A}\,| = \begin{vmatrix} a & h & g \\ h & b & f \\ g & f & c \end{vmatrix} \neq 0,$$

called the *discriminant* of the conic (6). When $|\mathbf{A}| = 0$, the conic is known to degenerate to a pair of lines, i.e. (6) factorises into two linear equations. *Note: Under a collineation, a conic transforms into a conic,* for, letting the collineation have equation

$$\mathbf{X} = \mathbf{P}\mathbf{X}' \qquad (|\mathbf{P}| \neq 0)$$

we find that the conic (6′) is transformed into

$$(\mathbf{X}')^T\mathbf{P}^T\mathbf{A}\mathbf{P}\mathbf{X}' = \mathbf{O} \qquad (|\mathbf{A}| \neq 0)$$

i.e. $$(\mathbf{X}')^T\mathbf{B}\mathbf{X}' = \mathbf{O}$$

where $$\mathbf{B} = \mathbf{P}^T\mathbf{A}\mathbf{P}$$

i.e. a conic, which is non-singular, since

$$\begin{aligned} |\mathbf{B}| &= |\mathbf{P}^T| \;\; |\mathbf{A}| \;\; |\mathbf{P}| \\ &= |\mathbf{P}| \;\; |\mathbf{A}| \;\; |\mathbf{P}| \\ &= |\mathbf{P}|^2 \;\; |\mathbf{A}| \\ &\neq 0. \end{aligned}$$

Thus, *conic is a projective invariant.*

§ 27.2 Equation of the Line-Conic.

EXAMPLE 32: Dually the line-equation of the conic Σ is

$$\boxed{\Sigma \equiv Al^2 + Bm^2 + Cn^2 + 2Fmn + 2Gnl + 2Hlm = 0} \qquad \text{(6a)}$$

where A, B, C, F, G, H are the cofactors of a, b, c, f, g, h respectively in $|\mathbf{A}|$. *Solution:* Getting this equation (6a) is equivalent to finding the condition for a line to touch the conic. Anticipating (9′) and taking the equation of the (tangent) line to be (1′), we have, on equating coefficients of x, y, z and allowing (x_1, y_1, z_1) to be the point of contact of the tangent,

$$\left\{\begin{aligned} ax_1 + hy_1 + gz_1 &= kl \\ hx_1 + by_1 + fz_1 &= km \\ gx_1 + fy_1 + cz_1 &= kn \\ lx_1 + my_1 + nz_1 &= 0 \end{aligned}\right. \qquad (k \neq 0)$$

where k is a factor of proportionality. Bring the terms kl, km, kn to the left-hand sides of their equations. Eliminate $x_1 : y_1 : z_1 : -k$.

$$\therefore \begin{vmatrix} a & h & g & l \\ h & b & f & m \\ g & f & c & n \\ l & m & n & 0 \end{vmatrix} = 0$$

i.e. $Al^2 + Bm^2 + Cn^2 + 2Fmn + 2Gnl + 2Hlm = 0$, on calculation.

§ **27.21 Jacobi's Theorem.** (Carl Jacobi, 1804–1851, German.) When the line-equation (6a) of the conic is given and it is desired to obtain the point-equation $\sigma = 0$, the aid is invoked of Jacobi's Theorem on determinants, namely

$$\frac{A^*}{a} = \frac{B^*}{b} = \frac{C^*}{c} = \frac{F^*}{f} = \frac{G^*}{g} = \frac{H^*}{h} = \Delta$$

where $A^*, B^*, \ldots, H^*$ are the cofactors of the cofactors of $a, b, \ldots, h$ in $\Delta = |\mathbf{A}|$, i.e. $A^*, B^*, \ldots, H^*$ are the cofactors of $A, B, \ldots, H$ in $\begin{vmatrix} A & H & G \\ H & B & F \\ G & F & C \end{vmatrix}$.

Then $\sigma = \Delta S$ by Jacobi's Theorem and (6). Since $\Delta \neq 0$, it follows that $\sigma = 0$, $S = 0$ represent the same locus, as we expect.

§ **27.3 Pole and Polar.** Proceeding, let P_1, P_2 be two distinct points having respective co-ordinate vectors $\mathbf{X}_1, \mathbf{X}_2$ where $\mathbf{X}_1^T = [x_1, y_1, z_1]$ and $\mathbf{X}_2^T = [x_2, y_2, z_2]$. It follows that the point with co-ordinate vector $\lambda\mathbf{X}_1 + \mu\mathbf{X}_2$ (on P_1P_2) lies on the conic if

$$(\lambda\mathbf{X}_1^T + \mu\mathbf{X}_2^T)\,\mathbf{A}\,(\lambda\mathbf{X}_1 + \mu\mathbf{X}_2) = \mathbf{O} \qquad \text{(i)}$$

Because $\mathbf{A}$ is symmetric, $\mathbf{X}_1^T\mathbf{A}\mathbf{X}_2 = \mathbf{X}_2^T\mathbf{A}\mathbf{X}_1$, both matrices being of size 1 (i.e. 1×1 matrices) and therefore regarded as scalars. Equation (i), which is called *Joachimsthal's ratio equation* (Ferdinand Joachimsthal, 1818–1861, German) may be written

$$\boxed{\lambda^2 f_{11} + 2\lambda\mu f_{12} + \mu^2 f_{22} = 0} \qquad (7)$$

where $\boxed{f_{11} \equiv \mathbf{X}_1^T\mathbf{A}\mathbf{X}_1, f_{12} \equiv \mathbf{X}_1^T\mathbf{A}\mathbf{X}_2 = \mathbf{X}_2^T\mathbf{A}\mathbf{X}_1 \equiv f_{21}, f_{22} \equiv \mathbf{X}_2^T\mathbf{A}\mathbf{X}_2} \qquad (8)$

You should write out in full the expressions involving matrices in (8). Roots of (7) in $\frac{\mu}{\lambda}$ determine the two points of intersection of P_1P_2 and the conic. For these two intersections to be harmonic conjugates w.r.t. P_1, P_2, i.e. expressible as $\mathbf{X}_1 \pm \frac{\mu}{\lambda}\mathbf{X}_2$, we must have that the sum of the roots is zero, i.e. $f_{12} = 0$. (See § 19 VI(a).) Now, since points conjugate to P_1 lie on the polar of P_1, it follows immediately that the equation of the polar is

$$\boxed{f_1 \equiv \mathbf{X}_1^T\mathbf{A}\mathbf{X} = \mathbf{O}} \qquad (9)$$

When P_1 lies on the conic, this polar becomes the tangent, i.e. the equation of the tangent at (x_1, y_1, z_1) is $f_1 = 0$. Fully written out, $f_1 = 0$ (equation of tangent at, or polar of, $P_1(x_1, y_1, z_1)$) is

$$(ax_1 + hy_1 + gz_1)x + (hx_1 + by_1 + fz_1)y + (gx_1 + fy_1 + cz_1)z = 0 \quad (9')$$

i.e. in terms of partial derivatives

$$x\left(\frac{\partial f}{\partial x}\right)_{P_1} + y\left(\frac{\partial f}{\partial y}\right)_{P_1} + z\left(\frac{\partial f}{\partial z}\right)_{P_1} = 0 \quad (9'')$$

where $\left(\frac{\partial f}{\partial x}\right)_{P_1}$, say, means the value of $\frac{\partial f}{\partial x}$ at $P_1(x_1, y_1, z_1)$.

Next we determine the equation of a pair of tangents to the conic. To do this, observe that the requirement for a tangent is that the roots of Joachimsthal's equation (7) be equal, i.e. $f_{12}^2 = f_{11}f_{22}$. Consequently, as P_2 varies, its co-ordinates satisfy the equation

$$f_1^2 = f_{11}f \quad (10)$$

This is an equation of the second degree in x, y, z and must, from the nature of its derivation, represent two lines. That is, (10) is the equation of the tangent-pair from P_1 to the conic.

Diagrams illustrating the above theory are so elementary that they are omitted.

§ 28 SPECIAL FORMS OF THE CONIC

By suitable choice of the triangle of reference, the equation of the conic may be made to take simple, special forms.

I *Conic Circumscribing Triangle of Reference*

Since $X \equiv (1, 0, 0)$ lies on the conic, we have, after substitution in (6), that $a = 0$.

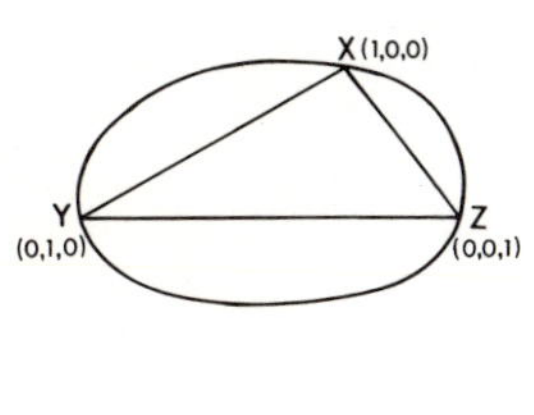

FIG. 54. (a) A circum-conic of the triangle of reference

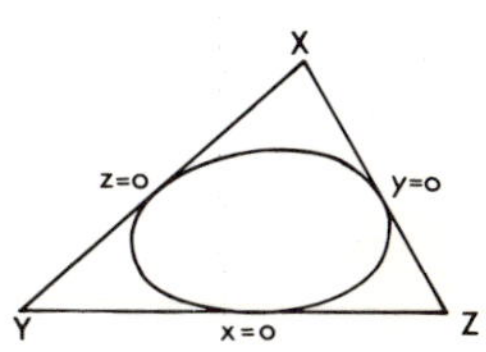

FIG. 54. (b) An in-conic of the triangle of reference

Likewise $b = c = 0$ (Fig. 54(a)) since $Y \equiv (0, 1, 0)$ and $Z \equiv (0, 0, 1)$ lie on the conic. So the equation of the conic is

$$\boxed{fyz + gzx + hxy = 0} \tag{11}$$

II *Conic Inscribed in Triangle of Reference*

Dually, from (6a) the line-equation of a conic inscribed in the triangle of reference (Fig. 54(b)) is

$$\boxed{Fmn + Gnl + Hlm = 0} \tag{11$'$}$$

III *Conic Referred to a Self-Polar Triangle*

By (9′) the polar of $X \equiv (1, 0, 0)$ is $ax + hy + gz = 0$.

But since the triangle of reference is self-polar this is the same as YZ, i.e. $x = 0$.

Comparison of the two equations yields $h = g = 0$. Similarly, we have $h = f = 0$, $g = f = 0$. Thus the equation of the conic is

$$\boxed{ax^2 + by^2 + cz^2 = 0} \tag{12}$$

Next make the transformation

$$\begin{bmatrix} \xi \\ \eta \\ \zeta \end{bmatrix} = \begin{bmatrix} \sqrt{a} & \cdot & \cdot \\ \cdot & \sqrt{b} & \cdot \\ \cdot & \cdot & \sqrt{c} \end{bmatrix} \begin{bmatrix} x \\ y \\ z \end{bmatrix},$$

i.e. the triangle of reference is unchanged, but the point $\left(\frac{1}{\sqrt{a}}, \frac{1}{\sqrt{b}}, \frac{1}{\sqrt{c}}\right)$ is transformed into the unit point. Then (12) becomes

$$\boxed{x^2 + y^2 + z^2 = 0} \tag{12a}$$

on replacing ξ, η, ζ by x, y, z. Zeros are represented by dots in the matrix of the transformation. Observe that the algebraic effect of making the appropriate selections of triangle of reference and unit point is to reduce the homogeneous quadratic form in (6) to a sum of squares (12a).

IV *Two Conics*

As triangle of reference, take the common self-polar diagonal triangle of the complete quadrangle formed by the four points of intersection of the two conics (Fig. 51). (To prove that two conics *do* meet in four points, solve two quadratic equations like (6) for, say, the ratios $\frac{x}{z}, \frac{y}{z}$ and we obtain a quartic equation, i.e. an equation with four roots, each root corresponding to a common point of the two conics.) With suitable choice of unit point, the conics may be written

$$\boxed{\begin{aligned} ax^2 + by^2 + cz^2 &= 0 \\ x^2 + y^2 + z^2 &= 0 \end{aligned}} \qquad (12b)$$

Usually this is the most convenient way of writing down the equations of two conics of general position. Certainly, as will become evident later in this Chapter, (12b) simplifies considerably the algebraic calculations associated with a pair of conics.

EXAMPLE 33: Obtain the equation of the Hessian line.

Solution: Having regard to Fig. 49, we take ABC as triangle of reference, so that the equation of the conic is, by (11), $fyz + gzx + hxy = 0$.

The equation of the tangent at A is $hy + gz = 0$, i.e. $\frac{y}{g} + \frac{z}{h} = 0$.

This meets $x = 0$ on the line

$$\frac{x}{f} + \frac{y}{g} + \frac{z}{h} = 0.$$

$\left(\text{Why } \frac{x}{f}\text{? Why not some other multiple of } x\text{?}\right)$

By symmetry, the two points obtained similarly on CA, AB also lie on this line, i.e. this equation is that of the Hessian line.

Compare this equation with the equation of the Desargues line in Exercise 3.19. Have you any comments to make?

V Choose the triangle of reference as XYZ shown in Fig. 55. Then, since X, Z lie on the conic, whose equation initially is (6), we have $a = c = 0$, i.e. the equation reduces to

$$by^2 + 2fyz + 2gzx + 2hxy = 0.$$

Now the equation of the polar of $Y(0, 1, 0)$ w.r.t. this conic is

$$hx + by + fz = 0$$

i.e.

$$y = 0,$$

whence $h = f = 0$. Consequently, the equation reduces to $y^2 + \frac{2g}{b}zx = 0$ which, by suitable choice of unit point, further simplifies to

$$\boxed{y^2 - zx = 0} \qquad (13)$$

Any point on the conic, with parameter θ, has co-ordinates $(\theta^2, \theta, 1)$ as may easily be verified. Obviously the unit point arises from the value $\theta = 1$. When $\theta = 0, \infty$ we obtain Z, X respectively.

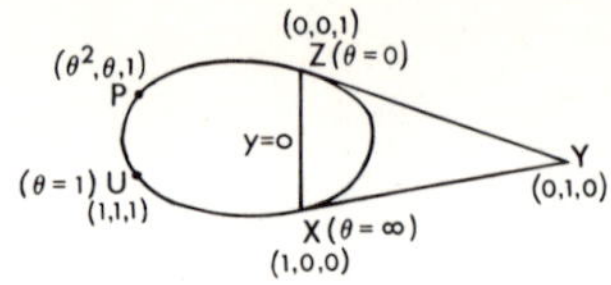

FIG. 55.

Determination of the latter fact is as follows: An arbitrary point may be expressed as $\left(1, \frac{1}{\theta}, \frac{1}{\theta^2}\right)$ on dividing $(\theta^2, \theta, 1)$ by θ^2, which is possible since $\theta = 0$, giving Z, is excluded. Then $\theta = \infty$ leads immediately to X.

§29 INVARIANTS OF A PAIR OF CONICS

Rather surprisingly, we find that quite a lot of pretty and powerful Mathematics emerges from the study of a pair of conics considered as a single object. Since a projective transformation does not alter the mutual relationships of a pair of conics, we may expect such geometrical properties to be expressible by algebraic invariant relations.

Only a fairly brief outline of the theory can be permitted here. Texts such as Loney [136] and Sommerville [105] provide further information. Some further references of a metrical nature will be made in the next Chapter.

§ 29.1 The Four Invariants $\Delta, \Delta', \Theta, \Theta'$. Consider the two conics S, S' (Fig. 56) whose equations in matrix form are

$$\boxed{\begin{aligned} S &\equiv \mathbf{X}^T\mathbf{A}\mathbf{X} = \mathbf{O} \\ S' &\equiv \mathbf{X}^T\mathbf{B}\mathbf{X} = \mathbf{O} \end{aligned}} \tag{14}$$

where $\mathbf{X}^T$ is the row co-ordinate matrix $[x, y, z]$, $\mathbf{O}$ is the zero 1×1 matrix, and

$$\mathbf{A} = \begin{bmatrix} a_{11} & a_{12} & a_{13} \\ a_{21} & a_{22} & a_{23} \\ a_{31} & a_{32} & a_{33} \end{bmatrix}, \quad \mathbf{B} = \begin{bmatrix} b_{11} & b_{12} & b_{13} \\ b_{21} & b_{22} & b_{23} \\ b_{31} & b_{32} & b_{33} \end{bmatrix} \quad (|\mathbf{A}| \neq 0, |\mathbf{B}| \neq 0)$$

where $\qquad a_{ij} = a_{ji}, b_{ij} = b_{ji} \; (i, j = 1, 2, 3),$

i.e. $\mathbf{A}$, $\mathbf{B}$ are both symmetrical. There should be no confusion when we speak of the conic with equation $S = 0$ as 'the conic S'.

Take the pencil of conics (through A, B, C, D)

$$\boxed{\lambda S + S' \equiv \mathbf{X}^T(\lambda\mathbf{A} + \mathbf{B})\mathbf{X} = \mathbf{O}} \tag{15}$$

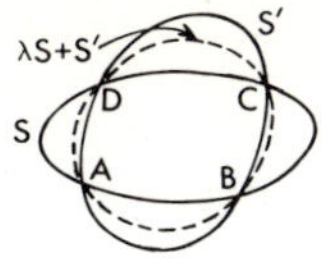

FIG. 56. Pencil of conics

Degenerate conics of this pencil (namely the three pairs of common secants AB, CD; AC, BD; AD, BC) are given by the three values of λ for which

$$|\lambda\mathbf{A} + \mathbf{B}| = 0,$$

i.e. for which

$$\begin{vmatrix} \lambda a_{11} + b_{11} & \lambda a_{12} + b_{12} & \lambda a_{13} + b_{13} \\ \lambda a_{21} + b_{21} & \lambda a_{22} + b_{22} & \lambda a_{23} + b_{23} \\ \lambda a_{31} + b_{31} & \lambda a_{32} + b_{32} & \lambda a_{33} + b_{33} \end{vmatrix} = 0.$$

Expanding this 3×3 determinant (which ought to be calculated in full by the reader), we derive the discriminating cubic,

$$\boxed{\Delta\lambda^3 + \Theta\lambda^2 + \Theta'\lambda + \Delta' = 0} \tag{16}$$

where

$$\boxed{\begin{cases} \Delta = |\mathbf{A}| \\ \Delta' = |\mathbf{B}| \\ \Theta = \Sigma b_{ij} A_{ij} \\ \Theta' = \Sigma a_{ij} B_{ij} \end{cases}} \tag{17}$$

A_{ij}, B_{ij} being the cofactors of a_{ij}, b_{ij}, in $|\mathbf{A}|$, $|\mathbf{B}|$ respectively.

Observe that

$$\begin{cases} \Theta = \text{trace } \mathbf{BA}^*, \\ \Theta' = \text{trace } \mathbf{AB}^*, \\ 3\Delta = \text{trace } \mathbf{AA}^*, \\ 3\Delta' = \text{trace } \mathbf{BB}^* \end{cases}$$

where $\mathbf{A}^*$, $\mathbf{B}^*$ are the adjoint matrices of $\mathbf{A}$, $\mathbf{B}$ respectively. The two Θ-expressions are not symmetrical relations between the conics S, S', because each results from the point-equation of one and the line-equation of the other.

As we shall shortly see, Δ, Δ', Θ, Θ' are invariants of projective transformations. They may be referred to as the *four fundamental invariants.*

EXAMPLE 34: Calculate the fundamental invariants Δ, Δ', Θ, Θ' for the conics $S \equiv b^2x^2 + a^2y^2 - a^2b^2z^2 = 0$, $S' \equiv (x - \alpha z)^2 + (y - \beta z)^2 - r^2z^2 = 0$.
Solution: The invariants of the conics are

$$|\mathbf{A}| = \begin{vmatrix} b^2 & . & . \\ . & a^2 & . \\ . & . & -a^2b^2 \end{vmatrix}, \quad |\mathbf{B}| = \begin{vmatrix} 1 & . & -\alpha \\ . & 1 & -\beta \\ -\alpha & -\beta & (\alpha^2 + \beta^2 - r^2) \end{vmatrix}$$

$$= -a^4b^4 \qquad\qquad = -r^2,$$

$$\Theta = 1.(-a^4b^2) + 1.(-a^2b^4) + (\alpha^2+\beta^2-r^2).a^2b^2 + (-\beta).0 + (-\alpha).0 + 0.0$$
$$= a^2b^2(\alpha^2+\beta^2-r^2-a^2-b^2),$$
$$\Theta' = b^2.(\alpha^2+\beta^2-r^2-\beta^2) + a^2(\alpha^2+\beta^2-r^2-\alpha^2) - a^2b^2.1 + 0.\beta + 0.\alpha + 0.\alpha\beta$$
$$= a^2\beta^2 + b^2\alpha^2 - r^2(a^2+b^2) - a^2b^2.$$

EXAMPLE 35: Evaluate the fundamental invariants of the following pairs of conics:

(i) $S \equiv y^2 - 4axz = 0$, $S' \equiv (x-\alpha z)^2 + (y-\beta z)^2 - r^2z^2 = 0$;
(ii) $S \equiv f^2x^2 + g^2y^2 + h^2z^2 - 2ghyz - 2hfzx - 2fgxy = 0$ (in-conic);
$S' \equiv 2f'yz + 2g'zx + 2h'xy = 0$ (circum-conic);
(iii) $S \equiv f^2x^2 + g^2y^2 + h^2z^2 - 2ghyz - 2hfzx - 2fgxy = 0$ (in-conic)
$S' \equiv ax^2 + by^2 + cz^2 = 0$ (self-polar conic);
(iv) $S \equiv 2fyz + 2gzx + 2hxy = 0$ (circum-conic)
$S' \equiv ax^2 + by^2 + cz^2 = 0$ (self-polar conic).

(By a 'self-polar conic' we mean a conic w.r.t. which the triangle of reference is self-polar.)

Solution: These are left as exercises (Exercises 7.7–7.10). For subsequent reference, we present the answers in tabular form thus:

	Δ	Δ'	Θ	Θ'
(i)	$-4a^2$	$-r^2$	$-4a(a+\alpha)$	$\beta^2 - 4a\alpha - r^2$
(ii)	$-4f^2g^2h^2$	$2f'g'h'$	$4fgh(ff'+gg'+hh')$	$-(ff'+gg'+hh')^2$
(iii)	$-4f^2g^2h^2$	$a\ b\ c$	0	$f^2bc + g^2ca + h^2ab$
(iv)	$2fgh$	$a\ b\ c$	$-(af^2+bg^2+ch^2)$	0

Theorem 27: Δ, Δ', Θ, Θ' are projective invariants.

Proof: Make the (non-singular) projective transformation of co-ordinates

$$\mathbf{X} = \mathbf{KY} \ (|\mathbf{K}| \neq 0)$$
$$\therefore \quad \mathbf{X}^T(\lambda\mathbf{A}+\mathbf{B})\mathbf{X} = \mathbf{Y}^T\{\mathbf{K}^T(\lambda\mathbf{A}+\mathbf{B})\mathbf{K}\}\mathbf{Y}$$

Now $|\mathbf{K}^T(\lambda\mathbf{A}+\mathbf{B})\mathbf{K}| = |\mathbf{K}|^2\,|\lambda\mathbf{A}+\mathbf{B}|$

$$= k^2(\Delta\lambda^3 + \Theta\lambda^2 + \Theta'\lambda + \Delta') \text{ where } k = |\mathbf{K}|$$
$$= \Delta_1\lambda^3 + \Theta_1\lambda^2 + \Theta_1'\lambda + \Delta_1' \text{ where } \Delta_1 = k^2\Delta, \text{ etc.}$$

Under a projective transformation, each quantity Δ, Δ', Θ, Θ' is reproduced except for the constant factor k^2. Because of the presence of k^2, we refer to the four fundamental quantities Δ, Δ', Θ, Θ' as *relative* (not absolute) *projective invariants*. (If $k^2 = 1$, the invariants are *absolute*.)

Note that any homogeneous relation among the invariants is invariant under any projective transformation. Such an invariant relation must—and we so assume—be homogeneous when Δ, Θ, Θ', Δ' (heed the order) are considered to be of

(i) the same 'dimension' (nothing to do with the geometrical meaning of dimension);
(ii) dimensions 0, 1, 2, 3 respectively;
(iii) dimensions 3, 2, 1, 0 respectively.

Consequently, $\Theta = \Theta'$ is meaningless unless both are zero since both requirements (ii) and (iii) are violated. Assuming dimensions are added, we see that $\Theta^2 = 4\Delta\,\Theta'$ is a valid relation whereas $\Theta\Delta = \Theta'\Delta'$ is not. For more details, see Loney [136].

Analogous to the theory developed above for the pencil $\lambda S + S' = 0$ there is a corresponding theory for the dual system, i.e. for the system of line-conics $\lambda\Sigma + \Sigma' = 0$.

§ 29.2 Triangular Inscription and Circumscription. From Example 35(ii) we derive the result (due to Cayley):

Theorem 28: If S is inscribed in and S' is circumscribed about a triangle, then

$$\Theta^2 = 4\Delta\Theta'.$$

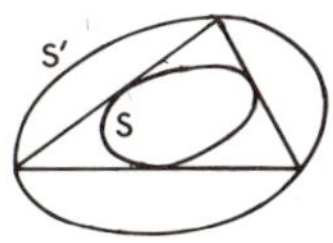

FIG. 57. Triangular inscription and circumscription

Under these circumstances, S is *triangularly inscribed* in S' or, equivalently, S' *triangularly circumscribes* S (Fig. 57). It may be shown that this is a *porism* (or *poristic property*), i.e. there is not just one such triangle but an infinite number of such triangles.

§ 29.3 Simplified Forms of the Invariants. Quite often, and especially in this section on two conics, much algebraic manipulation is avoided by choosing as triangle of reference the common self-polar triangle of the complete quadrangle formed from the four points of intersection of the two conics. Then their equations can be cast in the homogeneous form (refer to 12(b))

$$S \equiv ax^2 + by^2 + cz^2 = 0$$
$$S' \equiv \ \ x^2 + \ \ y^2 + \ \ z^2 = 0.$$

Consequently $\mathbf{A}$ = diag. (a, b, c) and $\mathbf{B}$ = diag. $(1, 1, 1)$ and the now simplified fundamental invariants of the conic-pair S, S' are

$$\boxed{\begin{cases} \Delta \ = a\,bc \\ \Delta' = 1 \\ \Theta \ = ab + bc + ca \\ \Theta' = a + b + c \end{cases}} \qquad (17')$$

Clearly, $\mathbf{A}^*$ = diag. (bc, ca, ab), $\mathbf{B}^*$ = diag. $(1, 1, 1)$ so that the duals of S, S' have equations

$$\boxed{\begin{array}{l} \Sigma \equiv bcl^2 + cam^2 + abn^2 = 0 \\ \Sigma' \equiv l^2 + m^2 + n^2 = 0 \end{array}} \qquad (18)$$

§ 30 APOLARITY

From Example 35(iii) we deduce

Theorem 29: $\Theta = 0$ means that there is a triangle circumscribing Σ and self-polar w.r.t. S'.

Under these circumstances, Σ is said to be *inpolar* to S'.

From Example 35(iv) we deduce further that $\Theta' = 0$ means that there is a triangle inscribed in S and self-polar w.r.t. S'. Interchanging dashed and undashed symbols, we obtain a second interpretation of $\Theta = 0$, namely, $\Theta = 0$ signifies that there is a triangle inscribed in S' and self-polar w.r.t. Σ, i.e. S' is *outpolar* to Σ. Also see Sommerville [105].

Briefly, we summarise the situation thus:

$\Theta = 0$ means (i) S' is outpolar to Σ, and
(ii) Σ is inpolar to S'.

In either case, we say S', Σ are *apolar* (Fig. 58).

Similar results apply to $\Theta' = 0$.

An alternative and suggestive notation is to write

$$\Theta \equiv \Theta_{S'\Sigma},\ \Theta' \equiv \Theta'_{S\Sigma'}.$$

Accordingly, in this notation the invariant relation in Theorem 28 is expressible as

$$\{\Theta_{S'\Sigma}\}^2 = 4\Delta_S \Theta'_{S\Sigma'}.$$

Two conics are *mutually apolar* if

$$\Theta = 0 = \Theta'.$$

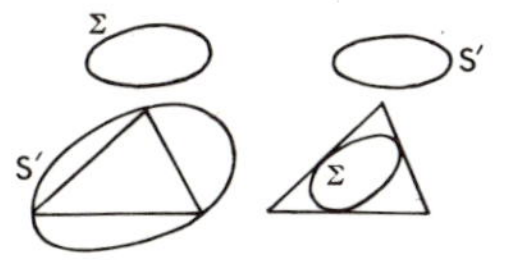

FIG. 58. Apolarity ($\Theta = 0$)

Some of the most interesting problems in the theory of two conics relate to the special conics (harmonic conics) which we now introduce.

§ 31 HARMONIC CONICS

Basic questions which would naturally be asked in a serious study of a pair of conics S, S' are:

(i) What is the locus of points from which the tangents to S, S' form a harmonic pencil (Fig. 60)?
(ii) What is the envelope of lines cut harmonically by S, S' (Fig. 59)?

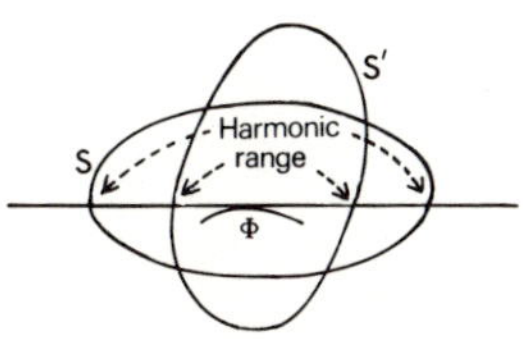

FIG. 59. Generation of the Φ-conic

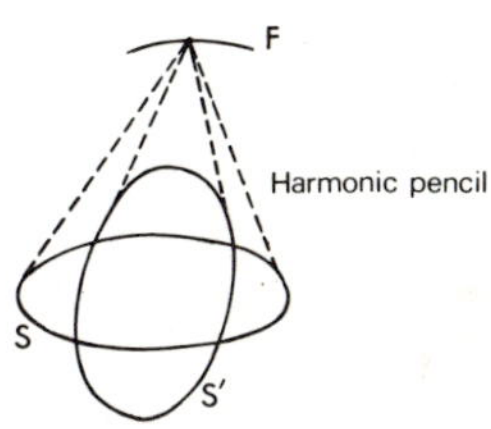

FIG. 60. Generation of the F (Salmon)-conic

Theorem 30: The envelope of lines cut harmonically by two conics is a conic (the *harmonic conic-envelope*)

$$\Phi \equiv (b + c)\, l^2 + (c + a)\, m^2 + (a + b)\, n^2 = 0.$$

Proof: No proof is given here, but the interested reader is referred to Todd [111] where the locus is derived by matrix methods based on the equations (17′), or Faulkner [38] where the locus is treated synthetically, or Maxwell [69] who uses straightforward analytical techniques.

By a similar argument it may be shown that

Theorem 31: The locus of points from which tangents to two conics form a harmonic pencil is a conic (the *harmonic conic-locus*)

$$F \equiv a(b + c)\,x^2 + b\,(c + a)\,y^2 + c\,(a + b)\,z^2 = 0.$$

Credit for the discovery of the F-conic is probably due to George Salmon who wrote in 1879: 'I believe I was the first to direct attention to the importance of this conic in the theory of two conics.' One may therefore appropriately refer to the F-conic as the *Salmon conic* of S, S'.

Note:

(i) Theorem 30 is the dual statement of Theorem 31, but the Φ-conic is not the line-equation of the F-conic. Compare their equations.

(ii) S, S', F, Φ all have the same self-polar triangle (since their equations are similar).

(iii) Working with the simplified forms (12b) of S, S' requires us, where necessary, to make an invariant equation homogeneous by putting $\Delta' = 1$. Had we used the more general equations of S, S', i.e. were the triangle of reference not specially selected, then Δ ($\neq 1$ under those circumstances) would of course occur in the equation.

EXAMPLE 36: If F is outpolar to Φ then $\Theta\Theta' + 3\Delta\Delta' = 0$.

Solution: The required condition is $\Theta'_{F\Phi} = 0$ where F is in the point-form and Φ in the line-form. That is,

$$a(b + c).(b + c) + b\,(c + a).(c + a) + c\,(a + b).(a + b) = 0$$
$$\text{i.e. } \Sigma a^2 b + 3abc + 3abc = 0$$
$$\text{i.e. } \Theta\Theta' + 3\Delta\Delta' = 0.$$

[Try to obtain a corresponding relation when Φ is outpolar to F.]

Theorem 32: The Φ-conic touches the eight tangents from the points of intersection of S, S' (Fig. 61).

Proof: Solve $S = 0$, $S' = 0$ in (12b). Co-ordinates of A, B, C, D are then $(\pm\sqrt{b - c}, \pm\sqrt{c - a}, \pm\sqrt{a - b})$. Tangents to S at these four points have, by (9′), equations

$$\pm a\sqrt{b - c}\,x \pm b\sqrt{c - a}\,y \pm c\sqrt{a - b}\,z = 0 \qquad \text{(i)}$$

Tangents to S' at these four points have, by (9′), equations

$$\pm\sqrt{b-c}\,x \pm\sqrt{c-a}\,y \pm\sqrt{a-b}\,z = 0 \qquad \text{(ii)}$$

The line-co-ordinates of these eight tangents (i) and (ii) all satisfy $\Phi = 0$.

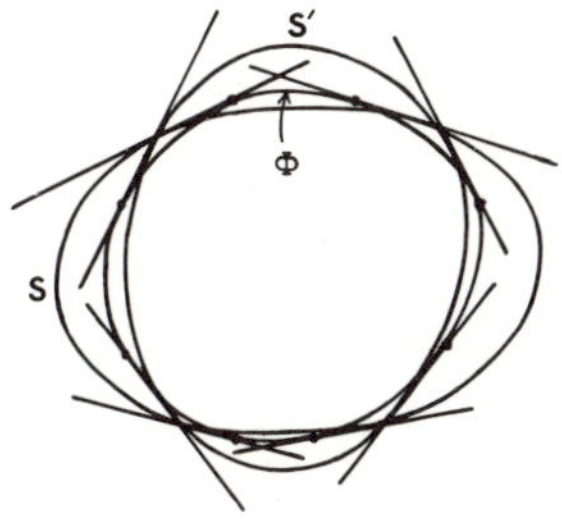

FIG. 61. 8 tangents to Φ

Arguing in a similar fashion, we have the dual result:

Theorem 33: The F-conic passes through the eight points of contact of the common tangents of S, S' (Fig. 62).

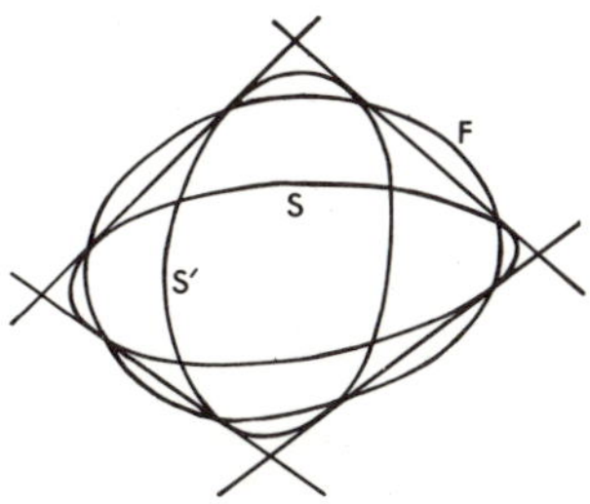

FIG. 62. 8 points on F

Proof: This should be supplied by the reader as an exercise.

Finally, notice that on solving $\Sigma = 0$, $\Sigma' = 0$ we discover the line-co-ordinates of the four common tangents to be $(\pm\sqrt{a(b-c)}, \pm\sqrt{b(c-a)}, \pm\sqrt{c(a-b)})$ and hence the equations of the four common tangents to be

$$\pm\sqrt{a(b-c)}\,x \pm\sqrt{b(c-a)}\,y \pm\sqrt{c(a-b)}\,z = 0.$$

All four equations may be combined into the quartic equation in x, y, z:

$$F^2 = 4\Delta\Delta' SS'.$$

Likewise, the four common points A, B, C, D are expressed by the quartic equation in l, m, n:

$$\Phi^2 = 4\Sigma\Sigma'.$$

Aspects of the theory of a pair of conics will be encountered in the next Chapter in a metrical setting.

Suggested topics for further study

(1) Deeper study of apolarity and harmonic conics.

(2) Higher plane curves (Primrose [89]).

(3) Weierstrassian elliptic functions and the general plane cubic curve (Karl Weierstrass 1815–1897, German) (Primrose [89]).

EXERCISES 7

1. What is the line-equation of the conic $y^2 - zx = 0$?

2. Justify the statement that the point-equation of a conic inscribed in the triangle of reference is expressible as $x^{\frac{1}{2}} \pm y^{\frac{1}{2}} \pm z^{\frac{1}{2}} = 0$.

3. Verify that the line-equation of a conic circumscribing the triangle of reference may be written in the form $l^{\frac{1}{2}} \pm m^{\frac{1}{2}} \pm n^{\frac{1}{2}} = 0$.

4. Determine the equation of the line joining two points with parameters θ, θ' on the conic $y^2 - zx = 0$. Also determine the equation of the tangent at θ.

5. Obtain the co-ordinates of the fixed point through which pass the secants joining pairs of points in involution on a conic.

6. Show that the equation of two conics having osculating (= three-point) contact may be written $y^2 - zx = 0$ and $y^2 - zx + kzy = 0$. (Here, the osculating contact is at X, i.e. there are three infinite values of θ. Had the point of osculating contact been at Z, how different would the equation of the second conic be?)

7–10. Perform the calculations of Example 35(i)–(iv).

11. Eliminate λ between (16) and the pencil $\lambda S + S' = 0$ to obtain the equation $\Delta S'^3 - \Theta S'^2 S + \Theta' S' S^2 - \Delta' S^3 = 0$ (a sextic in the co-ordinates).

Interpret this as the equations of the three pairs of common secants (= degenerate conics) of the pencil.

12. Evaluate the discriminants of the F- and Θ-conics. [Answers: $\Delta\Delta'(\Theta\Theta' - \Delta\Delta')$, $(\Theta\Theta' - \Delta\Delta')^2$ respectively, i.e. F degenerates if Θ degenerates (when $\Theta\Theta' = \Delta\Delta'$).]

13. Express the point-equation of Φ as $S'\Theta + S\Theta' - F = 0$.

14. Mutual apolarity of S, S' implies that their F and Θ-conics are identical. Prove. [Note that the result follows immediately from Exercise 7.13.]

15. If S is outpolar to Σ' prove that

(i) Σ is inpolar to F;

(ii) S' is inpolar to Θ.

16. Establish that the envelope of the polar of a variable point on S w.r.t. S' is a conic (the *polar reciprocal*)

$S_o \equiv \Theta S' - F = 0.$

[Try to discover further properties of this locus.]

17. If S, S' are mutually apolar, what is the 'F-conic' (Salmon conic) of F and S'?

18. Triangles ABC, $A'B'C'$ are inscribed in a conic and are in perspective from the unit point U. Tangents at A', B', C' meet BC, CA, AB at A'', B'', C'' respectively. Show that, with ABC as triangle of reference,

(i) $A' \equiv \left(\dfrac{-f}{g+h}, 1, 1\right)$;

(ii) the equation of the tangent at A' is $\dfrac{(g+h)^2}{f}x + gy + hz = 0$;

(iii) A'', B'', C'' are collinear on the line $fx + gy + hz = 0$;

(iv) the equation of the Desargues line of the perspective triangles is

$$\frac{x}{f+h} + \frac{y}{g+h} + \frac{z}{f+g} = 0;$$

(v) the equation of the Pascal line of the hexagon $AA'BB'CC'$ is

$$gx + fy + hz = 0.$$

19. The three-cusped quartic curve (actually the *cardioid*) $x^2y^2 + y^2z^2 + z^2x^2 - 2xyz(x + y + z) = 0$ is known to have cuspidal tangents at the vertices X, Y, Z of the triangle of reference having equations $y = z$, $z = x$, $x = y$ respectively so that they are concurrent at the unit point.

These cuspidal tangents at X, Y, Z meet the cardioid again at L, M, N respectively. Prove that the equation of the tangent at L is

$$-8x + y + z = 0$$

and that the tangents at L, M, N form a triangle which is in perspective with the triangle XYZ. What is the equation of the Desargues line of these perspective triangles?

20. In $[n]$, a general non-linear locus, called a *primal* of order r, is given by the equation $f(x_1, x_2, \ldots, x_{n+1}) = 0$ where f is a polynomial of degree r in the homogeneous co-ordinates $x_1, \ldots, x_{n+1}$. Determine the number of coefficients k in this equation (and hence the number of points, $k-1$, necessary to define this primal uniquely). What do these numbers become in the case of the conic and the general plane cubic? [This problem could lead in to the study of algebraic objects called *homogeneous products* which are related to the theory of combinations.]

CHAPTER 8

AFFINE AND EUCLIDEAN GEOMETRY

> 'Given any group of transformations . . . which includes the principal group (Euclidean and similarity transformations, reflections) as a subgroup, then the invariant theory of this group gives a definite kind of geometry, and every possible geometry can be obtained in this way. Thus each geometry is characterised by its group, which therefore assumes the leading place in our considerations.'
>
> (F. Klein [57] p. 133)

REVIEWING briefly the progress which we have made in the first seven Chapters, we see how, from a given abstract axiomatic basis, the chief synthetic and analytic features of an entirely non-metrical geometry may be developed. While these processes could be extended to provide greater geometrical and algebraic detail, our wish is to modify Projective Geometry to allow for the entry of concepts such as parallelism and perpendicularity. Ultimately, plane Euclidean Geometry is the goal, and this requires the introduction into the projective plane of definitions of linear and angular measurement.

What does the ordinary concept of parallelism involve?

'Looking out on the world around us,' says Robinson [92], 'we have recognised the line as a significant contour; but in supposing that any two lines of projective geometry have a point in common, we have adopted a simplification which is suggested by the visual process. Two "coplanar" lines in physical space may be so situated that, though they appear to "approach" one another, yet no point of intersection can be found. It was this property of lines in physical space which was transferred by the Greeks to the geometry which they constructed in their minds: two such lines were called "parallel".'

These observations suggest that we should pick out certain elements in the projective plane π in terms of which parallel lines can be defined. Choose *any* projective line and call it the *ideal line* or *line at infinity*, l_∞. Though l_∞ is selected arbitrarily, once chosen it remains the only l_∞ for a particular problem. As the situation (or problem) varies, so may l_∞. By *PP*1 any (non-ideal) line l in π meets l_∞ in just one point, namely the *ideal point* or *point at infinity* P_∞ $(= l \cap l_\infty)$ on l. Expressed otherwise, l_∞ is the locus of all the points at infinity in the plane, one point at infinity for each line.

Let us now concentrate for a few moments on the *real* projective plane. In the real projective plane, geometry w.r.t. l_∞, i.e. the relations of other points and lines to the points of l_∞, is called *Affine Geometry*, which is thus a modification of Projective Geometry. (Let us note in passing that Affine Geometry can be developed axiomatically quite independently of Projective Geometry, just as Euclidean Geometry can be.) As we shall see, Affine Geometry occupies an intermediate stage between Real Projective Geometry and Euclidean Geometry. Later, it will be established that in terms of Klein's Erlangen Programme, Affine Geometry is the set of properties remaining invariant under a group of affine transformations (the *affine group*), a special feature of which—as we shall see—is that its elements always leave l_∞ invariant while permuting amongst themselves points of the plane not on l_∞; points (at infinity) on l_∞ are permuted among themselves, i.e. l_∞ is invariant as a whole, not point-by-point (Theorem 35).

Label the affine plane π_a, i.e. the real projective plane with l_∞ removed. (This means that an affine line is not closed, whereas a projective line is closed.) Affine points are the real projective points which do not lie on l_∞. Affine lines are real projective lines (excluding l_∞) minus the point at infinity on each line. Symbolically

$$\begin{aligned} l_a + P_\infty &= l \\ \pi_a + l_\infty &= \pi \end{aligned}$$

where l_a, π_a stand for affine line and plane, while l, π represent the corresponding augmented affine (real projective) line and plane. Accordingly, we can move freely from π_a to π or from π to π_a by the adjunction or deletion of l_∞. (Suitable manipulation of the affine plane gives rise to the Minkowski plane of *Minkowski Geometry* (Hermann Minkowski 1864–1909, Russian-German) which is the geometrical basis for the Special Theory of Relativity.)

§ 32 PARALLELISM, PERPENDICULARITY AND BISECTION

Furthermore, on l_∞ in the projective plane (not necessarily real) choose two *arbitrary* points I, J which we will call the *circular points at infinity*. Preferential treatment accorded these two special points will be seen as the key to Euclidean Geometry. The stage is now set for the first break-down of the projective concepts.

§ 32.1 Definitions: (i) Lines having a point in common on l_∞ are *parallel* (Fig. 63(a)).

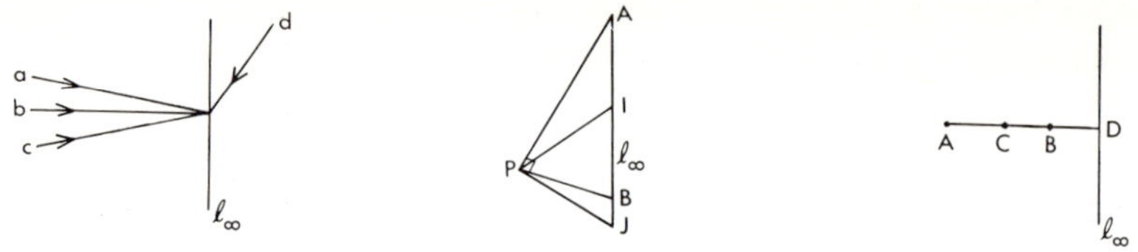

FIG. 63. (a) Parallelism $a\|b\|c\|d$ (b) Perpendicularity $P(AB, IJ) = -1$ (c) Mid-point $(AB, CD) = -1$

Obviously, according to this definition, there is just one line through a given point of the plane parallel to a given line (see *PP*1). (As this deduction is not valid for parallelism in the non-Euclidean geometries, we must conclude that they are not sub-geometries of Affine Geometry, i.e. that they are derived from Projective Geometry independently of Affine Geometry.)

Is there a line parallel to l_∞? (No. Why?)

(ii) Two lines are *perpendicular* (or at *right angles*) if they meet l_∞ in two points harmonically separating I, J (Fig. 63(b)). Lines through I, J (excluding IJ itself), called *isotropic lines*, are said to be self-perpendicular.

Can a line be perpendicular to l_∞? (No. Why?)

Pairs of perpendicular lines through a point (each pair thus forming on l_∞ an harmonic range with I, J) constitute an involution whose double lines are the isotropic lines through the point. Such an involution is called the *orthogonality involution*. On l_∞ there is thus an involution having as double points I, J (which are complex points, as we shall see later). Such an involution, which originated with von Staudt, is called the *absolute involution*. Symbols $\|$, $\perp$ may be employed to represent parallelism and perpendicularity respectively.

(iii) C is the *mid-point* of AB (i.e. AB is *bisected* at C), meaning C is the mid-point of the points A and B, if AB meets l_∞ in the point which is the harmonic conjugate of C w.r.t. A, B (Fig. 63(c)). Bisection clearly cannot refer to l_∞ itself, e.g. there is no mid-point of IJ.

In the above definitions of perpendicularity and bisection, there is no suggestion whatever of metric ideas at this stage. Phrases like 'at right angles' and 'mid-point of' are here used only for convenience, as they simplify the language when additional modifications (e.g. introduction of measurement) are made.

§ 32.2 Order in the Real Projective and Affine Planes. We pause for a while to discuss the real projective plane. Earlier, in § 13.5, we mentioned that order could be introduced into the real projective plane by means of an undefined relation called separation which could be shown to be a projective invariant.

Assuming that the projective Axioms of Order can be explained in terms of cross-ratio (Robinson [92]), we define separation as follows:

Given four distinct points A, B, C, D on the projective line l, we say that

C, D *separate* A, B if and only if $(AB, CD) = \lambda < 0$.

(In particular, $\lambda = -1$ gives the familiar harmonic separation.)

Specifying D as D_∞ (and then omitting D_∞ to make l an affine line), we say that

C is *between* A, B if and only if $(AB, CD_\infty) = \lambda < 0$.

(In particular, $\lambda = -1$ yields the mid-point of AB.)

Since cross-ratio is a projective invariant, it follows that separation, as defined, is a projective invariant. Betweenness is an affine invariant for, as we shall see (Theorem 35), affine transformations permute the points on l_∞, i.e. under an affine transformation, $(AB, CD_\infty) = (A'B', C'D'_\infty) = \lambda < 0$.

§ 32.3 Segments in the Real Projective and Affine Planes. On the projective line l, we define the concept of (projective) segment in this way:

Given three distinct points A, B, C, D on l, then

the (projective) *segment* AB consists of all points C for which C, D separate A, B.

On the affine line,

the (affine) *segment* AB consists of all points C between A and B.

Remarks:

(1) Both the projective and affine segments are non-metrical.

(2) The projective segment AB cannot contain D since $(AB, DD) = 1 (\nless 0)$

and in any case the four points were specified as being distinct.

(3) A, B are called the *end-points* of the segment AB (which may be taken to include A, B as well as the points between A, B).

(4) Though we use the same symbol AB both for the name of the line determined by A, B and for the segment of which the end-points are A, B, there will be no confusion as the meaning will always be clear from the context.

(5) Corresponding to the concepts of separation and betweenness of collinear points (§ 32.2) there are dual concepts of separation and betweenness of concurrent lines.

(6) Just as we have the non-metrical concept of segment AB, so we may have a non-metrical concept of the *angular region ab* between two lines a, b.

§ 32.4 Congruence in the Affine Plane. A non-metrical definition of congruence may be introduced into the affine plane by means of parallel lines. If $BB'C'C$ is a parallelogram (i.e. a quadrangle with two pairs of opposite sides parallel), we say that the affine segments BB', CC' are *congruent by translation*, written $BB' \equiv CC'$. Thus, in Fig. 82, $AB \equiv DC$, $AD \equiv BC$. As parallelism is an affine invariant (Theorem 36), it follows that congruence, as defined, is an affine invariant.

Looking at the special affine case of Desargues' Theorem shown in Fig. 113(b), we see that

$$BB' \equiv CC',\ CC' \equiv AA' \Rightarrow BB' \equiv AA'.$$

This concept of transitivity of congruent segments on parallel lines may be extended to congruent segments on the same line. To avoid drawing a fresh figure, imagine that the metrical Fig. 80(a) exists in the (non-metrical) affine plane. Then

$$OB \equiv PK,\ PK \equiv AC \Rightarrow OB \equiv AC.$$

If $B = A$, then A is the mid-point of OC. (Check to convince yourself that this accords with the previous definition of mid-point, namely, that the harmonic conjugate of A w.r.t. O, C is on l_∞.)

As an exercise, the student should refer to Fig. 15(c) and imagine it drawn in the affine plane with $HB'P_\infty$ as l_∞.

Nothing in our assumptions about the affine plane enables us to compare segments on distinct, non-parallel lines or to compare angles. These problems require the introduction of measurement (length of segment, measure of angular region) into the affine plane by means of an arbitrarily chosen elliptic involution called (§ 32.1) the absolute involution. (See § 38.7.) The metric plane thus created, namely the affine plane endowed with a metric, is then the familiar Euclidean plane. Automatically, affine results become (metrical) Euclidean results. (It may be shown (e.g. Fishback [40]) with a different approach, that all of Hilbert's axioms for the Euclidean plane, except the congruence axioms, are valid in an affine plane. Accordingly, many Euclidean results are also affine results, since not all Euclidean results are concerned with measurement.)

Logically, some affine results (Theorems 35, 36) have been used before they have been proved, but we believe there are pedagogical advantages in the procedure adopted. Full details of some affine properties which are also valid in Euclidean Geometry (e.g. order) are not provided here because they would add too much to the content of our material. To this extent, our treatment of the Euclidean plane will be incomplete, though our derivation of it will give us its essential features.

On account of what we have just said, the properties of conics which we develop for the real affine plane are also true in the metrical context of the Euclidean plane.

Leaving the real projective plane, we return to the more general plane.

§ 33 THE CIRCLE

Definitions: A conic through I, J is a *circle*. (Therefore, there exists just one circle through three points other than I, J (Theorem 19). If these three points are collinear, the circle degenerates into this line of collinearity and $IJ (= l_\infty)$. A circle may, however, degenerate into two lines OI, OJ (Fig. 64(a)).

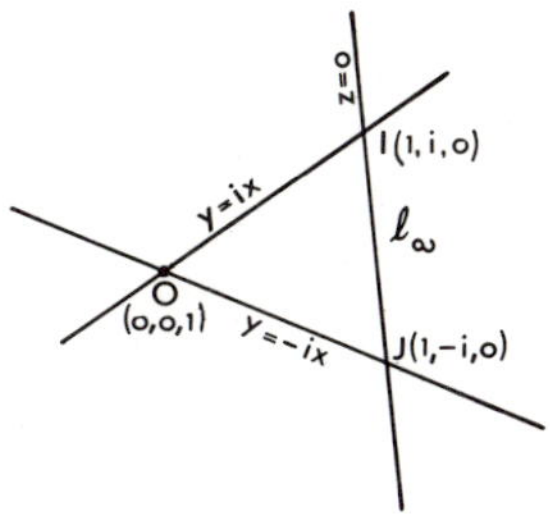

FIG. 64. (a) Degenerate circle $x^2 + y^2 = 0$

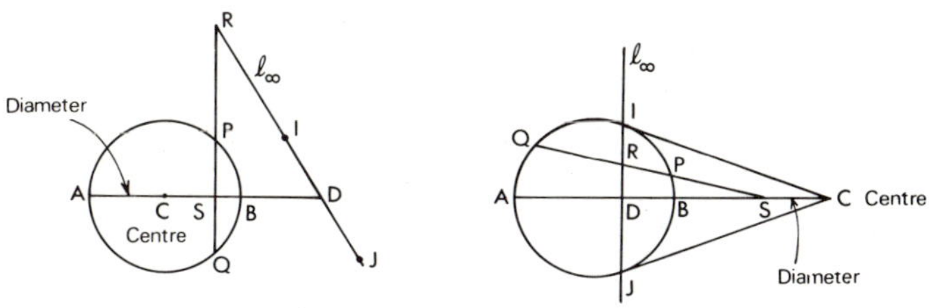

FIG. 64. Centre and diameter of a circle

The *centre* of the circle is the pole of l_∞.

Any secant through the centre is a *diameter*. (No measure is associated with this concept.) The two points of intersection of a diameter and a circle are called the *ends* of the diameter.

In constructing diagrams involving the centre of a circle, we have the choice of (a) taking the centre C 'inside' the circle so that the tangents from C cannot be physically drawn, or (b) taking C 'outside' the circle, in which case the tangents can be drawn and the polar is then represented according to Fig. 50, i.e. in the usual way. (See Fig. 64.) Often, depending on the nature of the problem, it may be unnecessary to indicate I, J in the diagram, e.g. in Theorem 34, Fig. 65.

It is well to convince ourselves that this definition does in fact lead to familiar properties of a circle. For example, since C, D are harmonic conjugates of A, B (Fig. 64) and D is on l_∞, it follows that $(AB, CD) = -1$ and C is the mid-point of AB. That is

the centre is the mid-point of any diameter AB.

Moreover

$-1 = (AB, CD)$	given
$= I(AB, CD)$	building up the range of collinear points into a pencil
$= I(AB, IJ)$	bringing the base points of the pencil from the line to the conic
$= P(AB, IJ)$	by Chasles' Theorem
$\therefore PA \perp PB$	by § 32, definition (ii).

[This is the counterpart of the Euclidean Theorem: The angle in a semi-circle is a right angle (attributed to Thales (6th century B.C., Greek)).]

A particularly elegant result which follows without much advanced knowledge is Miquel's Theorem, discovered in 1838 ('*Liouville's Journal de Mathématiques*', 3 (1838) 485–487). Our proof is due to R. J. Lyons.

Theorem 34 (*Miquel's Theorem*)*:* If D, E, F are points on the sides BC, CA, AB respectively of the triangle ABC, then the circumcircles of the triangles AEF, BFD, CDE have a point in common (*the Miquel Point*).

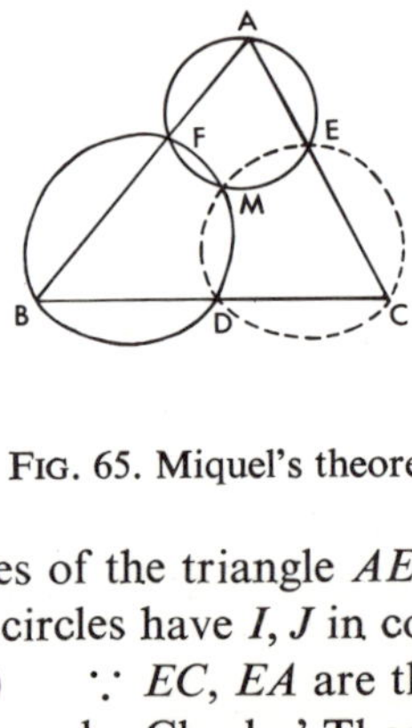

FIG. 65. Miquel's theorem

Proof: Let the circumcircles of the triangle AEF, BFD meet at M (Fig. 65).
We use the fact that all circles have I, J in common.

$E(MC, IJ) = E(MA, IJ)$	$\because$ EC, EA are the same line
$= F(MA, IJ)$	by Chasles' Theorem
$= F(MB, IJ)$	$\because$ FA, FB are the same line
$= D(MB, IJ)$	by Chasles' Theorem
$= D(MC, IJ)$	$\because$ DB, DC are the same line.

$\therefore$ by the Converse of Chasles' Theorem, E, D, M, C, I, J lie on a conic.
i.e. E, D, M, C lie on a circle.

$\therefore$ circumcircles of triangles AEF, BFD, CDE meet at M (*Miquel Point*).

Question: What happens when D, E, F are collinear? (We then obtain the Miquel point of a quadrilateral.)

Subsequently, we shall need the idea of concentric circles. How should they be defined? Recalling the definition of the centre as the pole of l_∞ we make the following assertion:

Definition: Concentric circles are conics of a system having double contact at I, J (Fig. 66).

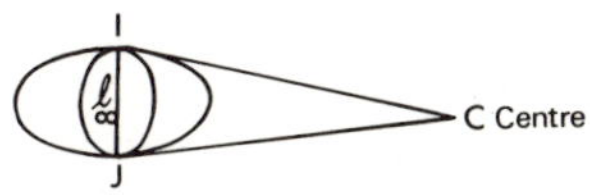

FIG. 66. Concentric circles

§ 34 THE CONIC

Conics may be specialised according to their behaviour in relation to l_∞. Following is the classification (Fig. 67(c), drawn in the real affine plane):

§ 34.1 Definitions: An *ellipse* is a conic cutting l_∞ in two complex points (i.e. in no real points).

An *hyperbola* is a conic cutting l_∞ in two real points.

A *rectangular hyperbola* is a conic cutting l_∞ in two real points which harmonically separate I, J (Fig. 67(a)).

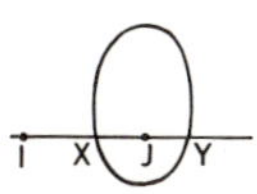

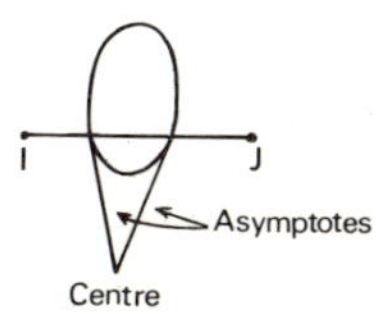

FIG. 67. (a) Rectangular hyperbola $(XY, IJ) = -1$

(b) Asymptotes and centre of an hyperbola

A *parabola* is a conic touching l_∞.

Centre and *diameter* of a conic are defined as for the circle. As with the circle, the centre is the mid-point of every diameter (except in the case of the parabola). Any non-parabolic conic is called a *central conic*, since it has a centre.

The tangents to a conic at its points of intersection with l_∞ are the *asymptotes* (Fig. 67(b)). (Consequently, asymptotes of a parabola coincide with l_∞, and asymptotes of a rectangular hyperbola are perpendicular.)

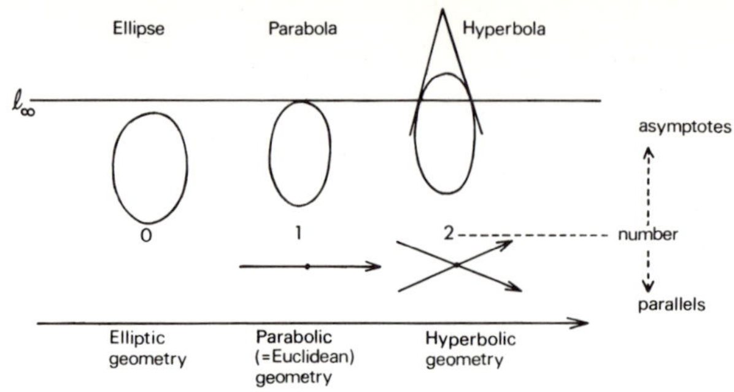

FIG. 67. (c) Klein's nomenclature for geometries

As a representative problem which is not only of interest *per se* but which also incorporates many aspects of the theory of conics, we choose the example hereunder.

EXAMPLE 37: Every conic through the four points of intersection of two rectangular hyperbolas H_1, H_2 is itself a rectangular hyperbola and the four points are the vertices of a triangle and its *orthocentre* (i.e. the point of concurrency of lines from the vertices perpendicular to the opposite sides).

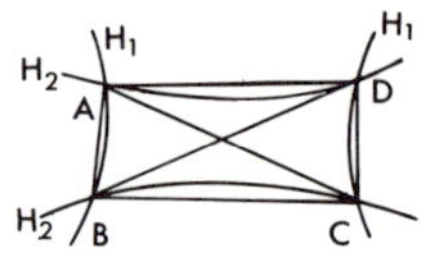

FIG. 68. Orthocentre of a triangle

Solution: By Desargues' Theorem (Theorem 23), the conics of the pencil (through the four points A, B, C, D) cut l_∞ in pairs of points in involution (Fig. 68). But $H_1 \cap l_\infty$, $H_2 \cap l_\infty$ are points harmonically dividing I, J by definition

$\therefore$ the involution consists of pairs of points harmonically separating I, J

$\therefore$ every conic of the pencil is a rectangular hyperbola.

Included in the pencil are the three degenerate rectangular hyperbolas AB, CD; AC, BD; AD, BC which are thus pairs of perpendicular lines

$\therefore$ D is the orthocentre of the triangle ABC.

Similarly for the points A, B, C.

§ 34.2 Foci and Axes of a Conic. Definitions: A *focus* of a conic is the intersection of two isotropic tangents, i.e. tangents from I, J (a concept due to Poncelet). Associated with each focus is its *directrix*, viz. the polar of the focus w.r.t. the conic. It follows that a conic generally has four foci and four directrices. However, a parabola has just one focus and a corresponding directrix. (Two of the foci of the central conic become I, J in the case of the parabola and the remaining focus coincides with the point of contact of the parabola and l_∞. By formal analogy with our concept of centre, we may then refer to this point of contact as the 'centre' of the parabola. Strictly speaking, however, the parabola has no centre.)

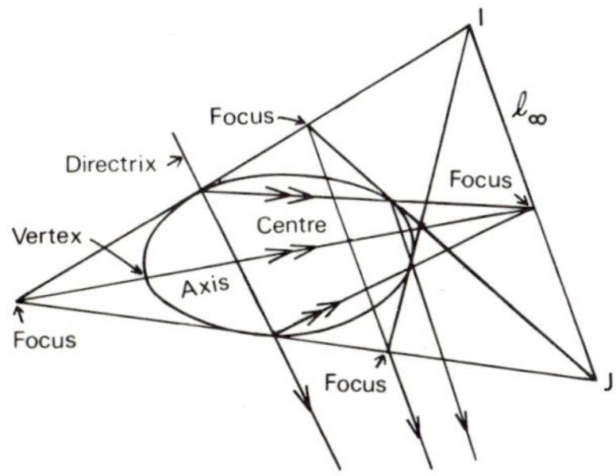

FIG. 69. (a) Basic features of a central conic

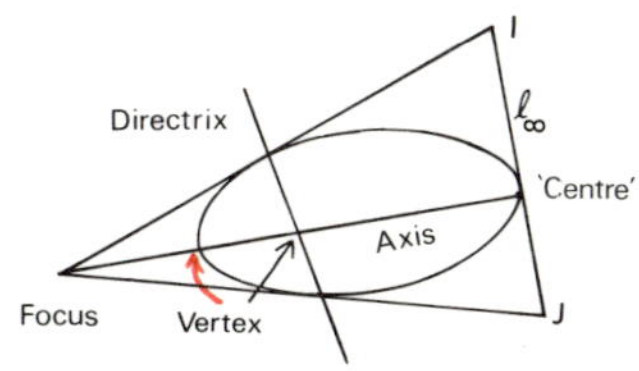

(b) Basic features of a parabola

Axis is defined to be a line joining two foci so that, as shown in Fig. 69(a) a central conic has only two axes, meeting in the centre. Parabolas Fig. 69(b) have but a single axis. *Vertex* is defined to be the intersection of the conic and an axis. Accordingly, a central conic has four vertices and a parabola has one proper vertex in addition to the 'centre' on l_∞. Historically, it is worth mentioning that this 'centre' is none other than Kepler's 'blind focus' (Fig. XII) defined in 1604 in his '*Paralipomena in Vitellionem*', at which lines parallel to the axis of the parabola intersect.

One statement above must not pass uncontested. The centre of a conic is defined to be the intersection of two axes, but we need to reassure ourselves that this intersection is indeed the same point as the centre as previously defined, namely the pole of l_∞. Justification of this fact is straightforward: Labelling pairs of opposite foci F_1 and F_2, F_3 and F_4 we see that the complete quadrilateral has a self-polar diagonal triangle formed by the lines F_1F_2, F_3F_4, IJ so that $F_1F_2 \cap F_3F_4 =$ centre C *is*, in actual fact, the pole of l_∞.

Elementary properties of the conics which should be verified as exercises are:

I Axes are perpendicular lines (since CF_1F_2, CF_3F_4 harmonically separate CI, CJ).

II Axes are the unique pair of perpendicular conjugate diameters (refer to Exercise 8.6 for the definition of conjugate diameters).

III A directrix is perpendicular to one axis and parallel to the other.

IV For the parabola, the directrix is perpendicular to the axis.

§ 34.3 Miscellaneous Examples on Conics.

EXAMPLE 38: A, B, C, D are four points on a parabola. Lines are drawn through C, D parallel to the axis of the parabola to meet AD, BC in H and K respectively. Prove that HK is parallel to AB (Fig. 70).

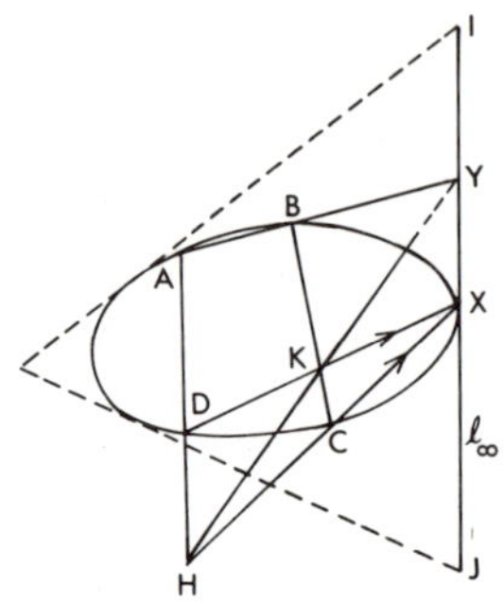

FIG. 70.

Solution: Consider the degenerate Pascal hexagon $ADXXCB$. Its Pascal line is HKY where $Y = AB \cap XX$ (i.e. $AB \cap IJ$)

$\therefore AB \cap HK = Y$

i.e. $AB \| HK$.

EXAMPLE 39: Every conic through the four foci of a conic is a rectangular hyperbola.

Solution: Use Fig. 69(a) and consider the involution on l_∞ ($= IJ$) w.r.t. the pencil of conics through the four loci. Label the foci F_1, F_2, F_3, F_4:

$$I = F_1F_3 \cap F_2F_4, \qquad J = F_2F_3 \cap F_1F_4.$$

Since F_1F_3, F_2F_4 and F_2F_3, F_1F_4 are degenerate conics of the pencil, it follows that I, J are the double points of the involution. Hence the result, keeping in mind the definition of a rectangular hyperbola.

EXAMPLE 40: An asymptote to a hyperbola is cut by the three pairs of opposite sides of a complete quadrangle inscribed in the hyperbola in three segments with a common middle point. Prove.

Solution: Let the asymptote meet the pairs of opposite sides of the complete quadrangle in the points A, A'; B, B'; C, C' which, by the Corollary to Desargues' Theorem, must be pairs of points in an involution (Fig. 71).

Double points of this involution are given by the two points in which two conics of the system through A, B, C, D touch the asymptote.

One of these is E, by the definition of a hyperbola. Suppose the other is F

$\therefore (AA', EF) = (BB', EF) = (CC', EF) = -1$ by § 20 I

$\therefore F$ is the common mid-point of AA', BB' and CC' ($\because E$ is on l_∞).

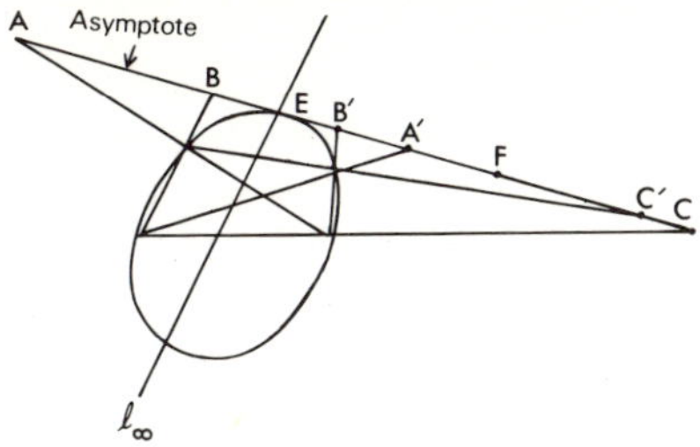

FIG. 71.

EXAMPLE 41: Through the points of contact of two tangents to a parabola, which intersect at O, lines parallel to these tangents are drawn to intersect at R. Prove that OR is a diameter of the parabola (i.e. a line through the 'centre' C of the parabola) and that it is bisected by the line joining the points of contact.

Choosing this line joining the points of contact as the line at infinity, interpret this result as a statement about a rectangular hyperbola.

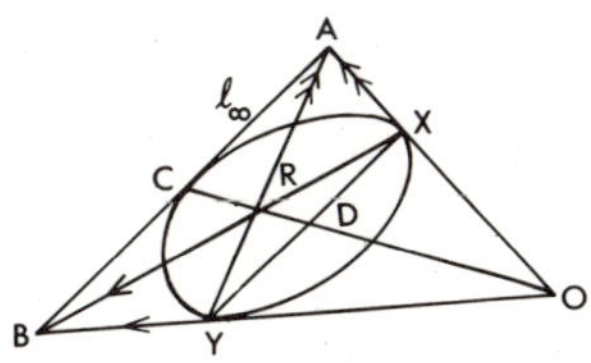

FIG. 72. (a)

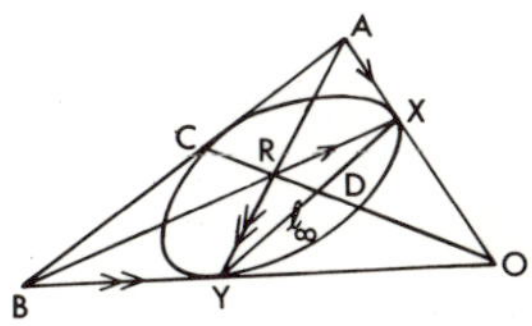

FIG. 72. (b)

Solution: R is the Hessian point of the triangle ABO circumscribing the conic (Fig. 72(a))

$\therefore$ ORC is a line

i.e. OR is a diameter of the parabola.

From the complete quadrangle based on A, B, X, Y we have

$(OR, DC) = -1$ and C is on l_∞

$\therefore$ D is the mid-point of OR.

Instead of choosing AB as l_∞, choose XY as l_∞ (Fig. 72(b)).

As previously $(OR, DC) = -1$ but now D is on l_∞

$\therefore$ C is the mid-point of OR.

The interpretation of the result is now as follows: If, through A, B, where tangent to any hyperbola meets the asymptotes, lines parallel to the symptotes are drawn to intersect at R and O is the centre of the hyperbola, then OR passes through the point of contact C of the tangent and is bisected at C.

§ 35 PROJECTIVE METRIC

§ 35.1 Euclidean Metrics. Ultimately Euclidean Geometry, which we desire, is based on measurement and, therefore, in modifying the projective plane π to make it Euclidean, we must introduce into π projective analogues of the Euclidean metrics, namely measures of line-segment and angle.

These projective definitions must be such that they lead eventually into the corresponding Euclidean definitions. Now, in Euclidean Geometry, both distance and angle are for the moment taken as intuitively understood. In order to generalise these notions to the projective plane, we need to know what fundamental properties characterise them. A careful study might lead us to accept the following properties as basic, where we represent the distance between two points P, Q by the symbol (PQ), and the angle between two lines p, q by (pq):

(1) $(PP) = 0$;
(2) $(PQ) = -(QP)$;
(3) $(PQ) + (QR) + (RP) = 0$ for three collinear points P, Q, R;
(1′) $(pp) = 2n\pi$ where n is any integer (including zero);
(2′) $(pq) = (-qp) + 2n\pi$ where n is an integer (including zero);
(3′) $(pq) + (qr) + (rp) = 2n\pi$ for three concurrent lines p, q, r, where n is any integer (including zero).

In what follows all previously established results for the augmented plane retain their validity.

§ 35.2 Projective Metrics. Historically, the problem was solved independently by Arthur Cayley [20] 2, p. 583 (*A Sixth Memoir on Quantics*) and Felix Klein ('*Ueber die sogenannte Nicht-Euklidische Geometrie*', *Mathematische Annalen* **4**, (1871) 573–632, though the basic idea goes back to 1853 to Edmond Laguerre ('*Note sur la théorie des foyers*', *Nouvelles Annales de Mathématique* (1853) 57–66), who, however, did not develop it sufficiently. (Also see Busemann and Kelly [17].)

Cayley and Klein make the following definition of 'distance' on a line and 'angle' at a point in the projective plane, where O, U are fixed reference points and o, u are fixed reference lines, and k is the arbitrary *scale-constant*.

Length		Angle	
$(PQ) = k \log (PQ, OU)$	duals $\longleftrightarrow$	$(pq) = k \log (pq, ou)$	(19

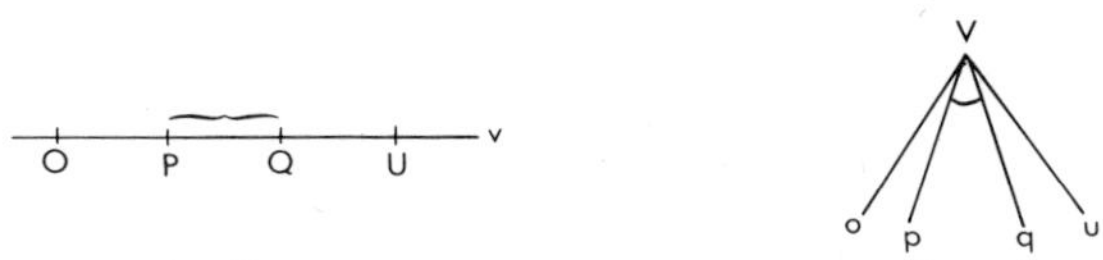

FIG. 73. Projective metrics

It is not necessary for one or both of P, Q (p, q) to be 'between' O, U (o, u) as shown in Figs. 73(a), (b). (As used in (19), the symbol U is unrelated to its previous designation for the unit point. Likewise, O has an independent meaning.)

These definitions will have to be modified later to meet the requirements of Euclidean Geometry, but it is worth checking at this stage to see how closely they reproduce the properties (1), (2), (3) and (1′), (2′), (3′). Using the facts

$$\begin{cases} (PP, OU) & = 1 \text{ (two points coincident)} \\ (PQ, OU) & = \dfrac{1}{(QP, OU)} \\ (PQ, OU)\,(QR, OU)\,(RP, OU) & = 1 \end{cases}$$

we readily verify the validity of (1), (2), (3) so that the strange-looking definition (19) is in accord with the ordinary notion of distance. (Note that $(PQ, OU) + (QR, OU) \neq (PR, OU)$.) Similar remarks apply to (1)′, (2)′, (3)′. In passing, it should be emphasised that the deductions from (19) lead to the presence of a term $2n\pi ki$ added on to the right hand side of (1), (2), (3) and (1′), (2′), (3′). This ambiguity derives from the periodicity of the logarithmic function for complex variables. Otherwise the definitions (19) satisfy the requirements of 'distance' on a line and 'angle' at a point.

§ **35.3 Absolute conic.** For the whole projective plane π, the definitions (19) must be true for distances on every line of π and for angles at every point of π. Says Cayley [20] 2,p. 586: 'Passing now to geometry of two dimensions, we have here to consider a certain conic, which I shall call the Absolute.' *There is nothing absolute whatever about Cayley's 'Absolute'. It is quite arbitrarily chosen.* However, once chosen (assumed non-degenerate) it is fixed. While a warning has been sounded about the possible misleading nature of the term, we shall continue in accord with common practice to use it, though some writers refer to it as the *gauge-conic* (which is undoubtedly a more suggestive name). Label the *Absolute conic* $\mathscr{A}$. Since cross-ratio is invariant under a projective transformation, the group of such transformations which leaves $\mathscr{A}$ invariant must also preserve the invariance of distance and angle in (19).

FIG. 74. Absolute conic

Distance and angle, in relation to $\mathscr{A}$, are now defined precisely as before, i.e. by (19). Evidently (Figs. 74(a), (b)), $\mathscr{A}$ is the locus of all points O, U (absolute points) as the line v varies, while dually it is also the envelope of all lines o, u (absolute lines) as the vertex V varies.

Special cases arise in the following ways (and here reference ought to be made to § 19 II):

(a) P lies on $\mathscr{A}$ (p is a tangent to $\mathscr{A}$)

$\therefore (PQ, OU) = 0, \infty$ according as $P = O, U$. (Similarly for $p = o, u$)

$\therefore$ distance and angle are undefined.

(b) v is a tangent to $\mathscr{A}$ (V lies on $\mathscr{A}$), i.e. $O = U$ ($o = u$)

$\therefore (PQ, OU) = (pq, ou) = 1 = e^{2n\pi i}$ $(n = \ldots, -1, 0, 1, \ldots)$

$\therefore$ from (19), $(PQ) = (pq) = 2nk\pi i \left(= n\pi \text{ if } k = \dfrac{1}{2i}\right)$

$\therefore$ distance and angle are undefined.

[Obviously, we could have operated with Q and q instead of P and p.]

Summing up:

Distance in Projective Geometry is not an inherent property of two points (as it is in Euclidean Geometry), but is merely a property of these points w.r.t. arbitrarily chosen $\mathscr{A}$.

Similar remarks apply to the projective concept of angle. Absolute points and lines are not intrinsically special but enjoy a special privilege purely by virtue of the choice of $\mathscr{A}$.

Quotations from Cayley [20] 2, p. 592, written in 1859, are appropriate at this stage:

'. . . . the theory in effect is, that the metrical properties of a figure are not the properties of the figure considered *per se* apart from anything else, but its properties when considered in connection with another figure, viz. the conic called the absolute'. [Klein, at the beginning of his *Erlanger Programm*, makes a similar statement.]

'Metrical geometry is thus a part of descriptive [= projective, in Cayley's usage] geometry, and descriptive geometry is *all* geometry, and reciprocally . . .'

Next, consider the metrical theory algebraically. In Fig. 74, let $P \equiv (x_1, y_1, z_1)$ and $Q \equiv (x_2, y_2, z_2)$ be real. Take the equation of $\mathscr{A}$ to be quite general, i.e. $f(x, y, z) = 0$ as in §27 (6). Then $\mathbf{O} = \mathbf{P} + \mu_1\mathbf{Q}$, $\mathbf{U} = \mathbf{P} + \mu_2\mathbf{Q}$.

Note the variation of technique here. Ordinarily, we should want to express P, Q in terms of O, U. Using (7), with $\lambda = 1$, we have Joachimsthal's equation in the form

$$\boxed{\mu^2 f_{22} + 2\mu f_{12} + f_{11} = 0} \qquad (7')$$

of which the roots μ_1, μ_2 are $\dfrac{-f_{12} \pm \sqrt{f_{12}^2 - f_{11}f_{22}}}{f_{22}}$. Parameters of P, Q, O, U are consequently $0, \infty, \mu_1, \mu_2$

$$\therefore (PQ, OU) = \frac{\mu_1}{\mu_2} \text{ by § 19 VI}$$

$$\therefore (PQ) = k \log\left(\frac{\mu_1}{\mu_2}\right) \text{ using (19)}$$

$$\text{i.e. } (PQ) = k \log\left\{\frac{f_{12} + \sqrt{f_{12}^2 - f_{11}f_{22}}}{f_{12} - \sqrt{f_{12}^2 - f_{11}f_{12}}}\right\}$$

$$\text{or } \boxed{(PQ) = k \log\left\{\frac{\dfrac{f_{12}}{\sqrt{f_{11}f_{22}}} + \sqrt{\dfrac{f_{12}^2}{f_{11}f_{22}} - 1}}{\dfrac{f_{12}}{\sqrt{f_{12}f_{22}}} - \sqrt{\dfrac{f_{12}^2}{f_{11}f_{22}} - 1}}\right\}} \qquad (20)$$

Having introduced a projective analogue of measurement, we now proceed to interpret it and so initiate the theory of metrical Euclidean Geometry. The value of (19) is clearly dependent on the nature of $\mathscr{A}$. Three cases may arise:

(i) If $\mathscr{A}$ *is real*, i.e. consists of at least one real point, the resulting metrical geometry is called *Hyperbolic Geometry* or *Lobatschewsky-Bolyai Geometry* after its originators. Such an $\mathscr{A}$ is the conic $-x^2 + y^2 + z^2 = 0$. Ordinary points of the hyperbolic plane are defined to be points of the real projective plane for which $-x^2 + y^2 + z^2 < 0$.
(Do not be overcome by the phrase 'at least one'—it applies only if $\mathscr{A}$ degenerates into two complex lines which must meet in a real point. (Can you prove it?) Otherwise, there is an infinite number of real points on a non-degenerate $\mathscr{A}$.)

(ii) If $\mathscr{A}$ *is complex*, i.e. consists entirely of complex points, the resulting metrical geometry is called *Elliptic Geometry* or *Riemannian Geometry* after its founder (though the Riemannian Geometry of Relativity Theory is more general). A form of $\mathscr{A}$ generating this geometry is $x^2 + y^2 + z^2 = 0$. Elliptic and Hyperbolic Geometries are together known as *Non-Euclidean Geometry*, or we may say that each is a Non-Euclidean Geometry. Points of the elliptic plane are defined to be points in the real projective plane which are not on $\mathscr{A}$, i.e. all the points in the real projective plane.

(iii) If $\mathscr{A}$ *is degenerate* and consists of two complex points, the resulting metrical geometry is a limiting case of Non-Euclidean Geometry. When these complex points are the circular points at infinity I, J we have *Euclidean Geometry* (sometimes, following the analogy, called *Parabolic Geometry*). Why this conic-phraseology? Klein, who gave the geometries these names in 1871, later wrote: 'I proposed for these three geometries the names *hyperbolic*, *elliptic*, and *parabolic*, since the existence of two real, two imaginary, or two coincident parallels corresponds precisely to the behaviour of the asymptotes of these three conics, respectively' (Fig. 67(c)). Points of the Euclidean plane are defined to be the points of the real projective plane which are not on $\mathscr{A}$ $(= l_\infty)$, i.e. the points of the real affine plane.

Attention is now focused on this third case, which is the ultimate objective of this book.

§ 36 EUCLIDEAN GEOMETRY

§ 36.1 Circular Points at Infinity. As $\mathscr{A}$, choose the conic

$$\boxed{\lambda(x^2 + y^2) + z^2 = 0} \tag{21}$$

of which the line-equation is

$$\boxed{l^2 + m^2 + \lambda n^2 = 0} \tag{21'}$$

where λ is quite arbitrary. (Observe that the discriminant of $\mathscr{A}$ is λ^2.)

If $\lambda > 0$, $\mathscr{A}$ is complex and $\therefore$ Elliptic Geometry results.

If $\lambda < 0$, $\mathscr{A}$ is real and $\therefore$ Hyperbolic Geometry results.

If $\lambda = 0$, $\mathscr{A}$ has equation $z^2 = 0$, or $l^2 + m^2 = 0 = (l + m)(l - im)$ i.e. the line $z = 0$ repeated, or the point-pair (in the standard notation)

$$\boxed{\begin{aligned} I &\equiv (1, i, 0) \\ J &\equiv (1, -i, 0) \end{aligned}} \tag{22}$$

the *circular points at infinity* (forming a degenerate line-circle) on $z = 0$. Clearly, the equations of I, J are, respectively, $l + im = 0$, $l - im = 0$. Vectorially,

$$\mathbf{I} = \begin{bmatrix} 1 \\ i \\ 0 \end{bmatrix}, \qquad \mathbf{J} = \begin{bmatrix} 1 \\ -i \\ 0 \end{bmatrix}.$$

The essential unity of Euclidean and Non-Euclidean Geometries, demonstrated thus by Klein in 1871, is an exquisitely beautiful mathematical concept.

§ 36.2 Euclidean Angle Formulas. Now, in § 35 we saw that the general distance function was

$$(PQ) = k \log \left(\frac{\mu_1}{\mu_2}\right), \text{ i.e. if } k = \frac{1}{2i},$$

$$e^{2i(PQ)} = \frac{\mu_1}{\mu_2}$$

whence $\cos^2(PQ) = \left\{\dfrac{e^{i(PQ)} + e^{-i(PQ)}}{2}\right\}^2$ by the ordinary complex definition of the cosine function

$$= \tfrac{1}{4}\left(e^{2i(PQ)} + 2 + e^{-2i(PQ)}\right)$$

$$= \tfrac{1}{4}\left(\frac{\mu_1}{\mu_2} + 2 + \frac{\mu_2}{\mu_1}\right) \text{ substituting for } e^{2i(PQ)}$$

$$= \frac{(\mu_1 + \mu_2)^2}{4\mu_1\mu_2} \text{ on simplifying}$$

$$= \frac{f_{12}^2}{f_{11}f_{22}} \text{ applying the elementary theory of equations to Joachimsthal's equation } (7')$$

$$\therefore \quad \boxed{\cos^2(PQ) = \frac{f_{12}^2}{f_{11}f_{22}}} \qquad (23)$$

Dually, if the line-equation of $\mathscr{A}$ is $F(l, m, n) = 0$, the angle between the lines $p(l_1, m_1, n_1)$ and $q(l_2, m_2, n_2)$ is

$$\boxed{\cos^2(pq) = \frac{F_{12}^2}{F_{11}F_{22}}} \qquad (23')$$

Use (21′) and let $\lambda \to 0$

$$\therefore \cos^2(pq) = \frac{(l_1l_2 + m_1m_2)^2}{(l_1^2 + m_1^2)(l_2^2 + m_2^2)}$$

since, by (8) in dual form, $F_{12} = l_1l_2 + m_1m_2 + \lambda n_1n_2$ with $\lambda \to 0$.

But $(l_1l_2 + m_1m_2)^2 + (l_1m_2 - l_2m_1)^2 = (l_1^2 + m_1^2)(l_2^2 + m_2^2)$

$$\therefore \sin^2(pq) = 1 - \cos^2(pq) = \frac{(l_1m_2 - l_2m_1)^2}{(l_1^2 + m_1^2)(l_2^2 + m_2^2)}.$$

Ignoring signs we thus have the usual Euclidean expressions for an angle:

$$\boxed{\cos(pq) = \frac{l_1 l_2 + m_1 m_2}{\sqrt{(l_1^2 + m_1^2)(l_2^2 + m_2^2)}}, \quad \sin(pq) = \frac{l_1 m_2 - l_2 m_1}{\sqrt{(l_1^2 + m_1^2)(l_2^2 + m_2^2)}}} \quad (24)$$

§ **36.21 Laguerre's Formula.** Since $\mathscr{A} = \{I, J\}$ (as a line-conic) for Euclidean Geometry, (24) is equivalent to the Euclidean definition of an angle as

$$\boxed{(pq) = \frac{1}{2\sqrt{-1}} \log(pq, ij)} \quad (19a)$$

where, in the cross-ratio, i, j are symbols for the isotropic lines joining the circular points I, J to the vertex V of the angle, i.e. VI, VJ may be considered as tangents to the point-conic $z^2 = 0$. (See Fig. 73(b) and equation (19).) Equation 19(a) was discovered by Laguerre in 1853 and its importance was recognised by Klein.

It is an instructive exercise to show that (19a) yields π as the angle-sum of a Euclidean triangle. See Exercise 8.30.

As defined by (19a), the angle θ between the lines p, q is assigned no particular sense of rotation. There is nothing to indicate which sense of direction (clockwise or counter-clockwise) in the Euclidean plane is given by (19a), so we need to take a stand on this issue. Of course, it is always understood that the isotropic lines i, j are through I, J respectively. Reversing this correspondence amounts to reversing the sense of rotation.

Convention: Define θ to be positive when p rotates about O to coincide with q in an anti-clockwise fashion.

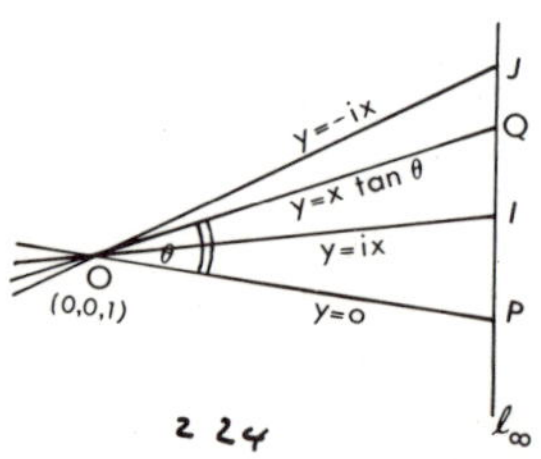

FIG. 75. Isotropic lines. Laguerre's definition of angle

Consider Fig. 75. Each of the lines OP, OQ, OI, OJ is given by an equation of the form $y = cx$ where c is some number. (See (25′) for OI, OJ.) Accordingly we may take c to be the parameter which characterises each line.

$$\therefore\ O(QP, IJ) = (\tan\theta, 0; i, -i) \qquad (i = \sqrt{-1})$$

$$= \frac{(\tan\theta - i)(0 - (-i))}{(\tan\theta + i)(0 - i)}$$

$$= \frac{(\tan\theta - i).i}{(\tan\theta + i).-i}$$

$$= \frac{\cos\theta + i\sin\theta}{\cos\theta - i\sin\theta}$$

$$= \frac{e^{i\theta}}{e^{-i\theta}} \qquad \text{by Euler's formula}$$

$$= e^{2i\theta} \qquad (\alpha)$$

and $O(PQ, IJ) = (0, \tan\theta; i, -i) = \dfrac{1}{O(QP, IJ)} = e^{-2i\theta}$ using (4) and (α)

$$= e^{-2i\theta} \text{ on actual calculation of the cross-ratio}$$

$$= e^{2i(\pi-\theta)} \quad \because e^{2\pi i} = 1$$

i.e. if $\frac{1}{2i}\log\{O(QP, IJ)\}$ defines the angle θ, then $\frac{1}{2i}\log\{O(PQ, IJ)\}$ defines the supplementary angle $\pi-\theta$ (measured in the clockwise direction from OP to OQ). See Hopkins and Hails [52] pp. 247–248, and Maxwell [69] pp. 185–186.

Interchange of I and J will, of course, reverse the sense of rotation in the Euclidean plane.

§ 36.3 Euclidean Distance Formula. Dealing next with linear measure, we ask: How must we define (PQ) in Euclidean Geometry where $P \equiv (x_1, y_1, z_1)$ and $Q \equiv (x_2, y_2, z_2)$? As we saw, $\mathscr{A}$ degenerates to l_∞ repeated and this is met by the line PQ in two coincident points. Consequently, the distance (PQ) is undefined in accordance with § 35.3, (b). Hence, Euclidean distance must be approached by a limiting process, i.e. it may be taken as (21) with $\lambda \to 0$.

Choose $k = \dfrac{1}{2i\sqrt{\lambda}}$ $\left(\text{instead of } k = \dfrac{1}{2i}\right)$ and parallel the calculations which follow (22). Write $(PQ) = d$. Applying (23), we then have

$$\sin^2(d\sqrt{\lambda}) = \frac{f_{11}f_{22} - f_{12}^2}{f_{11}f_{22}} \quad \text{where, for instance, } f_{12} = \lambda(x_1x_2 + y_1y_2) + z_1z_2$$

$$= \frac{\begin{pmatrix}\lambda^2(x_1^2 + y_1^2)(x_2^2 + y_2^2) + \lambda\{(x_1^2 + y_1^2)z_2^2 + (x_2^2 + y_2^2)z_1^2\} + z_1^2z_2^2 \\ -\lambda^2(x_1^2x_2^2 + y_1^2y_2^2 + 2x_1x_2y_1y_2) - 2\lambda(x_1x_2z_1z_2 + y_1y_2z_1z_2) - z_1^2z_2^2\end{pmatrix}}{\lambda^2(x_1^2 + y_1^2)(x_2^2 + y_2^2) + \lambda\{(x_1^2 + y_1^2)z_2^2 + (x_2^2 + y_2^2)z_1^2\} + z_1^2z_2^2}$$

$$\text{i.e. } \frac{\sin^2(d\sqrt{\lambda})}{\lambda} = \frac{(x_1^2 + y_1^2)z_2^2 + (x_2^2 + y_2^2)z_1^2 - 2(x_1x_2z_1z_2 + y_1y_2z_1z_2)}{z_1^2z_2^2}$$

letting $\lambda \to 0$ on the R.H.S.

$$= \frac{(z_1x_2 - z_2x_1)^2 + (z_1y_2 - z_2y_1)^2}{z_1^2z_2^2}.$$

Furthermore, P and Q do not lie on l_∞, i.e. $z_1 \neq 0$, $z_2 \neq 0$ (so length on l_∞ is not defined). Thus we may take $z_1 = z_2 = 1$. So

$$\lim_{\lambda \to 0} \left\{ \frac{\sin^2(d\sqrt{\lambda})}{\lambda} \right\} = (x_2 - x_1)^2 + (y_2 - y_1)^2.$$

$$\begin{aligned} \text{Now, the L.H.S.} &= \lim_{\lambda \to 0} \left\{ d^2 . \frac{\sin^2(d\sqrt{\lambda})}{d^2\lambda} \right\} \\ &= \lim_{\lambda \to 0} \left\{ d^2 . \left(\frac{\sin(d\sqrt{\lambda})}{d\sqrt{\lambda}} \right)^2 \right\} \\ &= \lim_{\lambda \to 0} \quad d^2 . 1^2 \\ &= d^2. \end{aligned}$$

Hence
$$\boxed{d = \pm\sqrt{(x_2 - x_1)^2 + (y_2 - y_1)^2}} \tag{24'}$$

which is easily recognised as the ordinary Euclidean distance formula, the co-ordinate form of Pythagoras' Theorem. Variations of this proof (from O'Hara and Ward [79]) occur in Faulkner [38], and in Hopkins and Hails [52]. Co-ordinates are taken over the field of real numbers in order to make the Euclidean distance (24′) real. (If (24′) yields d real in the case of two complex points, we will still speak of d as the distance between them.) We call the distance d the length or measure of the *segment PQ* (portion of the line PQ). Congruent segments are then segments having the same length.

Summarising
we may characterise Euclidean Geometry as a metrical geometry whose absolute conic is degenerate and consists of two conjugate complex points $(1, i, 0)$, $(1, -i, 0)$ w.r.t. the co-ordinate system.

Points of the (real) Euclidean plane are all points of the real projective plane not on $z = 0$. Lines of the Euclidean plane are all lines of the real projective plane excluding $z = 0$. (Isotropic lines, i.e. lines through I, J—see § 36.11—being complex, cannot be lines of the (real) Euclidean plane.)

§ 36.4 Triangle of Reference. Rectangular Cartesian Co-ordinates. It yet remains for us to elucidate further the essential nature of the triangle of reference XYZ.

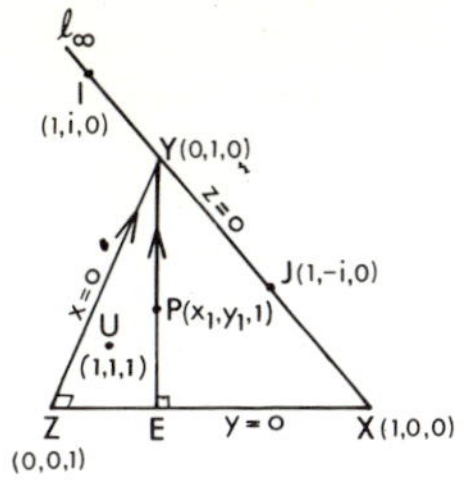

FIG. 76. (a) Metrical meaning of the triangle of reference

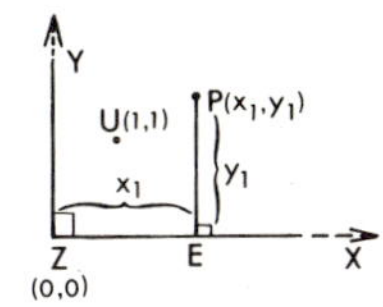

FIG. 76. (b) Rectangular cartesian co-ordinates

In Fig. 76(a), $\mathbf{I} = \mathbf{X} + i\mathbf{Y}$

$\mathbf{J} = \mathbf{X} - i\mathbf{Y}$

$\therefore$ I, J harmonically separate X, Y (by Example 5)

$$\therefore Z(XY, IJ) = -1$$

$$\therefore ZX \perp ZY$$

$\therefore$ *choosing* (1, i, 0), (1, $-i$, 0) *as co-ordinates of* I, J *is equivalent to taking a triangle of reference to consist of two perpendicular lines and* l_∞;
Z is called the *origin*.

If $\mathbf{P} = x_1\mathbf{X} + y_1\mathbf{Y} + \mathbf{Z} \equiv (x_1, y_1, 1)$ and $PY \cap ZX = E$, then
$\mathbf{E} = x_1\mathbf{X} + \mathbf{Z} \equiv (x_1, 0, 1)$ on calculation
$\therefore$ by (24′), $\begin{cases} ZE = x_1 \\ EP = y_1. \end{cases}$
Moreover, $EP \perp ZE$ since $ZY \cap EP = Y$ on l_∞.
We call $\begin{cases} x_1, y_1 \text{ the } \textit{rectangular Cartesian co-ordinates of } P \\ ZX, ZY \text{ the } \textit{rectangular Cartesian axes } (x\text{-}, y\text{-axes respectively}). \end{cases}$
(The homogeneous co-ordinates giving rise to these co-ordinates and in which we may put $z = 1$ are called *homogeneous Cartesian co-ordinates.* The unit point (1, 1, 1) corresponds to $x_1 = 1 = y_1$.)
Fig. 76(b) is then the undisguised and familiar form of Fig. 76(a). Thus, *selection of* (1, i, 0), (1, $-i$, 0) *for* I, J *is equivalent to having a triangle of reference as an ordinary rectangular Cartesian co-ordinate system* (*with* l_∞ *adjoined*).

§ 36.5 Metrical Cross Ratio. Furthermore, the non-metrical symbols of § 19, by which cross-ratio was defined, may now be revealed in their metrical form: with O as origin, we readily deduce from (24′) that $OA = a$, $OB = b$, etc., so that $a - c = \pm AC$. Metrically, therefore, (4) becomes

$$(AB, CD) = \frac{AC.BD}{AD.BC}.$$

(*Note:* We take $AB = -BA$.) Harmonic cross-ratio, in a metrical and physical sense, is associated with musical notes emitted from a stretched string. Coxeter [28] p. 23 gives a numerical illustration of this. From $(AB, CD) = -1$, we have $\dfrac{AC}{CB} = \dfrac{AD}{BD}$, i.e., C, D divide AB internally and externally in the same ratio.

§ 36.6 Angle-Bisection and Involution. An interesting application of the theory is the definition of angle-bisection. In Fig. 77, let X, Y be the double points of the involution determined by $P, Q; I, J$. Then, we assert, OX, OY are the bisectors of $P\hat{O}Q$. For, from the involution, we have

$$
\begin{aligned}
O(XP, IJ) &= O(XQ, JI) \\
&= O(QX, IJ) \text{ by § 19 III.}
\end{aligned}
$$

i.e. $P\hat{O}X = X\hat{O}Q$ by (19a)

so that OX bisects $P\hat{O}Q$ internally.

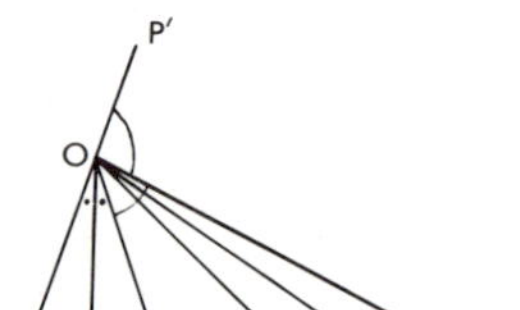

FIG. 77. Metrical meaning of an involution pencil

Can you show, in this manner, that OY is the external bisector of $P\hat{O}Q$?

Alternatively, from the involution we have, by § 20 I,

$$(PQ, XY) = -1$$

i.e. $$\frac{PX.QY}{PY.QX} = -1$$

i.e. $$\frac{PX}{XQ} = \frac{PY}{QY}\left(= \frac{PO}{QO}\right)$$ having due regard to signs of segments.

In Euclidean Geometry this is known to mean that OX, OY are internal and external bisectors of $P\hat{O}Q$.

Hence, our projective definition of angle-bisection is justified in Euclidean Geometry.

§ 36.7 Metrical Involution. Again we ask: How do we derive the metrical notion of involution in Euclidean Geometry from the projective definition?

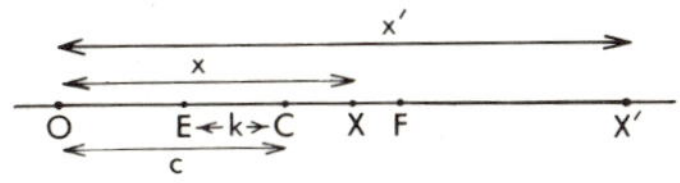

FIG. 78. Metrical meaning of involution

In Fig. 78, let E, F be the double points of the involution having X, X' as a pair and let C be the mid-point of EF. From § 20 I, we have

$$(EF, XX') = -1$$

i.e. $EX.FX' = -EX'.FX$

i.e. $(EC + CX)(CX' - CF) = -(EC + CX')(CX - CF)$.
$(\because FX = -XF)$

Thence $$CX.CX' = CE^2 = CF^2 = k^2$$
which is the metrical form for the involution, centre C. (Actually, this is Desargues' definition of an involution.) W.r.t. C as origin, we have

$$(x - c)(x' - c) = k^2$$

i.e. $xx' - c(x + x') + d = 0 \qquad (d = c^2 - k^2)$

which is the metrical equivalent of the projective equation (5). If C is at O, i.e. $c = 0$, the equation of the involution reduces to $xx' = k$. (Refer to § 20 for this equation.) Double points are given by $x' = x$, i.e. by $x = \pm k$.

§ 36.8 Equivalence of Definitions of a Circle.

EXAMPLE 42: Show that the definition of the circle given in § 33 is consistent with the usual definition in Euclidean Geometry.

Solution: In the equation (6) of the general conic, substitute the co-ordinates $(1, i, 0)$, $(1, -i, 0)$ for I, J. We find $a = b$, $h = 0$. Choose $Z \equiv (0, 0, 1)$ as the centre of the circle. Then by (9'), the equation of the polar of Z is

$$gx + fy + cz = 0.$$

But the polar of Z is l_∞, i.e.

$$z = 0$$

Comparing these equations, we have $f = g = 0$. That is, the equation of the circle reduces to $x^2 + y^2 + cz^2 = 0$, or, with $z = 1$ and $c = -r^2$, to

$$x^2 + y^2 = r^2$$

$\therefore$ by (24') any point $(x, y, 1)$ on the circle is distant r (constant) from the centre $(0, 0, 1)$. This is the definition of a circle in Euclidean Geometry.

§ 36.9 Parabola.

EXAMPLE 43: Show that the conic $y^2 - zx = 0$ is a parabola. Determine the co-ordinates of its focus and the equation of its directrix.

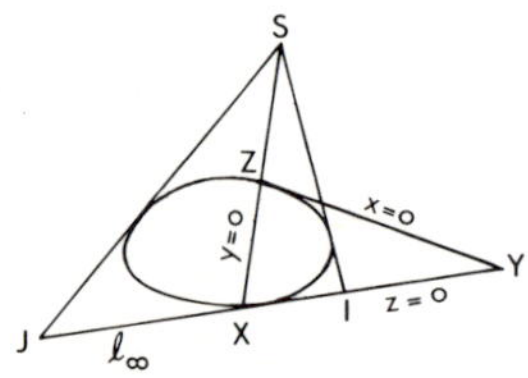

FIG. 79. Parabola

Solution: Now, $y^2 - zx = 0$ touches $l_\infty(z = 0)$, so it is a parabola. By (10), the equations of the isotropic tangents from I are (Fig. 79)

$$(iy - \tfrac{1}{2}z)^2 = -1.(y^2 - zx)$$

$$\text{i.e. } z\left(-iy + \frac{z}{4} - x\right) = 0.$$

$\therefore$ equation of IS is $-iy + \dfrac{z}{4} - x = 0$.

Similarly, equation of JS is $iy + \dfrac{z}{4} - x = 0$.

Solve and we find the focus is

$$S \equiv (\tfrac{1}{4}, 0, 1) \equiv (1, 0, 4).$$

Since the directrix is the polar of the focus, its equation is, by (9′),

$$x.-\tfrac{1}{2}.4 + y.1.0 + z.-\tfrac{1}{2}.1 = 0,$$

i.e. $$4x + z = 0$$

i.e. $$x = -\tfrac{1}{4}z$$

$\therefore$ we have the result, in ordinary analytic geometry, that the focus of the parabola $y^2 = x$ is the point $(\tfrac{1}{4}, 0)$ and the directrix has equation $x = -\tfrac{1}{4}$, as we know.

Note: (1) W.r.t. the conic (12): $ax^2 + by^2 + cz^2 = 0$, apply (10) and find the equation of the tangent pair. Putting $y = 0$, $x = 0$ in succession, we find the foci are $(\pm k, 0, 1)$, $(0, \pm ik, 1)$ where $k = \sqrt{\dfrac{(a-b)c}{ab}}$, i.e. if a, b, c are real, then two foci are real and two are complex.

(2) Using (10) on (12) again, we find the tangents from Z (i.e. the asymptotes) are $y = \pm\sqrt{\dfrac{a}{b}}ix$. But ZI, ZJ divide these harmonically. Parametrically

then, the cross-ratio $\left(\sqrt{\frac{a}{b}}, -\sqrt{\frac{a}{b}}; 1, -1\right) = -1$ whence $a + b = 0$, i.e. the conic is a rectangular hyperbola if $a + b = 0$ (and the asymptotes are then $y = \pm x$).

§ 36.10 Metrical Addition and Multiplication of Points. One can now appreciate the metrical significance of the various constructions used in the symbolism of § 13. Figs. 80(a), (b) represent Fig. 15 (addition of points) and Fig. 16 (multiplication of points) transformed into the ordinary diagrams of Euclidean Geometry. Observe in Fig. 80(b) that, by similar triangles,

$$\frac{OU}{OA} = \frac{OP}{OQ} \quad \text{where } OU = 1$$

$$= \frac{OB}{OC} \quad \text{again by similar triangles}$$

$$\therefore OC = OA.OB$$

$$= ab.$$

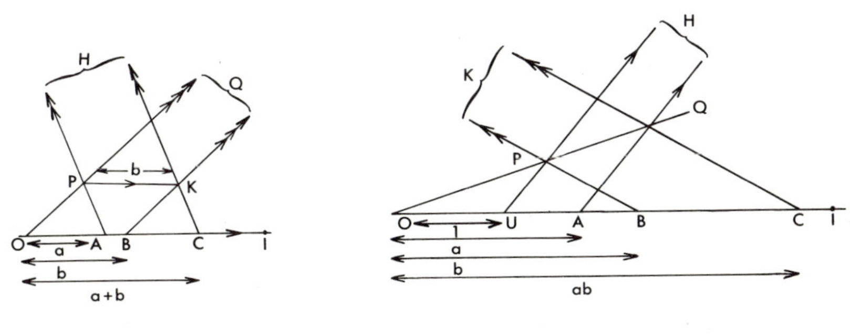

FIG. 80.

(a) Metrical meaning of addition of points

(b) Metrical meaning of multiplication of points

Let us look at this result from a different viewpoint.

Recall that the product § 13.3 (and sum, § 13.1) of two points could be expressed in terms of a quadrangular set which, by Desargues' Involution Theorem (Theorem 23), involves an involution. In the case of the product, the pairs of corresponding points of the involution have co-ordinates $1, c$; a, b; $0, \infty$ though the symbol ∞ has no meaning for the Euclidean plane. By § 36.7,

$$1.c = a.b$$

$$c = ab.$$

Points O, U, I determine a *scale* on the line such that U is at unit distance (arbitrarily chosen) from the origin O, and I is at infinity.

§ **36.11 Isotropic lines** (so named by Laguerre). General isotropic lines PI, PJ where $P \equiv (x_1, y_1, 1)$ have, by (1′), the equations

$$\boxed{y = \pm ix + k} \tag{25}$$

(where $k = -ix_1 + y_1$).

Isotropic lines OI, OJ through the origin consequently have equations

$$\boxed{y = \pm ix} \tag{25'}$$

the + sign referring to OI. (See Fig. 64(a).)

EXAMPLE 44(a): The distance between any two ordinary points on an isotropic line is zero. (Ordinary points are points not on l_∞.)

Solution: This is a necessary and sufficient condition.

(a) *Necessary:* Given the two points $P_1(x_1, y_1, 1)$, $P_2(x_2, y_2, 1)$ on $y = ix + k$ (say), we have immediately

$$y_1 = ix_1 + k,\ y_2 = ix_2 + k$$

whence $y_2 - y_1 = i(x_2 - x_1)$ implying $(y_2 - y_1)^2 = -(x_2 - x_1)^2$

$$\therefore P_1P_2 = \sqrt{(x_2 - x_1)^2 + (y_2 - y_1)^2}$$
$$= 0.$$

(b) *Sufficient:* Given $P_1P_2 = 0$ $(P_1 \neq P_2)$, we suppose the line has equation

$$y = mx + c$$
$$\therefore y_1 = mx_1 + c,\ y_2 = mx_2 + c$$
$$\therefore y_2 - y_1 = m(x_2 - x_1) \quad \ldots\ldots\ldots\ldots \quad (\alpha)$$
$$\therefore (P_1P_2)^2 = (x_2 - x_1)^2 + (y_2 - y_1)^2 \text{ by } (24')$$
$$= (x_2 - x_1)^2 (1 + m^2) \text{ substituting for } y_2 - y_1 \text{ in } (\alpha)$$
$$= 0 \because P_1P_2 = 0 \text{ (given)}$$
$$\therefore 1 + m^2 = 0 \text{ since } x_2 \neq x_1, \text{ i.e. } P_2 \neq P_1$$
$$\therefore m = \pm i$$

$\therefore$ the line *is* isotropic.

The word 'isotropic' is derived from two Greek words (*isos* ≡ 'equal', *tropein* ≡ 'to turn') because of the following property. Consider an ordinary line with gradient m and an isotropic line with gradient i. Suppose θ is the angle between these lines. Then, if $t = \tan\theta$, (24) gives

$$t = \frac{m - i}{1 + mi}.$$

Two cases arise:

Case (i): $m \neq i$ implies $t = \dfrac{m - i}{i(m - i)} = \dfrac{1}{i} = -i.$

Case (ii): $m = i$ implies $t = \frac{0}{0}$ which is indeterminate.

The first result means that the isotropic line has a constant gradient w.r.t. any ordinary line, whereas the second result means that the angle between two isotropic lines with the same gradient is indeterminate.

EXAMPLE 44(b): The distance between an ordinary point and a circular point at infinity is indeterminate.

Solution: Write equation (24′) in the homogeneous form used in its derivation, namely

$$d = \frac{\sqrt{(z_1x_2 - z_2x_1)^2 + (z_1y_2 - z_2y_1)^2}}{z_1z_2}.$$

For the points $O(0, 0, 1)$, $I(1, i, 0)$ we have

$$OI = \frac{\sqrt{1^2 + i^2}}{1.0}$$

$$= \frac{0}{0}$$

i.e. OI is indeterminate.

§ 36.12 Degenerate Circle. Refer now to Example 42. Put $r = 0$.

Then the equation of the circle with zero radius and centre at the origin O is

$$x^2 + y^2 = 0$$

i.e. $$(y - ix)(y + ix) = 0$$

i.e. the circle is degenerate and consists of the two isotropic lines

$$\begin{cases} OI & y - ix = 0 \\ OJ & y + ix = 0. \end{cases}$$

Being a circle, the degenerate locus must pass through I, J (Fig. 64(a)).

If $P(x, ix, 1)$ is any point other than I, we have

$$\begin{cases} OP = 0 & \text{(Example 44(a))} \\ OI \text{ is indeterminate} & \text{(Example 44(b)).} \end{cases}$$

Similar remarks apply to the isotropic line OJ.

Observe, therefore, that the two equivalent definitions of a circle stated in Example 42 are not valid for a degenerate circle since not all points on the circle are zero distance from O (namely I, J are not).

§ 37 EUCLIDEAN INTERPRETATION OF PROJECTIVE THEOREMS

Several basic projective concepts—e.g. involution and cross-ratio—have been given a metrical setting in the Euclidean plane. More detail would be required to show how all the features of Euclidean Geometry are derived from their projective analogues, but we do not propose to follow this lead. Interested readers may wish to consult, say, Heading [48]. Assume hereafter that Euclidean Geometry has in fact been fully derived from Projective Geometry. Words not previously defined will be understood to have their usual Euclidean meanings, e.g. 'chord'.

What we do wish to concentrate on now is the way in which well-known Euclidean theorems are derived from projective theorems. The particular interpretation of a projective theorem we want depends on the choice we make of l_∞, or more especially of I, J, in the extended Euclidean plane. When two points in π are chosen to play the roles of I, J we may speak of the process as *projecting to infinity* for this is what it is equivalent to in physical terms. (Refer to Part 1, § 3, though we may not assume the material there.) *Different choices of I, J will, of course, lead to different interpretations in the Euclidean plane. It is up to us to select I, J sensibly so that the result is meaningful in Euclidean Geometry.* From this it is clear that a theorem in Projective Geometry will have many diverse interpretations in Euclidean Geometry. It is important that we acquire the mental agility required to move freely from Projective Geometry to Euclidean Geometry. Hereunder is a fine illustration of the interrelation of our geometries.

EXAMPLE 45 (Euclidean interpretations of Example 29):

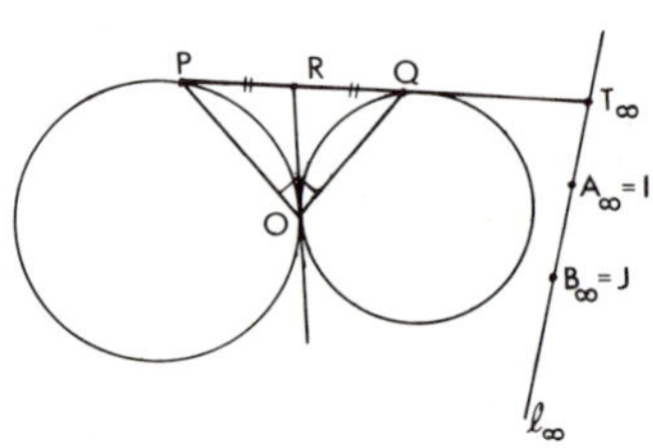

FIG. 81.

Solution: Referring to Example 29, Fig. 46, take the common chord as l_∞ and choose A, B as the circular points I, J (i.e. 'project A, B to infinity'), so that our diagram now becomes Fig. 81. From Example 29, we have $(RT, PQ) = -1$ which must still be valid under our choice of I, J

i.e. $(RT_\infty, PQ) = -1$

$\therefore$ R is the mid-point of PQ.

Similarly, $O(AB, PQ) = -1$ means $OP \perp OQ$.

(Both these results are simple to obtain in Euclidean Geometry.)

Instead of choosing AB as l_∞, suppose we select the common tangent OR. No longer does our Euclidean result concern a pair of circles; our conics are now parabolas. In this interpretation, $(RT, PQ) = -1$ means: the common chord of two parabolas with parallel axes bisects their common tangents.

Yet again, if we choose a common tangent (such as RB') as l_∞, a quite different result relating to a pair of parabolas is deduced.

Readers are advised to draw the diagrams in the Euclidean plane of these last two interpretations.

Does not all this cause us to reflect, in wonder, on the power which our projective thinking has unleashed? All these seemingly different and unconnected Euclidean theorems are, as we now see, but interpretations of one general (projective) theorem, i.e. they are related in a deeper and more profound way than we could have imagined. Of course, if I, J are so chosen that they do not lie on any of the lines of Fig. 46, the projective results of Example 29 carry over to Euclidean Geometry without alteration.

EXAMPLE 46 (Euclidean interpretation of Steiner's Theorem):

Solution: Assign to I, J the roles of the circular points in Theorem 21, Fig. 39. Steiner's Theorem is then enunciated in its familiar form: The circumcircle of a triangle circumscribed to a parabola passes through its focus.

Or we may restate Steiner's Theorem thus: The locus of the foci of the parabolas inscribed in a triangle is the circumcircle of the triangle.

[Using this Theorem, can you prove that the focus of a parabola is the Miquel Point (Theorem 34) of all quadrilaterals circumscribed to the parabola?]

Because this sort of work constitutes the chief purpose of this book, and because students invariably and naturally are attracted to it, it is worth our while to gain a little further experience in the technique of selecting the circular points. A few simple, yet interesting, specialisations are set out below in tabular form

Projective Theorem and Figure		Euclidean Interpretation
Complete quadrangle, Fig. 2(a)	Choose E, F as I, J	Parallelogram, Fig. 82
Harmonic property of complete quadrangle		Bisection properties of a parallelogram.
Chasles' Theorem (Theorem 20), Fig. 37(a)	Choose C, D as I, J	Fig. 83 Angles standing on the same arc are constant (equal).
Seydewitz's Theorem (Theorem 26), Fig. 42	Choose U, V as I, J.	Fig. 84 Angle in a semi-circle is a right-angle.
Exercise 6.7	Choose I, J as the circular points at infinity I, J	Fig. 85 $PV = VQ$.

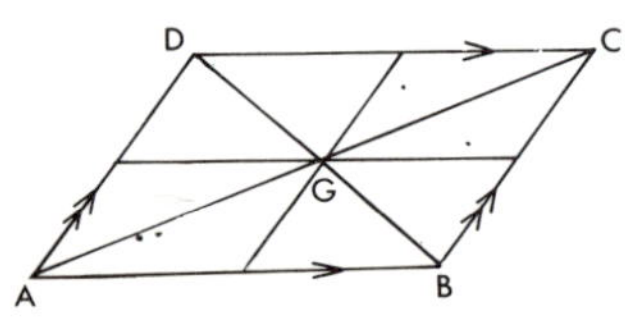

FIG. 82. Parallelogram properties

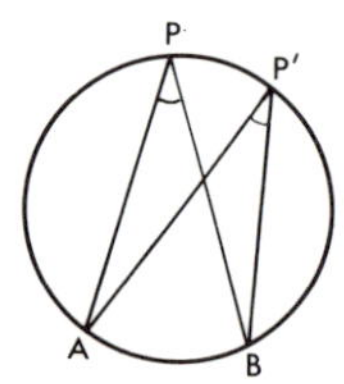

FIG. 83. Angles on the same arc

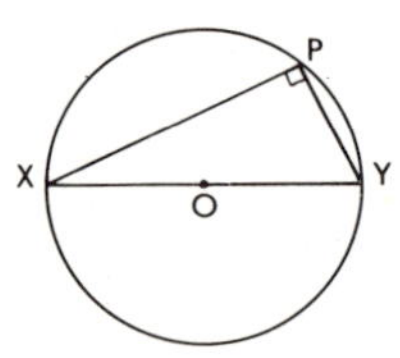

FIG. 84. Angles in a semi-circle

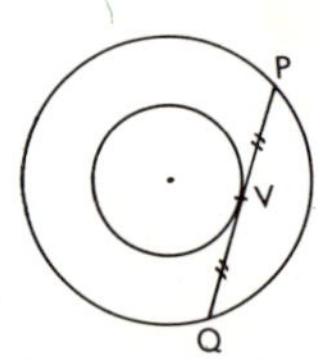

FIG. 85.

Much pleasure may be obtained by the reader who seeks further experience in interpreting projective relations in a Euclidean setting.

Historically, as indicated in Part 1, the process was the reverse, namely, scholars sought a better understanding of Euclidean Geometry by adjoining points to the plane and thus establishing a more logically rigorous system. Much can be learned from both processes.

Knowing the fact that the angle in a semi-circle is a right-angle, we might infer Seydewitz's Theorem (Theorem 26). Even more strikingly, Example 45 discloses how a simple theorem about two circles in the Euclidean plane may be conceived as a result in the projective plane, and then reinterpreted in the Euclidean plane as a theorem about two parabolas. Another chain of results, already mentioned in Part 1, involves Desargues' Involution Theorem (Theorem 23), Example 37 and Exercise 8.5 (iii). Have a look back at Example 41, also (Figs. 72(a), 72(b) and 86(a), 86(b)).

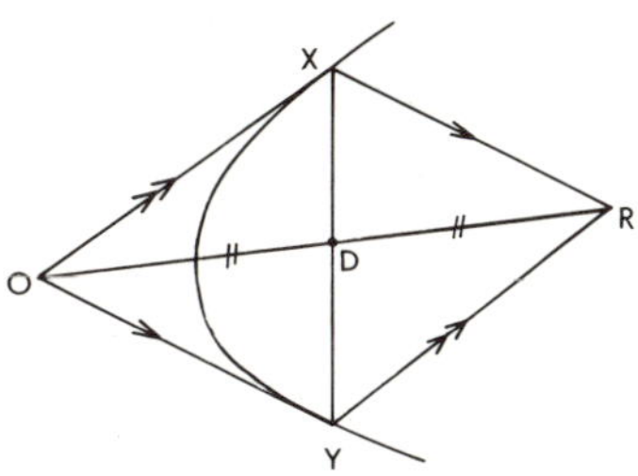

FIG. 86. (a)

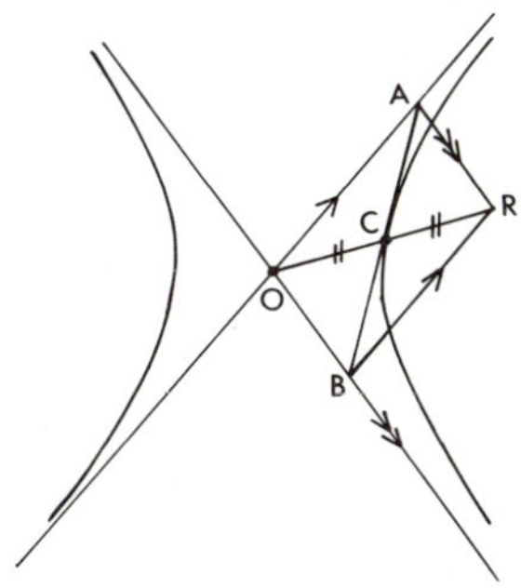

FIG. 86. (b)

Truly, these exciting sequences of ideas reveal the wonder, the power and the richness of Projective Geometry. Assuredly, the procedure of adjoining points to the Euclidean plane and exposing quite unexpected relations must belong among the real achievements in Mathematics.

Finally, to complete this Chapter, a quick survey is undertaken of important specialisations of projective transformations. This will appropriately conclude with a statement of Klein's view of a geometry as being determined by the behaviour of a certain group of transformations.

§ 38 REAL TRANSFORMATIONS IN THE EUCLIDEAN PLANE

In the remaining part of this Chapter we are concerned with linear transformations in the real affine and Euclidean planes. Elements in the matrices representing the transformations belong to the field R of real numbers.

§ 38.1 Affine Transformations. Previously, in Chapter 4, the subject of collineations, especially central collineations such as homologies and elations, was touched on, and it was demonstrated how different sets of these collineations had the group structure. Furthermore, it was proved in Theorem 11 that the non-singular collineations (= projective transformations) with matrices (2), i.e.

$$\begin{bmatrix} a_1 & a_2 & a_3 \\ b_1 & b_2 & b_3 \\ c_1 & c_2 & c_3 \end{bmatrix},$$

form a group, the projective linear group.

In making the first modification of the projective plane, we singled out a special line, l_∞, with equation $z = 0$. Consider now the matrix

$$\boxed{\begin{bmatrix} a_1 & a_2 & a_3 \\ b_1 & b_2 & b_3 \\ \cdot & \cdot & 1 \end{bmatrix}} \tag{26}$$

with the proviso

$$a_1 b_2 - b_1 a_2 \neq 0.$$

This is the matrix of an *affine transformation* and it is relatively straightforward to deduce that affine transformations form a group, the *affine group*, which is necessarily a subgroup of the projective linear group. Affine transformations clearly carry points into points and lines into lines.

Some characteristic properties of affine transformations are featured in the next three theorems.

Theorem 35: An affine transformation leaves l_∞ invariant as a whole.

Proof: Consider the effect of the transformation (26) on any point $(x, y, 0)$ on l_∞.

$$\begin{bmatrix} a_1 & a_2 & a_3 \\ b_1 & b_2 & b_3 \\ \cdot & \cdot & 1 \end{bmatrix} \begin{bmatrix} x \\ y \\ \cdot \end{bmatrix} = \begin{bmatrix} x' \\ y' \\ \cdot \end{bmatrix} \quad \text{where} \quad \begin{cases} x' = a_1 x + a_2 y \\ y' = b_1 x + b_2 y \end{cases}$$

i.e. the points of l_∞ are permuted among themselves by affine transtormations.

Theorem 36: An affine transformation takes parallel lines into parallel lines, i.e., *parallelism is an affine invariant.*

Proof: Consider the infinite set of parallel lines intersecting l_∞ $(z = 0)$ at P_∞. Under an affine transformation, P_∞ is taken into P'_∞ and the lines through P_∞ into lines through P'_∞. (Supply your own diagram.)

Hence the Theorem is proved.

Theorem 37: Under an affine transformation, the ratio of division of a segment by a point remains invariant, i.e., *segment-ratio is an affine invariant.*

Proof: Collinear points A, B, C, D_∞ are carried into collinear points A', B', C', D'_∞ by an affine transformation.

$$\text{Now } (AB, CD_\infty) = \frac{AC}{BC} \quad \text{on calculation}$$

$$\text{and } (A'B', C'D_\infty) = \frac{A'C'}{B'C'} \quad \text{on calculation.}$$

$$\text{But } (AB, CD_\infty) = (A'B', C'D'_\infty) \quad \text{by § 19 V}$$

$$\therefore \frac{AC}{BC} = \frac{A'C'}{B'C'}.$$

[This can be fairly simply done analytically, using Joachimsthal's section-formulas in Euclidean geometry.] Only the ratios of distance (on the same or parallel lines) are meaningful in Affine Geometry. Distance itself, as we have seen in Part 1, is not an affine property.

Affine transformations alter the distance between two points and the angle between two lines, as may be verified. It is not proposed here to examine all the special cases of affine transformations, but only three of them which are intimately connected with Euclidean geometry, namely rotations, translations, and reflections. That is, we now investigate some particular affine transformations.

In affine transformations, the restriction $a_1b_2 - b_1a_2 (= c) \neq 0$ has been stipulated. Now consider some cases when $c = \pm 1$.

§ 38.2 Euclidean Transformations ($c = +1$).

§ 38.21 Rotations. A counter-clockwise Euclidean *rotation* about O through an angle α (Fig. 87) is given by

$$\boxed{\begin{cases} x' = x\cos\alpha - y\sin\alpha \\ y' = x\sin\alpha + y\cos\alpha \\ z' = \qquad\qquad\qquad\quad z \end{cases}} \qquad (27)$$

with matrix

$$\begin{bmatrix} \cos\alpha & -\sin\alpha & \cdot \\ \sin\alpha & \cos\alpha & \cdot \\ \cdot & \cdot & 1 \end{bmatrix} \qquad (27')$$

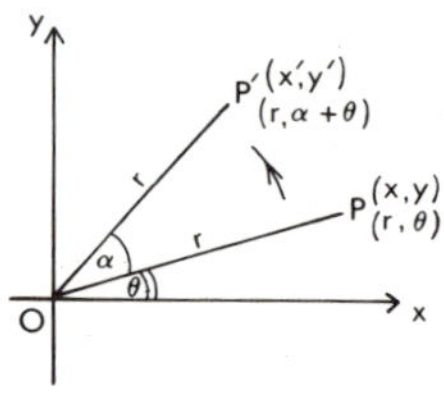

FIG. 87. Rotation

Clearly, these form an Abelian group under the operation of one rotation followed by another. (What is the identity rotation? What is an inverse rotation?) This fact may be established either geometrically or by consideration of the matrix form. Under this rotation, a point P with polar co-ordinates (r, θ) is transformed into the point $P'(r, \alpha + \theta)$. [Recall that the relationships between polar and rectangular Cartesian co-ordinates are $x = r\cos\theta$, $y = r\sin\theta$, $r = +\sqrt{x^2 + y^2}$, $\theta = \tan^{-1}\left(\frac{y}{x}\right)$.]

• § **38.22 Translation.** A Euclidean *translation* by which each point is moved a distance $\sqrt{h^2 + k^2}$ in the same direction is given by

$$\begin{cases} x' = x & & + hz \\ y' = & y & + kz \\ z' = & & z \end{cases} \qquad (28)$$

with matrix

$$\begin{bmatrix} 1 & \cdot & h \\ \cdot & 1 & k \\ \cdot & \cdot & 1 \end{bmatrix} \qquad (28')$$

Translations obviously form an Abelian group. [How are the identity and inverse translations formed?]

Both rotations and translations have $c = +1$.

Taken together, rotations and translations may be expressed by the equations

$$\begin{cases} x' = x\cos\alpha - y\sin\alpha + hz \\ y' = x\sin\alpha + y\cos\alpha + kz \\ z' = \qquad\qquad\qquad\qquad z \end{cases} \tag{29}$$

with matrices

$$\begin{bmatrix} \cos\alpha & -\sin\alpha & h \\ \sin\alpha & \cos\alpha & k \\ . & . & 1 \end{bmatrix} \tag{29'}$$

which may be proved to form a group, the *Euclidean group*, or *rigid motion group*. Rotations and translations together are said to be *rigid motions*, or *rigid motion transformations*, or *Euclidean transformations*. They possess the characteristic property that they leave invariant the distance between two given points and the angle between two given lines.

For example, putting $z = 1$ in (29) in order to deal with the familiar non-homogeneous co-ordinates, we have, if $P_1(x_1, y_1)$, $P_2(x_2, y_2)$ are given points initially, and $P'_1(x'_1, y'_1)$, $P'_2(x'_2, y'_2)$ are these points after the Euclidean transformation (29),

$$\begin{cases} x'_2 - x'_1 = (x_2 - x_1)\cos\alpha - (y_2 - y_1)\sin\alpha \\ y'_2 - y'_1 = (x_2 - x_1)\sin\alpha + (y_2 - y_1)\cos\alpha \end{cases}$$

whence

$$\{(x'_2 - x'_1)^2 + (y'_2 - y'_1)^2\}^{\frac{1}{2}} = \{(x_2 - x_1)^2 + (y_2 - y_1)^2\}^{\frac{1}{2}}$$

i.e. $$P'_1P'_2 = P_1P_2.$$

Verify the invariance of angle as an exercise.

A special feature of translations is expressed in the following result:

Theorem 38: Translations form an invariant Abelian subgroup of the affine group.

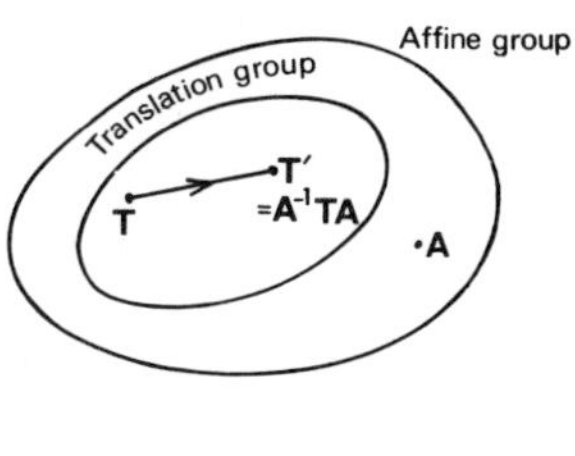

FIG. 88. (a) Translation group

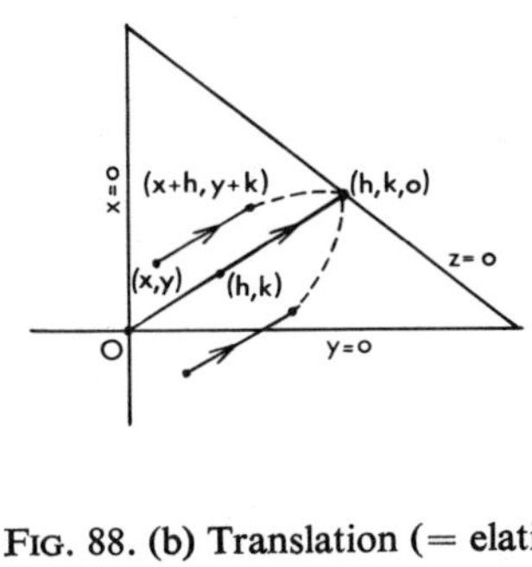

FIG. 88. (b) Translation (= elation)

Proof: Consider any affine transformation A and any translation T with corresponding matrices

$$\mathbf{A} = \begin{bmatrix} a & b & c \\ d & e & f \\ . & . & 1 \end{bmatrix}, \quad \begin{bmatrix} 1 & . & h \\ . & 1 & k \\ . & . & 1 \end{bmatrix} = \mathbf{T}$$

respectively.

$$\text{Now } \mathbf{A}^{-1}\mathbf{T}\mathbf{A} = \frac{1}{ae-bd}\begin{bmatrix} e & -b & bf-ce \\ -d & a & cd-af \\ . & . & ae-bd \end{bmatrix}\begin{bmatrix} 1 & . & h \\ . & 1 & k \\ . & . & 1 \end{bmatrix}\begin{bmatrix} a & b & c \\ d & e & f \\ . & . & 1 \end{bmatrix}$$

$$= \begin{bmatrix} 1 & . & h' \\ . & 1 & k' \\ . & . & 1 \end{bmatrix} \text{ after calculation (compute } h', k' \text{ yourself)}$$

$$= \mathbf{T}' \text{ the matrix of } T'$$

where T' is another translation. This completes the proof (Fig. 88(a)).

Note:

(i) Translations also form an invariant subgroup of the Euclidean group, as may be checked. This is the fundamental importance of the translation group. Thus we have the chain of subgroups:

Projective group →— Affine group →— Euclidean group →— Translation group.

(ii) Remember that (28′) is the matrix form of an elation, centre $(h, k, 0)$ and axis $z = 0$ (Fig. 88(b)). That is, a translation is an elation with centre on the line at infinity.

Emphasising again that rotations and translations have determinantal value $+1$, we now turn to reflections, for which the determinantal value is -1.

§ **38.3 Reflections.** A (*line*) *reflection* in the line $y = x \tan \alpha$ is expressible as

$$\begin{cases} x' = x \cos 2\alpha + y \sin 2\alpha \\ y' = x \sin 2\alpha - y \cos 2\alpha \\ z' = \qquad\qquad\qquad\qquad z \end{cases} \tag{30}$$

with corresponding matrix

$$\begin{bmatrix} \cos 2\alpha & \sin 2\alpha & . \\ \sin 2\alpha & -\cos 2\alpha & . \\ . & . & 1 \end{bmatrix}$$

in which $c = -1$. The line $y = x \tan \alpha$ is the perpendicular bisector of PP' (Fig. 89).

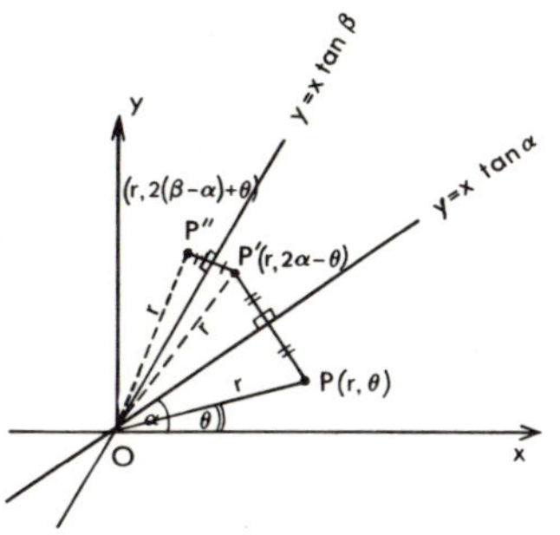

FIG. 89. (line) Reflection

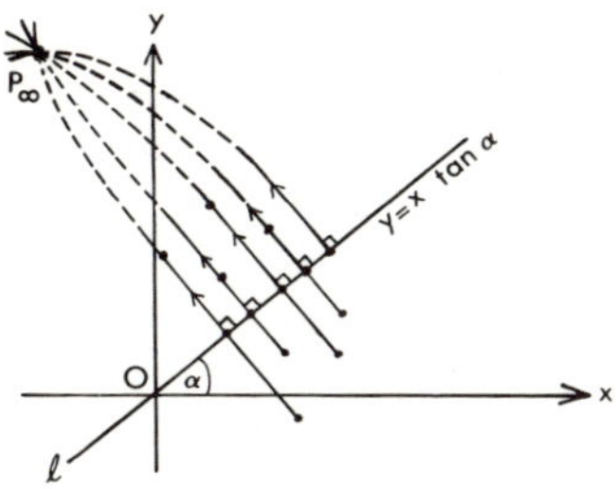

FIG. 89. (a) Line reflection as a harmonic homology

A further reflection in the line $y = x \tan \beta$ has the effect of taking $P'(r, 2\alpha - \theta)$ into the point $P''(r, 2\beta - (2\alpha - \theta)) \equiv (r, 2(\beta - \alpha) + \theta)$. But this is precisely the effect of a rotation through an angle $2(\beta - \alpha)$. Thus

the product of two reflections is a rotation.

Since closure is lacking, it follows that

reflections do not form a group.

Observe that a reflection in a line l may be regarded as a harmonic homology, of which the vertex is P_∞ on the line perpendicular to the reflecting line l (Fig. 89(a)).

Rotations, translations and reflections all carry points, lines and parallel lines into points, lines and parallel lines respectively and all preserve the invariance of angle-measurement and distance. Whereas rotations and translations preserve the sense of orientation of an angle, reflections reverse this orientation.

Some authors include reflections in the Euclidean transformations since they all retain the shape and size of figures. However, this custom is not followed here.

§ 38.31 Glide-reflection. An interesting combination of transformations is the *glide-reflection*, namely the product (in either order) of a line reflection and a translation along the axis of reflection. See Fig. 89(b).

Combining (28) ($k = 0$) and (30) ($\alpha = 0$), we obtain the equations for a glide-reflection in and along the x-axis

$$\boxed{\begin{cases} x' = x & & +hz \\ y' = & -y & \\ z' = & & z \end{cases}} \tag{30a}$$

with corresponding matrix

$$\begin{bmatrix} 1 & . & h \\ . & -1 & . \\ . & . & 1 \end{bmatrix} \tag{30'a}$$

i.e. a glide-reflection carries the point with Cartesian co-ordinates (x, y) into the point with Cartesian co-ordinates $(x + h, -y)$.

Glide-reflections have no invariant points.

A glide-reflection with identity translation (i.e. translation through zero distance) is an ordinary (line) reflection.

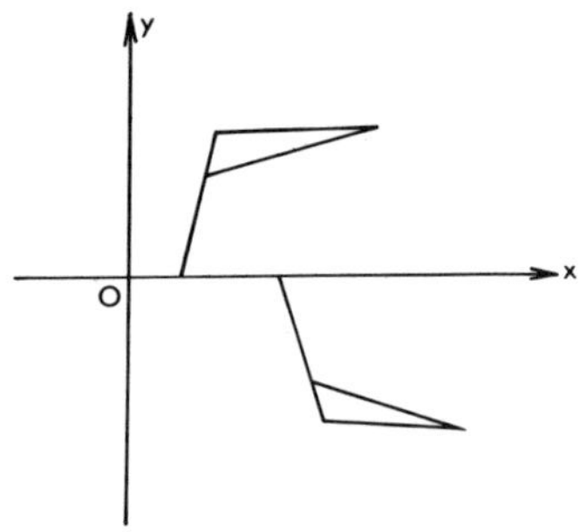

FIG. 89. (b) Glide-reflection

§ **38.32 Isometries.** An *isometry* (or *congruence*) is a transformation which preserves lengths of segments.

We have dealt with three types of isometries: Rotations, translations and reflections. Observe that a rotation leaves one point invariant, a translation has no invariant point—every point is moved—while a reflection leaves invariant points on the axis of reflection. Of course, the identity isometry leaves every point unchanged.

Isometries form a group.

As an example of a modern theorem on isometries, we quote without proof a result due to the Danish mathematician, Johannes Hjelmslev (1873–1950).

Hjelmslev's Theorem: If all the points P of one line are related by an isometry to all the points P' of another line, then the mid-points of the segments PP' are distinct and collinear, or they all coincide.

See Fig. 90 where $AB = A'B'$, $BC = B'C'$. For a proof, refer to Coxeter [26] pp. 46–47. The Theorem has applications in Hyperbolic Geometry.

A very recent publication on transformations which the reader may find useful is Gans, D: *Transformations and Geometries*, Appleton-Century-Crofts

(1969). Also see H. W. Guggenheim, *Plane Geometry and its Groups*, Holden-Day (1967).

FIG. 90. Hjelmslev's theorem

§ 38.4 Singular Transformations. Only non-singular transformations are of concern to us since they allow the existence of an inverse transformation and hence the possibility of group structure. None the less, it is worth pausing for a moment to see what happens in the singular case.

Take non-homogeneous co-ordinates and consider the singular transformation whose matrix is

$$\begin{bmatrix} a & b \\ c & d \end{bmatrix}$$

with $ad - bc = 0$, i.e. $\frac{c}{a} = \frac{d}{b} = k$, i.e. $c = ka, d = kb$, i.e. the matrix is really

$$\begin{bmatrix} a & b \\ ka & kb \end{bmatrix}.$$

The effect of this singular transformation on any point (ξ, η) of the plane is as follows:

$$\begin{bmatrix} a & b \\ ka & kb \end{bmatrix}\begin{bmatrix} \xi \\ \eta \end{bmatrix} = \begin{bmatrix} a\xi + b\eta \\ k(a\xi + b\eta) \end{bmatrix} = \begin{bmatrix} x \\ kx = y \end{bmatrix}$$

i.e. *every point of the plane is transformed into the line* $y = kx$. (What values of k relate to $x = 0, y = 0$?)

Alternately, this interesting result may be expressed as

the dimension of the plane is reduced

or equivalently

every plane region has its area reduced to zero.

Hereafter (and, indeed, heretofore) we are concerned purely with non-singular transformations.

Let us now pause for a few moments to review some of the basic numerical invariants which distinguish projective, affine and Euclidean geometries from each other.

§ 38.5 Fundamental Numerical Invariants. Associated with each of the main non-finite geometries there is a simple, fundamental numerical invariant related to sets of collinear points. These invariants may be tabulated thus

Geometry	Invariant	No. of points in collinear set
Projective	cross-ratio	4
Affine	ratio of segments	3
Euclidean Non-Euclidean }	distance	2

Having emphasised these features, we now proceed to examine the effect of the Euclidean group of transformations on the circular points at infinity. As the Euclidean group is a subgroup of the affine group, it follows from Theorem 35 that I, J must remain on l_∞ under the Euclidean transformations.

§ 38.6 Euclidean Group and I, J. Next, consider the effect of a typical transformation of the Euclidean group on the circular point I. We shall need to refer to Euler's Theorem: $e^{i\theta} = \cos\theta + i\sin\theta$.

$$\begin{bmatrix} \cos\alpha & -\sin\alpha & h \\ \sin\alpha & \cos\alpha & k \\ \cdot & \cdot & 1 \end{bmatrix} \begin{bmatrix} 1 \\ i \\ \cdot \end{bmatrix} = \begin{bmatrix} \cos\alpha - i\sin\alpha \\ \sin\alpha + i\cos\alpha \\ \cdot \end{bmatrix}$$

$$= \begin{bmatrix} e^{-i\alpha} \\ ie^{-i\alpha} \\ \cdot \end{bmatrix} \text{ using Euler's Theorem}$$

$$= e^{-i\alpha} \begin{bmatrix} 1 \\ i \\ \cdot \end{bmatrix}$$

i.e. Euclidean transformations take $I(1, i, 0)$ into $e^{-i\alpha}(1, i, 0)$. Since we are dealing with homogeneous co-ordinates, the non-zero factor $e^{-i\alpha}$ may be discarded ($\alpha \neq \infty$).

Thus under Euclidean transformations $I \rightarrow\!\!- I$.

Similarly $J \rightarrow\!\!- J$.

Thus the Euclidean group is characterised by the fact that it leaves I, J invariant. (Rotations leave the isotropic lines $y = \pm ix$ invariant.)

On the other hand, it is simple to show that reflections have the following effects: $I \rightarrow\!\!- J$, $J \rightarrow\!\!- I$, i.e. I, J are interchanged. This has the corollary that the sense of orientation of an angle is reversed by a reflection.

§ 38.7 Absolute Involution $\mathscr{I}_\infty$. Previously, in § 32, we saw that our definition of perpendicularity introduced an orthogonality involution into the augmented affine, and hence augmented Euclidean, plane. The section of this involution by l_∞ is an elliptic involution, called the absolute involution $\mathscr{I}_\infty$, of which the complex double points are $I(1, i, 0)$ and $J(1, -i, 0)$ (Fig. 63(d)).

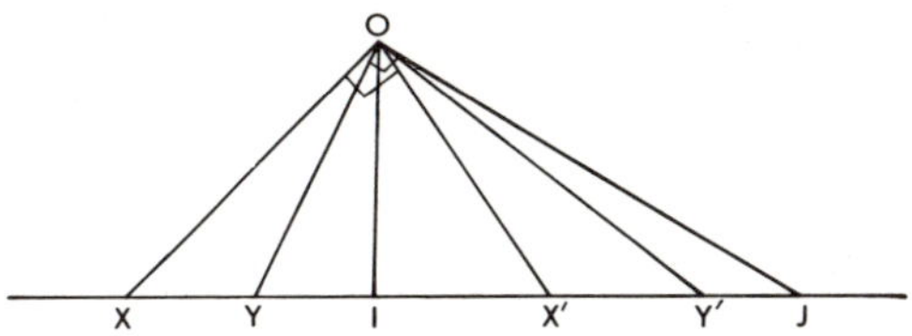

FIG. 63. (d) Orthogonality involution. Absolute involution

Suppose OX, OX' are a representative pair of perpendicular lines of the orthogonality involution, the vertex O being the origin. Let OX, OX' (equations $y = mx$, $y = m'x$ respectively) cut l_∞ in the points X, X', so that $X \equiv (1, m, 0)$, $X' \equiv (1, m', 0)$. Take the parameter to be the ratio of the second co-ordinate to the first co-ordinate. Hence, the parameters of X, X', I, J are $x(= m)$, $x'(= m')$, i, $-i$ respectively.

In the general equation (5) for an involution, use the fact that I is a double point of $\mathscr{I}_\infty$, i.e. that i corresponds to i

$\therefore p.-1 + q.2i + s = 0$

i.e. $\quad s - p + 2qi = 0.$

Equating real and imaginary parts we have

$$p = s, q = 0.$$

Thus the equation of $\mathscr{I}_\infty$ is

$$\boxed{xx' + 1 = 0} \tag{5''}$$

i.e.
$$x' = -\frac{1}{x}$$

(with double points given by $x^2 = -1$, i.e. $x = \pm i$, of course).

Remarks: (1) Equivalently, we may write (5″) as

$$mm' + 1 = 0$$

which is the familiar Euclidean expression for the perpendicularity of two lines with slopes m, m'. (Of course, this expression is also derived from $\{m, m'; i, -i\} = -1$. See Exercise 8.31.)

(2) Isotropic lines are self-perpendicular (as we already know) since they are double elements of the orthogonality involution (equivalently, since $i.i + 1 = 0$, i.e., $i = -\frac{1}{i}$).

(3) $\mathscr{I}_\infty$ was chosen to be an elliptic involution, rather than a hyperbolic involution, in order to exclude the possibility of an ordinary line being self-perpendicular.

(4) Choice of a hyperbolic involution as $\mathscr{I}_\infty$ leads to the two-dimensional case of Minkowski's geometry of space-time which Einstein used in his Special Theory of Relativity.

(5) *Euclidean transformations leave the absolute involution invariant.*

§ **38.8 Klein's Erlanger Programm.** Reference must now be made to § 15 where Klein's *Erlanger Programm* was first encountered. As previously explained, Klein's revolutionary thinking associated a geometry with a group. He defined a geometry to be the set of properties of figures which were left invariant by the transformations of this group (but not by the transformations of any subgroup).

Klein stated his philosophy in these words:

> *Klein's Erlanger Programm:* A geometry is defined by a group G of transformations when its definitions and theorems deal with properties of figures invariant under transformations of G but not invariant under transformations of any other group containing G.

Viewed in this light, our chain of sub-geometries and sub-groups may be summarised according to the simple table:

Geometry ⟵ Klein's *Programm*	⟶ *Group*
Real Projective Geometry	Projective Group
↓	↓
Affine Geometry	Affine Group
↓	↓
Euclidean Geometry	Euclidean Group

Originally, the statement of Klein's *Erlanger Programm* and its development appeared in '*Vergleichende Betrachtungen über neuere geometrische Forschungen*', printed in Erlangen in 1872, and was reprinted in *Mathematische Annalen*, 43 (1893) 63–100 and in Klein [56] 460–497. An English translation, by Haskell, is to be found in the *Bulletin of the New York Mathematical Society* 2 (1892–1893) 215–249 under the title: 'A comparative review of recent researches in geometry'.

Consult also Veblen and Young [113] 2, p. 71 for an exposition of the Kleinian concept of Geometry, and Levy [64] where the whole approach is conditioned by the *Erlanger Programm.*

Other types of geometry, not covered here under the title of Euclidean Geometry, are incorporated in Euclid's treatment. Indeed, the reader would normally think of these as being part of Euclidean Geometry. For instance, there is the theory of similar figures which are obtained from one another by magnification transformations, i.e. they preserve the shapes but alter the sizes of figures. These would leave I, J invariant too but they destroy the invariance of distance. Other transformations preserve sizes but alter the shapes of figures. Transformations which we have been mainly concerned with could be thought of as preserving the properties of congruent figures (i.e. both the shapes *and* the sizes of figures).

Fuller detail of transformation theory may be found in, say, Graustein [43], Levy [64], Meserve [73], Veblen and Young [113], Winger [120], Woods [122].

§ 38.9 Concluding Remarks on Euclidean Geometry. Most of the aims of this book, as stated in the Preface, have been achieved. Let us review briefly the main points of our progress.

Beginning with a small set of axioms, we developed the basic geometry of the projective plane and indicated how these axioms were wide enough to encompass a variety of geometries and flexible enough to extend the theory into higher dimensions. Associated with this synthetic approach, we introduced the analytical analogue of homogeneous co-ordinates, which allowed us to move freely between the geometrical and algebraic aspects of the work and to specify a geometry by the choice of field to which the co-ordinates belonged.

Taking as a major object the modification of the projective plane to derive the ordinary metric Euclidean plane, we dealt in turn with cross-ratio, involution and conics (defined in terms of projectivities). Ultimately, measurement was introduced into the projective plane by means of a cross-ratio associated with an arbitrarily chosen conic and many well-known Euclidean metric results were seen to be related to an elliptic involution on the line at infinity. In the process of creating this Euclidean specialisation of the projective plane we noted many features of the intermediary affine plane. Emphasis was placed on the unifying role of groups, namely on Klein's *Erlanger Programm* which asserted that a geometry was determined by a group of transformations leaving invariant certain properties of figures in that geometry.

We could stop here.

However, several other facets which have been mentioned along the way are deserving of further consideration. In the remaining portion of the book we deal with three of these important aspects of Projective Geometry according to the following plan:

(i) In Chapter 9 we consider some basic features of the fascinating non-Euclidean planes, preserving something of the spirit of their creators;
(ii) Chapter 10 deals with finite planes, one of the swiftly developing areas of research activity;
(iii) Another area of great mathematical creativity is the foundations of geometry, one aspect of which we discuss in Chapter 11 where we explore the geometrical implications of a ternary ring;
(iv) Finally and briefly, in Chapter 12, we glimpse (as through a glass, darkly) objects of future study.

As elsewhere in the book we shall seek in the final chapters to give something of the historical flavour of the geometrical objects being studied, to offer—where appropriate, though perhaps naively—some philosophical comment and to indicate some of their mathematical and scientific applications.

Suggested topics for further study

(1) Minkowski Geometry (Busemann and Kelly [17]).
(2) Hilbert Geometry (Busemann and Kelly [17]).
(3) Möbius transformation $z' = \dfrac{az + b}{cz + d}$ $(ad - bc \neq 0,\ c \neq 0)$ in the ordinary complex Argand (or Gauss) plane; it represents the product of several transformations (Hartshorne [132] p. 131).
(4) Application of linear transformations to Art and Biology (D'Arcy Thompson: *On Growth and Form*, C.U.P. (1961)).
(5) Similarity and equiareal transformations.
(6) The role of the Absolute in Geometry.

EXERCISES 8

1. Four coplanar lines a, b, c, d are such that $a \perp b$, $b \perp c$, $c \perp d$. Establish that $d \perp a$. (The quadrilateral so formed is called a *rectangle*.)

2. Prove that the line joining the mid-points of two sides of a triangle is parallel to the third side.

3. Show that the lines joining the vertices of a triangle to the mid-points of the opposite sides are concurrent (at the *centroid*).

4. An important result in the theory of higher plane curves is that called the *Theorem of eight associated points* which asserts that cubic curves passing through eight given points must pass through a ninth point. Considering a degenerate cubic curve to consist of either three lines or a conic and a line, prove:

(i) Pascal's Theorem;
(ii) Miquel's Theorem.

5. *Coaxal circles* are defined to be circles of a system passing through I, J and two other points P, Q. Label the following points: $PQ \cap IJ = R$, $IQ \cap JP = L'$, $IP \cap JQ = L$ (the complete quadrangle again!). Define L, L' as *limiting points*, LL' as the *line of centres* and PQ as the *radical axis*. Then prove:

(i) Centres of all circles of a system of coaxal circles lie on the line of centres (hence, of course, the name).
(ii) The radical axis is perpendicular to the line of centres.
(iii) A line is cut in involution by any system of coaxal circles. (This is Desargues' Involution Theorem for coaxal circles.)

6. *Conjugate diameters* are defined to be conjugate lines through the centre of a conic. Hence, show that

(i) The locus of mid-points of chords parallel to a diameter is the conjugate diameter.
(ii) Pairs of conjugate diameters are in involution, the asymptotes being the double lines of the involution.

7. Conjugate diameters of a circle are perpendicular. Prove.

8. The orthocentre of a triangle circumscribed to a parabola lies on the directrix. Prove. (A theorem due to Steiner.)

9. Directrices of all parabolas inscribed in a triangle are concurrent at the orthocentre. Prove.

10. Two parabolas and one rectangular hyperbola exist in every pencil of conics. Prove,

11. Demonstrate that the centre of a conic is the mid-point of the foci on either axis.

12. Where are the foci and directrices of a circle?

13. What is the locus of the point of intersection of perpendicular tangents to a conic? (The locus is the *director circle* or *circle of Monge* (Gaspard Monge, 1746–1818, French). See Faulkner [38] for a synthetic treatment of this, and the following exercise.) [An interesting illustration of this problem is that all rectangles circumscribing an ellipse have their vertices on the director circle.]

14. For the parabola, show that the locus of the point of intersection of perpendicular tangents is the directrix (and l_∞).

15. Tangents at the intersections of a line through the focus of a parabola and the parabola are perpendicular and meet on the directrix (attributed to de la Hire).

16. What does the director circle of a rectangular hyperbola become? [Answer: A degenerate point-circle.]

Confocal conics are conics having the same foci. Establish the results 17–21 (see Faulkner [38]):

17. Through any point in the plane there are two conics of a confocal system and these cut at right angles.

18. Director circles of a system of confocal conics are coaxal.

19. Director circles of a pencil of conics inscribed in a quadrilateral form a coaxal system (attributed to Plücker and Thomas Gaskin, about 1840). In this coaxal system, what are the radical axis and the limiting points in terms of special conics?

20. The three circles whose diameters are the joins of opposite vertices of a quadrilateral are coaxal.

21. Centres of all conics inscribed in a quadrilateral lie on the line containing the mid-points of the lines joining opposite vertices of the quadrilateral.

22. Fold a sheet of paper so that one corner lies on an opposite side. Determine the locus of the envelope of the line of the crease.

23. S is a point on the circumcircle of the triangle ABC; SL, SM, SN are perpendicular to SA, SB, SC respectively and meet BC, CA, AB in L, M, N. Prove that L, M, N are collinear with the circumcentre. (This line of collinearity is called the *Simson line* (Robert Simson, 1687–1768, English) or the *Wallace line* (William Wallace, 1768–1843).)

24. Discover what you can about the *hyperbola of Apollonius* (Faulkner [38]).

25. How might the measure of the angle (pq), when p, q intersect on the Absolute conic $\mathscr{A}$, suggest an analogue of parallelism in Euclidean Geometry? (Also see O'Hara and Ward [79] p. 262 for the relationship of different definitions of parallel lines to the metrical geometries.)

26. If a parallelogram is inscribed in a conic, prove that its diagonals are diameters.

27. Mid-points of a system of parallel chords of a parabola lie on a line parallel to the axis, i.e. a diameter of the parabola. Prove.

28. Given that the four points of intersection of two parabolas are concyclic, prove that their axes are perpendicular. (Normally, four such points are not concyclic.)

29. Suppose P, Q are conjugate w.r.t. $\mathscr{A}$ and p, q are conjugate w.r.t. $\mathscr{A}$. Evaluate (PQ) and (pq). Such points and lines are said to be *orthogonal*.

30. Angle-sum of a Euclidean triangle is π. Check the validity of this on the basis of equation (19a). [For a detailed solution, see Hopkins and Hails [52] pp. 221–222.]

31. Two lines $y = mx + cz$, $y = m'x + dz$ are perpendicular. Prove that $m' = -\dfrac{1}{m}$.

32. Deduce the condition that the two lines of Exercise 8.31 be parallel ($d \neq c$).

33. A useful result (*Gaskin's Theorem*, due to Thomas Gaskin (English) in 1852) states: A circle outpolar to a (central) conic envelope cuts the director circle of the conic orthogonally. Verify this Theorem in the case of the conic (ellipse) $b^2x^2 + a^2y^2 - a^2b^2 = 0$ and the circle $(x - \alpha)^2 + (y - \beta)^2 - r^2 = 0$.

34. Show that a circle and a rectangular hyperbola are mutually apolar if the centre of the hyperbola lies on the circle and if the centre of the circle lies on the rectangular hyperbola.

35. If a circle is outpolar to a parabola, show that the centre of the circle lies on the directrix of the parabola.

36. Two circles are mutually apolar. Establish that (i) their radii are equal, and (ii) they cut at an angle of $\dfrac{2\pi}{3}$.

37. Verify, for a circle S' and a parabola S, that if $\Theta^2 = 4\,\Delta\,\Theta'$ then the circle passes through the focus of the parabola. [Notice that this is really Steiner's Theorem (Example 46), viz: The circumcircle of a triangle circumscribing a parabola passes through the focus of the parabola.]

38. Prove that the Φ-conic of a circle and a rectangular hyperbola is a parabola. (Try a purely geometric proof.)

39. If S is a general conic and Σ' consists of the circular points I, J, find the F-conic.

40. Evaluate Θ and Θ' for the conics of Exercise 8.39. Hence deduce the condition that

(i) S should be outpolar to I, J, and
(ii) Σ should be inpolar to l_∞.

41. If the tangents to S, S' at A, B, C, D are perpendicular, prove that A, B, C, D are concyclic. (Hint: Use Theorem 32.)

42. Establish that the real foci of the Φ-conic of two circles are the centres of the circles.

43. Find the invariant equation for the condition that a quadrilateral can be drawn to circumscribe one conic and to be mutually inscribed in another—this is a poristic property. [One conic is then said to be *quadrilaterally inscribed* in the other.]

44. Using the ordinary Euclidean formula for the perpendicular distance from a given point to a given straight line, show that the distance from any finite point P to an isotropic line is infinite (if P does not lie on the line). However, if P does lie on the line, then show that this distance is indeterminate.

45. (*Frégier's Theorem.*) Chords joining the extremities of pairs of perpendicular chords meeting at a fixed point on a conic pass through a fixed point (the *Frégier Point*) on the normal at that point. Prove. (Gergonne's *Annales de Mathématique* 6 (1816) 229–234.)

46. Establish the result: The Frégier Point of a fixed point on a rectangular hyperbola lies on l_∞.

47. If ABC is a triangle, O is any point, $AO \cap BC = D$, $BO \cap CA = E$, $CO \cap AB = F$ and if, further, BC, CA, AB meet an arbitrary line in L, M, N respectively, then (Exercise 5.30) $(BC, DL) . (CA, EM) . (AB, FN) = -1$.

This is *Ceva's Theorem* (1678) in the projective plane. Interpret Ceva's Theorem in the Euclidean plane when LMN is l_∞.

48. Furthermore, if D, E, F are the mid-points of the sides BC, CA, AB respectively, verify that Ceva's Theorem becomes the well-known Euclidean theorem (Exercise 8.3): The medians of a triangle are concurrent.

49. *Menelaus' Theorem* (Menelaus, 1st century A.D., Greek) in the Euclidean plane states that if an arbitrary line cuts the sides BC, CA, AB respectively of a triangle ABC in the points D, E, F respectively, then $\frac{BD}{DC}\cdot\frac{CE}{EA}\cdot\frac{AF}{FB} = -1$. Assuming this result, verify that if D, E are mid-points of BC, CA, respectively, then Menelaus' Theorem becomes (Exercise 8.2): The line joining the mid-points of two sides of a triangle is parallel to the third side.

50. Show that a homology (or an elation) whose centre and axis are ordinary, transforms l_∞ into a line parallel to the axis.

51. Regarding Example 25, let the line l lie on the Euclidean plane such that the sides $B'C'$, $C'A'$, $A'B'$ of a triangle $A'B'C'$ meet l_∞ in the points A, B, C respectively. Interpret the result of Example 25 so as to determine the nature of the triangle $A'B'C'$. (Isosceles? Equilateral? Scalene?)

52. Redraw the diagram (Fig. 15(c)) for the harmonic sequence, taking $P_\infty H$ as l_∞.

53. Explain by means of diagrams what Desargues' Theorem (Figs. 5(a), (b) in [2]) becomes when (i) V is on l_∞; (ii) the Desargues line is l_∞; (iii) both (i) and (ii) simultaneously hold? [Case (ii) will lead to similar triangles with parallel sides and (iii) to congruent triangles with parallel sides. See Figs. 113(a), (b).]

54. Construct in the projective plane a figure illustrating the Euclidean theorem: Corresponding angles are equal.

55. Interpret Castelnuovo's Theorem (Exercise 6.6.) as a theorem about three circles in the Euclidean plane. Further, if the point of concurrency lies on the common chord, how must this theorem in the Euclidean plane be modified?

56. Ordinarily, Ceva's Theorem is not the dual of Menelaus' Theorem (Exercises 8.47, 8.49). However, in *pseudo-Euclidean Geometry*, they are exact duals if Ceva's Theorem is stated in terms of angles. Consult, as an exercise, C. M. Fulton, *Proceedings of the American Mathematical Society* 10 (1959) 304–306.

57. Show that the product of two reflections is non-commutative.

58. Establish that rotations, translations and reflections form a group. (Some authors speak of *this* group as the Euclidean group, as we have mentioned in § 38.3.)

59. Prove that the eigenvalues of a rotation matrix are in the ratio $1 : e^{i\theta} : e^{-i\theta}$. What points are left invariant by a rotation?

60. *A central (point) reflection*, centre $(h, k, 1)$ [$(0, 0, 1)$ if O is the centre], is a harmonic homology (§ 16) with matrix $\begin{bmatrix} -1 & . & h \\ . & -1 & k \\ . & . & 1 \end{bmatrix}$. Show that central reflections do not form a group. Show that the product of two central reflections is a translation and that every translation is the product of two central reflections. (See Busemann and Kelly [17] p. 85 for the proof and a diagram.) A central reflection is also known as a *half-turn*.

CHAPTER 9

NON-EUCLIDEAN GEOMETRY

'The study of Non-Euclidean Geometry is a fine, rare experience.'
(H. Wolfe)

'We have found a strange footprint on the shores of the unknown. . . . At last, we have succeeded in reconstructing the creature that made the footprint. And lo! it is our own.'
(Eddington)

RICH as has been the geometrical harvest so far gleaned, it would be incomplete without some reference to non-Euclidean Geometry. In our brief insight into this remarkable mathematical development we will discuss—fairly informally—some of the basic results of the classical research papers of Saccheri, Lobatschewsky and Bolyai, and Riemann. Following this, we will deal with certain models of the non-Eúclidean geometries and then consider their philosophical implications and their applications to the physical Universe.

§ 39 SACCHERI'S GEOMETRY

§ 39.1 Saccheri's Project. In 1733, the Jesuit priest (and Professor of Mathematics of the University of Pavia), Girolamo Saccheri, set out to free the 'most beauteous' *Elements* from 'certain blemishes'. Published in Milan, Saccheri's book, '*Euclides ab omni naevo vindicatus*', was destined to become the forerunner of a new type of mathematical thinking. Sub-titled 'A geometric endeavour in which are established the foundation principles of universal geometry' (*Conatus geometricus quo stabiliuntur prima ipsa universae geometriae principia*), it purported to demonstrate that Euclid's Postulate 5 could, in fact, be proved. Saccheri, in his own mind, had no doubt of the validity of Postulate 5. 'No one doubts the truth of this proposition,' he declared. ('*Porro nemo est, qui dubitet de veritate expositi Pronunciati.*')

To appreciate something of the spirit and expression of Saccheri's ideas, we present his first two theorems where he introduced the famous figure, the *Saccheri Quadrilateral* (Fig. 91(a) without the joining lines AD, CB). Sides

AB and CD of the Saccheri Quadrilateral are called the *base* and *summit* respectively of the Quadrilateral. Quotations are from Saccheri [95].

§ 39.2 Saccheri's Quadrilateral. *Saccheri's Proposition I:* 'If two equal straight lines *AC*, *BD*, make with the straight line *AB* angles equal toward the same parts: I say that the angles at the join *CD* will be mutually equal.

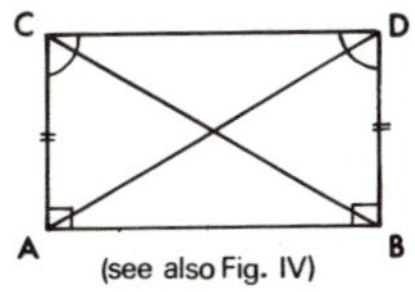

FIG. 91. (a) Saccheri quadrilateral

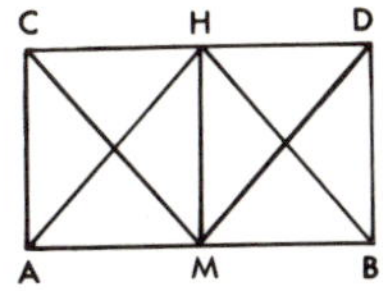

FIG. 91. (b)

Proof: Join *AD*, *CB*. Then consider the triangles *CAB*, *DBA*.

It follows (Euclid I, 4) that the bases *CB*, *AD* will be equal.

Then consider the triangles *ACD*, *BDC*. It follows (Euclid I, 8) that the angles *ACD*, *BDC* will be equal.'

'*Quod erat demonstrandum.*'

Saccheri's Proposition II: 'Retaining the uniform quadrilateral *ABCD*, bisect the sides *AB*, *CD* in the points *M* and *H*. I say the angles at the join *MH* will then be right angles [Fig. 91(b)].

Proof: Join *AH*, *BH*, and likewise *CM*, *DM*.

Because in this quadrilateral the angles *A* and *B* are taken equal and likewise (from the preceding proposition) the angles *C*, and *D* are equal; it follows (Euclid I, 4) (noting the equality of the sides) that in the triangles *CAM*, *DBM*, the bases *CM*, *DM* will be equal; and likewise, in the triangles *ACH*, *BDH*, the bases *AH*, *BH*.

Therefore, comparing the triangles *CHM*, *DHM*, and in turn the triangles *AMH*, *BMH*, it follows (Euclid I, 8) that we have mutually equal, and therefore right, the angles at the points *M*, and *H* .

'*Quod erat demonstrandum.*'

Query: Can you simplify Saccheri's proofs of Propositions I and II? Is it necessary to join *AD* in Proposition I, and *AH*, *BH* in Proposition II? (See Wolfe [121] p. 78.)

Saccheri's Proposition III: 'If two equal straight lines *AC*, *BD*, stand perpendicular to any straight line *AB*: I say the join *CD* will be equal to, or less, or greater than *AB*, according as the angles at *CD* are right, or obtuse, or acute.'

Here are stated the three famous Hypotheses of Saccheri on which depend his argument whereby he sought to demolish the assaults levelled against the fame of Euclid. The three possibilities

$$\text{summit} \gtreqless \text{base}$$

of the Saccheri Quadrilateral can be shown to correspond to

$$\left.\begin{array}{l}\text{Hyperbolic}\\ \text{Euclidean}\\ \text{Elliptic}\end{array}\right\}\text{Geometry}$$

respectively. That is, there exist three types of Saccheri Quadrilateral, one for each type of geometry. In each case the quadrilateral is characterised by the nature of its summit angles.

Saccheri's Propositions I and II, which are independent of the actual nature of the equal summit angles, are valid in all three geometries.

§ **39.3 Importance of Saccheri.** As we know, Saccheri failed in his endeavour to vindicate Euclid. But in actual fact, he achieved far more than he could have dared hope to accomplish.

Says Bonola [14] p. 43:

'Still, though it failed in its aim, Saccheri's work is of great importance. In it the most determined effort had been made on the Fifth Postulate; and the fact that he did not succeed in discovering any contradictions among the consequences of the Hypothesis of the Acute Angle, could not help suggesting the question, whether a consistent logical geometrical system could not be built upon this hypothesis, and the Euclidean Postulate be impossible of demonstration.'

Inherent in Saccheri's thinking is Euclid's tacit assumption that the straight line is infinite. Making use of this, he proved the impossibility of the Hypothesis of the Obtuse Angle, though he could not dispose of the Acute Angle Hypothesis so readily. This problem, he stated as

Saccheri's Proposition XXXIII: 'The hypothesis of acute angle is absolutely false; because repugnant to the nature of the straight line'

and took five supporting Lemmas to try to justify it.

Vailati finds Saccheri's position in the history of Mathematics as worthy of comparison with that of his great fellow-countryman, Columbus, who 'precisely in hopes of being able to reach by a new way regions already known, was led to the discovery of a new continent.'

A more earthy comment, from the editor of the '*Euclides . . . vindicatus*' [95], concludes our glimpse of Saccheri's mind at work: 'So flowered the beauteous body of a new geometry, mermaid-like, the latter portions somewhat fishy, but oh! the elegant torso.'(!)

§ 39.4 Lambert Quadrilateral. Allied to the Saccheri Quadrilateral is the trirectangular quadrilateral, or *Lambert Quadrilateral*, named after Johann Lambert (1728–1777, German) who described it in his *Theorie der Paralellinien* in 1766. It has three right angles, but no equal sides (Fig. 92). Lambert's Quadrilateral has many uses, e.g. it occurs in the establishment (§ 40.5) of Lobatschewskian co-ordinates in the Hyperbolic Plane. Shortly (Theorem 40), we will understand that a Lambert Quadrilateral may be regarded as 'half' a Saccheri Quadrilateral.

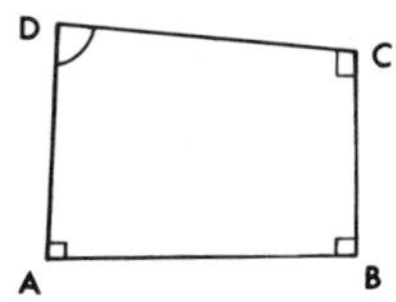

FIG. 92. Lambert quadrilateral

Historically, it is interesting to note again (refer to § 1.2, Fig. IV) that the Saccheri Quadrilateral originated with Nasîraddîn who asserted that the summit angles must be right angles.

§ 40 HYPERBOLIC GEOMETRY

§ 40.1 Gauss. During the next hundred years after Saccheri, many eminent mathematicians attempted to resolve the problem of the Parallel Postulate. One of these was Gauss, who thought deeply about the nature of Geometry. Said he:

'I keep coming closer to the conviction that the necessary truth of our geometry cannot be proved, at least *by* the human intellect *for* the human intellect. Perhaps in another life we shall arrive at other insights into the nature of space which at present we cannot reach. Until then, we must place geometry on an equal basis, not with arithmetic, which has a purely *a priori* foundation, but with mechanics.'

Because his thinking was ahead of his times, and conflicted with the prevailing philosophical doctrine of space and geometry asserted by the followers of Kant, Gauss published nothing of his meditations and doubts. He feared, as he said, 'the outcry of the Boeotians', a reference to an uncultured people of ancient Greece.

§ 40.2 Lobatschewsky. The first man to publish a system of geometry differing from Euclid's was the Russian, Nicolai Ivanovitsch Lobatschewsky, Professor in the University of Kazan. About his researches into the nature of Geometry, Lobatschewsky wrote: 'The fruitlessness of the attempts made, since Euclid's time, for the space of 2000 years, aroused in me the suspicion that the truth, which it was desired to prove, was not contained in the data themselves; to establish it the aid of experiment would be needed, for example, of astronomical observations, as in the case of other laws of nature. When I had finally convinced myself of the justice of my conjecture and believed that I had completely solved this difficult question, I wrote, in 1826, a memoir on this subject.'

After listing only those theorems of which a knowledge is necessary to appreciate his ideas, Lobatschewsky came to his basic idea, expressed in his Theorem 16 [66]:

'All straight lines which in a plane go out from a point can, with reference to a given straight line in the same plane, be divided into two classes—into *cutting* and *non-cutting*.

The *boundary lines* of the one and the other class of those lines will be called *parallel to the given line*.'

Thus, in the Lobatschewskian concept of parallelism, three things can happen to a pair of lines (Fig. 93, in which $AD \perp BC$, $AE \perp AD$):

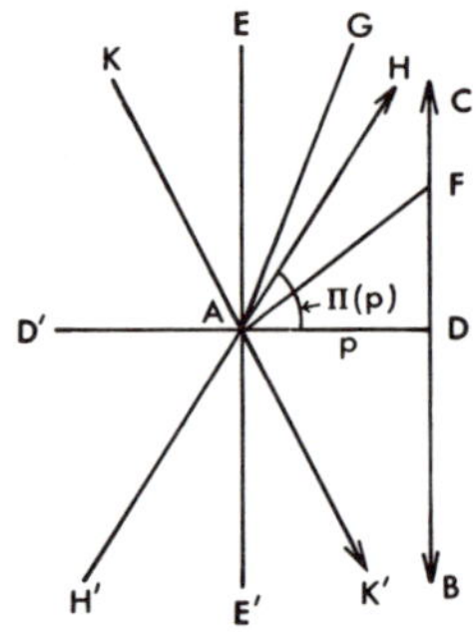

FIG. 93. Lobatschewsky's concept of parallelism

(i) they may intersect, as, e.g. AF and BC;
(ii) they may be parallel (as are the *Lobatschewskian parallels* AH and AK' with respect to BC);
(iii) they may be non-intersecting, as, e.g. all lines through A within the angle HAK with respect to BC.

One may imagine the lines through A as being various positions taken up by the rotating line AF.

We call $\hat{HAD}$ the *angle of parallelism* and represent it by the symbol $\Pi(p)$, a function of the distance $AD = p$. Sometimes the angle of parallelism is taken to be $\theta = 2\left(\frac{\pi}{2} - \Pi(p)\right)$.

The inequality $\Pi(p) < \frac{\pi}{2}$ implies the existence of two parallels, the *left-parallel* (say AK') and the *right-parallel* (say AH).

Two parallels may be said to meet in an *ideal point* (= *point at infinity*). Many useful results may be derived if no distinction is made in the hyperbolic plane between ordinary points and ideal points.

As a convention, we say that non-intersecting lines (which may be shown to have a unique common perpendicular) intersect in an *ultra-ideal point*.

Notice that three possibilities arise in relation to the rotating line *AF:*

I *AF* (*F* on *BC*) can never coincide with *AE*, however close it may get to it—this is the Euclidean assumption.

II *AF* may approach, but not reach, *AE*—this is the Lobatschewskian viewpoint, which may be shown to involve Saccheri's Hypothesis of the Acute Angle.

III *AF* and *AE* meet, i.e. every pair of lines in the plane intersect so that there is no such thing as parallelism—this is the Riemannian assumption, which may be shown to involve Saccheri's Hypothesis of the Obtuse Angle.

(Unwittingly, thus, Saccheri glimpsed the theory of parallels in its full generality.)

[Relying entirely on our finite intuition of space, we realise that there is always an angle $\theta = 2\left(\frac{\pi}{2} - \Pi(p)\right)$ which diminishes as our space is extended but which never vanishes and all lines falling within θ fail to intersect *BC*. However, we must be guided by logical reasoning rather than by sensuous images (physical diagrams drawn by a pencil) derived from the experience of our narrow portion of space.]

§ 40.3 Bolyai. Johann Bolyai's explanation of the same idea is given at the very beginning of his book [13] describing his Absolute Science of Space:

'If the ray *AM* is not cut by the ray *BN*, situated in the same plane, but is cut by every ray *BP* comprised in the angle *ABN*, we will call ray *BN parallel* to ray *AM* . . .' (Fig. 94).

It is altogether remarkable that these identical concepts should have developed so closely together in time. Lobatschewsky published his first paper 'On the principles of geometry' in 1829–30 in a local Kazan journal which, however, had limited distribution. Bolyai's work was published as '*Appendix. Scientiam spatii absolute veram exhibens* . . .' in 1832 as an

Appendix to the *Tentamen* , written by his father, Wolfgang Bolyai, himself a distinguished mathematician.

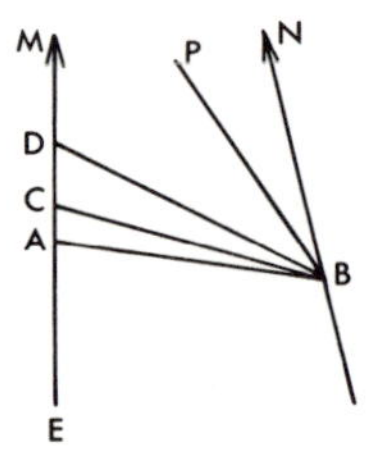

FIG. 94. Bolyai's concept of parallelism

Because the ideas of Lobatschewsky and Bolyai agreed so remarkably, their geometry will be called Lobatschewsky-Bolyai Geometry or Hyperbolic Geometry. But it should not be overlooked that the Russian had priority in publication, even though by only a short period. Looking at the two pioneering treatises, we get the impression that Lobatschewsky's is the better presented and therefore more readable.

When Bolyai's work was shown to the great Gauss for approval, Gauss commented that praising the work would be praising himself, as he was already familiar with its context from his own researches. This attitude caused Bolyai to proceed no further with mathematical research. On the other hand, Lobatschewsky struggled all his life against adversity, to bring the light of his many discoveries to the mathematical world. Perhaps it is for this reason that many writers refer simply to this creation as Lobatschewskian Geometry (which Lobatschewsky himself called Pangeometry or Imaginary Geometry ('*Voobražaemaya geometriya*')).

It is well to pause for a moment to emphasise the human factor by reflecting that major developments in Mathematics are often achieved only at the expense of great frustration and despair. Such an agony of mind and of spirit is eloquently expressed in Wolfgang Bolyai's impassioned entreaty to his son (quoted in Meschkowski [72] p. 21):

'You ought not to try the road of the parallels; I know the road to its end —I have passed through this bottomless night, every light and every joy of my life has been extinguished by it—I implore you for God's sake, leave the lesson of the parallels in peace. . . . I had purposed to sacrifice myself to the truth; I would have been prepared to be a martyr if only I could have delivered to the human race a geometry cleansed of this blot. I have performed dreadful, enormous labours; I have accomplished far more than was accomplished up until now; but never have I found complete satisfaction.

. . . When I discovered that the bottom of this night cannot be reached from the earth, I turned back without solace, pitying myself and the entire human race.'

§ 40.4 Hyperbolic Geometry. All properties of Euclidean Geometry which are independent of Postulate 5, e.g. the first 28 propositions of Euclid, carry over to Hyperbolic Geometry. Some of these are:

1. A straight line maintains its characteristic of parallelism at all its points.
2. Two lines are always mutually parallel in the same direction, i.e. Lobatschewskian parallels possess the property of symmetry.
3. Two straight lines which are parallel to a third are also parallel to each other (the transitivity property).
4. The exterior angle of a triangle is greater than the interior opposite angle.

An important figure which we need in the ensuing theory is an improper triangle of which two vertices C, D are ordinary points while the third vertex I is an ideal point (Fig. 95). Using this triangle, one may establish geometrically the following two basic results (which we assume)

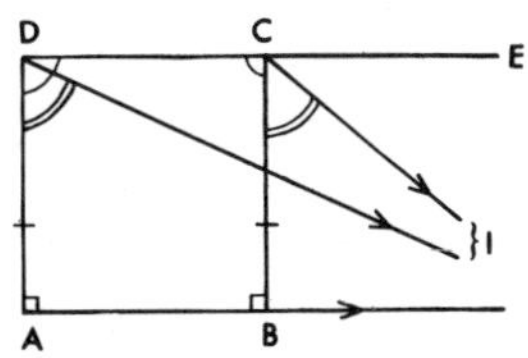

FIG. 95.

(a) the angle of parallelism increases (decreases) as the distance decreases (increases); and

(b) the angle of parallelism is constant for a given distance.

Next we establish some of the basic theorems of Hyperbolic Geometry. In doing so, we lean heavily on the clear presentation given in Wolfe [121] pp. 78–79, 81–83.

Theorem 39: Summit angles of a Saccheri Quadrilateral are acute.

Proof: Now (Fig. 95) the summit DC and the base AB of the Saccheri Quadrilateral $ABCD$ are non-intersecting (why?)

$\therefore$ CI, DI (parallels to AB) lie within the angles ADC, BCE respectively, where E is any point on DC produced through C.

Also $A\hat{D}I = B\hat{C}I$ (angles of parallelism for equal distances AD, BC).

In Δ *CDI*, exterior $E\hat{C}I >$ interior opposite $C\hat{D}I$

$\therefore B\hat{C}E > A\hat{D}C\ (= D\hat{C}B)$

$\therefore$ the equal summit angles are acute.

Therefore,
Saccheri's Hypothesis of the Acute Angle characterises Hyperbolic Geometry.

Theorem 40: In a Lambert Quadrilateral the fourth angle is acute.

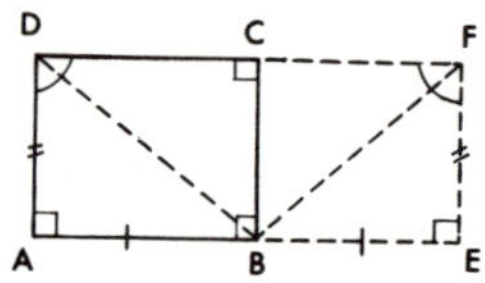

FIG. 96.

Proof: Suppose we are given the Lambert Quadrilateral *ABCD*.
Construct another 'equal' Lambert Quadrilateral *BEFC* (Fig. 96).

Join *BD*, *BF*.

Now Δ's *BAD*, *BEF* are congruent (α)

$\therefore BD = BF$

$\therefore \Delta$'s *BCD*, *BCF* are congruent (β)

It follows from (α), (β) that *AEFC* is a Saccheri Quadrilateral.

$\therefore A\hat{D}C$ is acute.

Thus we have the remarkable result:
Squares do not exist in Hyperbolic Geometry.

Theorem 41: Angle-sum of a triangle is < two right angles in Hyperbolic Geometry.

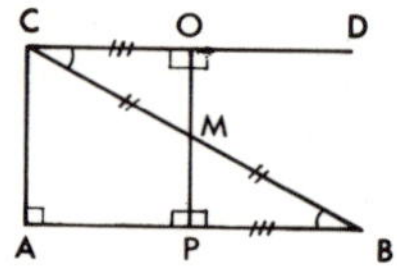

FIG. 97. (a)

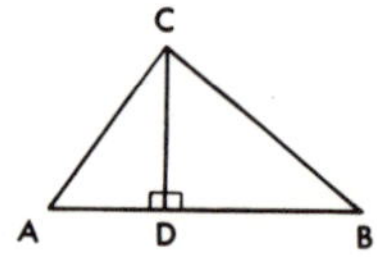

FIG. 97. (b)

Angle-sum of a triangle in hyperbolic geometry

Lemma: Angle-sum of a right-angled triangle is less than two right angles.

Proof: Let $C\hat{A}B = \frac{\pi}{2}$ in the given right-angled triangle ABC.

Now $A\hat{B}C + B\hat{C}A$ must be $< \pi$ (why?) i.e. $A\hat{B}C$, $B\hat{C}A$ are both acute.

Construct $B\hat{C}D = A\hat{B}C$.
From the mid-point M of CB draw $MP \perp AB$, so P lies between A and B.
On CD, cut off $CQ = PB$. Join MQ. (See Fig. 97(a).)
Then Δ's MBP, MCQ are congruent

$$\therefore C\hat{Q}M = \frac{\pi}{2}$$

$\therefore Q, M, P$ are collinear

$\therefore CAPQ$ is a Lambert Quadrilateral $\left(A\hat{C}Q < \frac{\pi}{2}\right)$

$\therefore$ sum of acute angles of right angled $\Delta\ ABC$ is $< \frac{\pi}{2}$

$\therefore$ angle-sum of a right-angled triangle is $<$ 2 right angles.

Proof of the Theorem: The Theorem follows by application of the Lemma to both the Δ's ADC, CDB (Fig. 97(b)).

Corollary: Angle-sum of a quadrilateral is $< 2\pi$.

Theorem 42: Similar triangles are congruent.

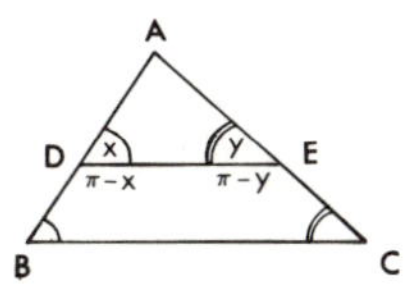

FIG. 98. Similar (congruent) triangles in hyperbolic geometry

Proof: Given the equality of angles as shown (Fig. 98), assume $AB > AD$. Quadrilateral $BCED$ has angle-sum $= 2\pi$, contrary to the Corollary

$\therefore AB \ngtr AD$.
Similarly, $AB \nless AD$.
$\therefore AB = AD,\ AC = AE$.

Expressed otherwise, we have the strange conclusion:
There are no similar triangles (or similar polygons) of different sizes in Hyperbolic Geometry.

Of this astounding property, E. B. Golos, *Foundations of Euclidean and Non-Euclidean Geometry*, Holt, Rinehart and Winston (1968) p. 209, has this to say: 'Thus,for a photograph to resemble its subject it would have to be a lifesize photograph: moreover, all the fields that depend upon similarity and proportion would be drastically altered. Consider what would happen to painting, designing, drafting, and architecture. Consider what would happen to the world around us. It would be a very strange new world.

Yet such a world is a possible one . . . it might even be that this geometry most closely describes the space world around us.'

§ **40.5 Analytical Hyperbolic Geometry** (after Shirokov [101] pp. 73–74). There are three types of co-ordinates which may be used to provide an analytic expression of results in the hyperbolic plane (Fig. 99, the Lambert Quadrilateral with non-right-angle at P):

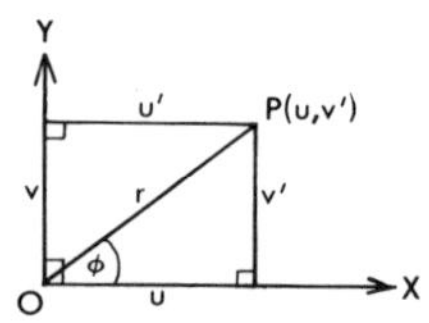

FIG. 99. Lobatschewskian, Beltrami and Weierstrass co-ordinates

I *Lobatschewskian co-ordinates:* $\begin{cases} x = u \\ y = v'. \end{cases}$

These are useful in dealing with geometrical problems relating to Calculus, but suffer the disadvantage that in them the line has a transcendental equation.

II *Beltrami co-ordinates:*

$$\begin{cases} x = \tanh \dfrac{u}{k} = \tanh \dfrac{r}{k} \cos \phi \\[2ex] y = \tanh \dfrac{v}{k} = \tanh \dfrac{r}{k} \sin \phi \end{cases}$$

which have the special value for projective problems that the equation of a line is linear.

III *Weierstrass co-ordinates:*

$$\begin{cases} x_1 = \sinh \dfrac{u'}{k} = \sinh \dfrac{r}{k} \cos \phi \\[2ex] x_2 = \sinh \dfrac{v'}{k} = \sinh \dfrac{r}{k} \sin \phi \\[2ex] x_3 = \cosh \dfrac{r}{k}. \end{cases}$$

While these have the advantages of the Beltrami system, they also express formulae for length and angle simply. They are convenient for investigating problems relating to Hyperbolic Geometry and Special Relativity.

Weierstrass co-ordinates satisfy the equation

$$x_3{}^2 - x_2{}^2 - x_1{}^2 = 1.$$

§ 40.6 Angle of Parallelism. Lobatschewsky and Bolyai discovered in 1823 the equation which is the key to all Hyperbolic Trigonometry

$$\boxed{\tan\left\{\frac{\Pi(p)}{2}\right\} = e^{-p/k}} \tag{31}$$

(k constant). Recall that every distance p has a corresponding angle of parallelism $\Pi(p)$, i.e. the angle of parallelism is a function of the distance.

Obviously, $p = k \log \cot\left(\frac{\Pi(p)}{2}\right)$, $\Pi(p) = 2 \tan^{-1}(e^{-p/k})$.

Deductions from equation (31):

(i) $\Pi(0) = \frac{\pi}{2}$;

(ii) $\Pi(\infty) = 0$;

(iii) $p_1 \gtrless p_2 \Rightarrow \Pi(p_1) \lessgtr \Pi(p_2)$.

When Hyperbolic Trigonometry is studied in detail it is found that Euclidean Trigonometry arises as a special case (Exercises 9.9, 9.11). That is

Euclidean formulas hold for 'infinitesimal' triangles.

Otherwise expressed

For small regions of space, Hyperbolic Geometry approximates to Euclidean Geometry.

§40.7 Unit of Length. Unit of length in Euclidean Geometry is measured in terms of some rigid body or its equivalent and is preserved by a bureau of standards or similar governmental organisation. This is not necessary for the unit of angle since a right angle may be constructed geometrically.

On the other hand, in Hyperbolic Geometry the unit of length is obtained by a geometrical construction (Wolfe [121] p. 134), at least theoretically (since it involves a transcendental number and consequently cannot be constructed by means of ruler and compasses). [In (31), $k = 1 \Rightarrow$ the unit of length = the distance corresponding to $2 \tan^{-1}(e^{-p})$ as the angle of parallelism.]

This fact is recognised by saying that in Hyperbolic Geometry there exist 'absolute' units of length, meaning that the units do not depend on the actual choice of prescribed segments.

Until recently, the unit of length in the metric system (the *standard metre*) was the distance between two fine lines engraved on gold plugs near the ends of a bar of a platinum-iridium alloy (at zero temperature, centigrade). In 1961, it was internationally agreed to adopt an atomic standard of length based on light which would keep both standards, old and new, as consistent as possible. One metre is now defined to be a certain number of wavelengths, in vacuum, of a particular orange radiation emitted by atoms of a particular isotope of krypton ($K_r{}^{86}$) in an electrical discharge. (See Halliday, D. and Resnick, R.: *Physics* , Wiley (1966).) Query: What advantages do you think the atomic standard has over the rigid bar standard? The suggestion that a light wave might be used to determine a standard of length was first advanced by the French physicist, Hippolyte Fizeau, in 1864.

§41 ELLIPTIC GEOMETRY

One of the turning points in 19th century Mathematics occurred on 10th June, 1854, when Riemann delivered his justly famous dissertation to the Philosophical Faculty at Göttingen. Entitled 'On the hypotheses which lie at the foundations of Geometry' ('*Über die Hypothesen welche der Geometrie zu Grunde liegen*'), Riemann's thesis propounded the bold and original idea that space (say, in particular, a line) need not be infinite, though it may still be unbounded. That is, Riemann emphasised the distinction between the concepts of *unboundedness* and *infinitude* in relation to spatial objects (and so opened the way for a study of Relativity).

Riemann [91] p. 284 asserted:

'In the extension of space-construction to the infinitely great, we must distinguish between *unboundedness* and *infinite extent*. . . . The unboundedness of space possesses . . . a greater empirical certainty than any external experience. But its infinite extent by no means follows from this [. . . '*Die unbegrenztheit des Raumes besitzt daher eine grössere empirische Gewissheit, als irgend eine äussere Erfahrung. Hieraus folgt aber die Unendlicheit keineswegs* . . .']; on the other hand, if we assume independence of bodies from position, and therefore ascribe to space constant curvature, it must necessarily be finite provided this curvature has ever so small a positive value.'

Thus, for Riemann,

a line is finite but unbounded.

Emphasis needs to be laid on the fact that Riemann's ideas were very largely general, rather than particular, and it was left to other people, notably Helmholtz, Lie, Klein and Beltrami, to work out the synthetic details of the system of plane non-Euclidean Geometry known as *Elliptic Geometry*, which complements that of the hyperbolic type. Here it might be borne in mind that the name *Riemannian Geometry* refers to a wider field of study involving differential geometry. Relativity Theory does, of course, make considerable use of this Geometry.

Replace Euclid's Postulate 5 by the following statement:

Two lines always intersect one another (and enclose an area).

This is the fundamental assumption of Elliptic Geometry.

As in the case of Hyperbolic Geometry, all Euclidean theorems which do not depend on his ideas of parallelism or the infinite extent of the straight line, are immediately available.

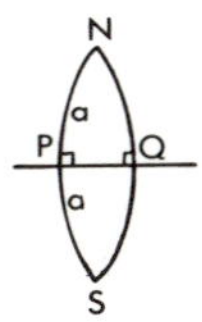

FIG. 100. Digon

By virtue of the fundamental assumption, a basic figure in Elliptic Geometry is the *digon* or *biangle* (Fig. 100) of which each side has length $2a$. Consequently, we may think of $2a$ as a natural unit of length in the elliptic plane.

§ 41.1 Some Basic Theorems of Elliptic Geometry. For the record, we append certain important theorems of Elliptic Geometry which correspond to basic theorems of Hyperbolic Geometry already enunciated.

Theorem 43: Summit Angles of a Saccheri Quadrilateral are equal and obtuse.

Therefore

Saccheri's Hypothesis of the Obtuse Angle characterises Elliptic Geometry.

Theorem 44: In a Lambert Quadrilateral the fourth angle is obtuse.

Hence

There are no squares in Elliptic Geometry.

Theorem 45: Angle-sum of a triangle is greater than two right-angles.

Note that this angle-sum decreases as the size of the triangle decreases, whereas in Hyperbolic Geometry this sum increases as the triangle diminishes. Refer to Figs. 101, 104.

Theorem 46: Similar triangles are congruent.

Otherwise expressed
There are no similar triangles (or similar polygons) of different sizes in Elliptic Geometry.

§42 MODELS OF NON-EUCLIDEAN GEOMETRY

§42.1 Hyperbolic Geometry.

§42.11 Beltrami Model. As stated in §1, Beltrami established the validity of non-Euclidean Geometry by associating it with a surface of Euclidean Geometry. For the Lobatschewsky-Bolyai variety it was found that the model required for the plane was one having constant negative curvature (like a saddleback surface). Such a surface is the *pseudosphere*, i.e. the surface formed by rotating a tractrix about its axis (Fig. 101). (The essential feature of a tractrix is that the length of a tangent intercepted by the axis is always constant ($= c$)). Correspondingly, Elliptic plane Geometry may be represented on a surface of constant positive curvature, e.g. a sphere—recall Riemann's comment on finiteness and unboundedness—while the Euclidean plane has (constant) zero curvature (and may thus be regarded as a sphere of infinite radius).

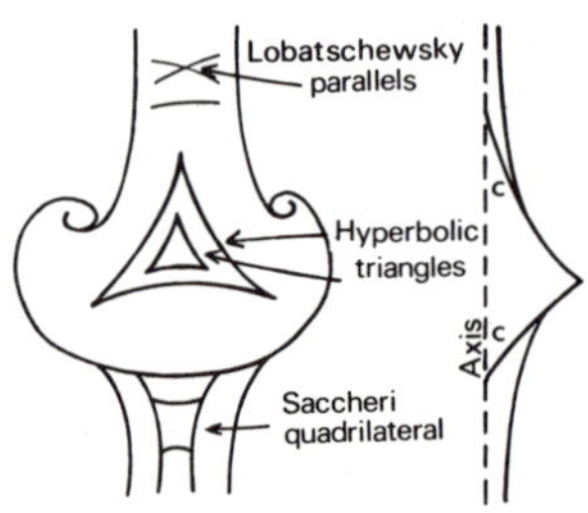

FIG. 101. Pseudosphere tractrix

Basic figures in Hyperbolic Geometry, such as the acute angled quadrilaterals of Saccheri and Lambert, ought to be drawn on the pseudospherical model for practice.

Beltrami's model is, however, not free from inadequacies. Shortcomings of his model are
(1) it is limited to a two-dimensional representation, and
(2) the pseudospherical surface does not represent all the non-Euclidean plane.

§ 42.12 Klein Model. These two weaknesses are absent from the simple plane Euclidean model of Hyperbolic Geometry devised by Klein (Fig. 102), also known as the Cayley-Klein model. Consider the conic S shown in the Figure —it is the (non-degenerate) real Absolute conic. Call points

inside	S *ordinary* (hyperbolic) points (= h-points)		(P)
on	S *ideal*		(A, B)
outside	S *ultra-ideal*		(C, D)

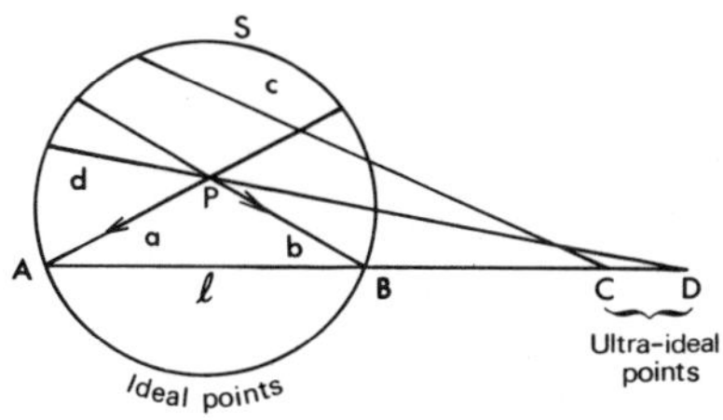

FIG. 102. Klein model of the hyperbolic plane

A hyperbolic line (h-line) is that portion of a line lying wholly within S. All the Euclidean Axioms are valid for these hyperbolic elements except the Parallel Postulate which is replaced by the axiom that two lines a, b separating the pencil of lines like d through $P\,(= a \cap b)$ and not cutting l are parallel (the Lobatschewskian parallels). In the Klein model, lines a, c are intersecting and lines c, l or d, l are non-intersecting.

Clearly, the hyperbolic plane consists of points and lines on the interior of S. Real S plays the role of Hyperbolic Geometry that l_∞ plays in Euclidean Geometry, e.g. two lines meeting on S are defined to be parallel. Ideal points (or points at infinity) on S are 'infinitely distant' from points inside S.

Two lines are defined to be perpendicular if they are conjugate w.r.t. S. Some theorems relating to perpendicularity are

I Parallel lines have no common perpendicular.

II Intersecting lines have no common perpendicular.

III Two lines intersecting in an ultra-ideal point have a common perpendicular.

§ 42.13 Poincaré Model (after Henri Poincaré). A beautiful, but more complicated Euclidean model, due to Poincaré,is obtained as a stereographic projection of the Northern Hemisphere from the North Pole on to the tangent plane at the South Pole. Points and orthogonal arcs to a fixed circle (which is the projection of the Equator from the North Pole) constitute the points and lines of the Hyperbolic Plane in the Poincaréan model. One of the special features of this model is that equality of angles is preserved, i.e. the angle between two hyperbolic lines equals the angle between the two corresponding orthogonal arcs. This angle-preserving quality is one of the basic properties of stereographic projection.

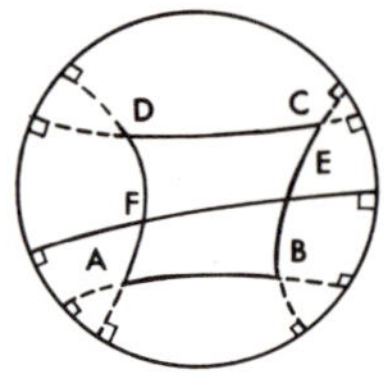

FIG. 103. Saccheri and Lambert quadrilaterals in the Poincaréan model

Details of the Poincaréan model (Fig. 103) and some applications of it are to be found in say Adler [2], Eves [36] 1, Meserve [73], Pedoe [84], and Wolfe [121]. For the equivalence of the Poincaréan and Klein models, consult Artzy [4] p. 251. Also consult H. Eves and V. E. Hoggatt, Jr.; 'Hyperbolic trigonometry derived from the Poincaré model', *Amer. Math. Monthly*, 58 (7) (1951) 469–474; Nguyêñ-cảnh-toàn (of Hanoi): '*Une méthode très simple pour établir les formules de la trigonométrie hyperbolique au moyen du modèle de Poincaré*', *Proceedings of the Second Hungarian Mathematical Congress I* (1960) 115–117.

§ **42.2 Elliptic Geometry.**

§ **42.21 Spherical Model** (a model in Euclidean space of three dimensions). Let us now interpret great circles on a sphere as lines in an elliptic plane. Such lines are indeed finite but unbounded. Now there is one blemish in the perfection of this correspondence, namely two great circles meet in diametrically opposite (= antipodal) points, i.e. two lines meet in two points. Remove this difficulty by identifying antipodal points, a bold stroke of imagination due to Klein. The surface of the sphere is thus a model of Elliptic Geometry. Without this identification of antipodal points the geometry of the sphere is called *Spherical* or *Doubly Elliptic Geometry* and this is probably what Riemann had in mind. A check ought to be made that the Axioms of Projective Geometry are valid in this model of Elliptic Geometry.

In this spherical model it is clear that Saccheri's Hypothesis of the Obtuse Angle (Fig. 104) is valid and that the angle-sum of a triangle is more than two right-angles. Look back to Fig. 5 also.

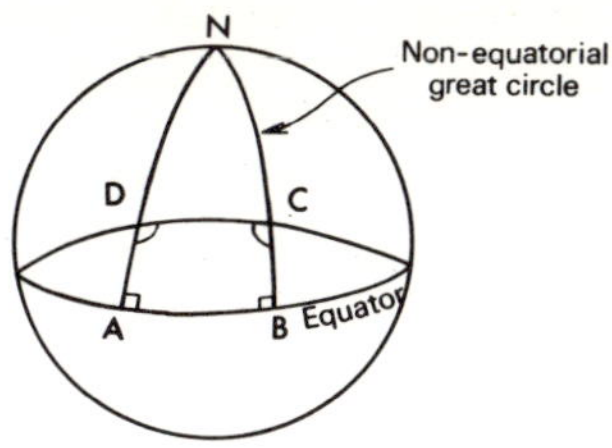

FIG. 104. Saccheri's quadrilateral in elliptic geometry

The elliptic plane may very simply be visualised by projecting the points of a hemisphere from the centre onto the tangent plane at the South Pole of the hemisphere. Antipodal points of the boundary of the hemisphere are identified, of course. See O'Hara and Ward [76] for the diagram, if necessary.

On the surface of the Earth (sphere), a great circle is always the path of shortest distance between two points. These are called *geodesics* (i.e. in the ordinary Euclidean plane a geodesic is a familiar straight line and in our elliptic geometry model, a geodesic is a great circle). Every air navigator knows that to get from one point on the Arctic Circle to another fairly distant point on it, it is quicker to travel over the North Pole, since the Arctic Circle is not a great circle, i.e. the geometry of great circle navigation is not Euclidean, so Euclid's is not the only geometry of human utility.

In Hyperbolic Geometry, geodesics are curves obeying Lobatchewsky's parallel postulates.

No parallel lines can exist in our elliptic model. You might well ask: But what about railway lines? Keeping in mind that two rails are equidistant, we remark that they cannot be parallel straight lines (which, on the sphere, must be great circles) since only one rail could actually be along a great circle, though in actual practice the other rail would differ negligibly from being part of a great circle.

For a survey of the properties of Elliptic Geometry, one may consult Prenowitz and Jordan [88].

§ 42.22 One-Sided Surface. Möbius Band. Essentially, the difference between spherical and elliptic planes is that the former is two-sided while the latter is one-sided, like the *Möbius Band.* To construct a Möbius Band, take an ordinary rectangular piece of paper and give it a half-twist, identifying points which were originally skew-opposite such as E, F and D, B (and A, C). This representation of the Möbius Band is called its *Poincaréan polygon* (Fig. 105).

A little thought will reveal that one can travel around the surface of the Möbius Band (Fig. 106) commencing at P and arrive at the point on the 'other' side of the Band without having perforated it, as would be the case for a sphere. Further, notice that the Möbius Band has only one boundary curve or edge.

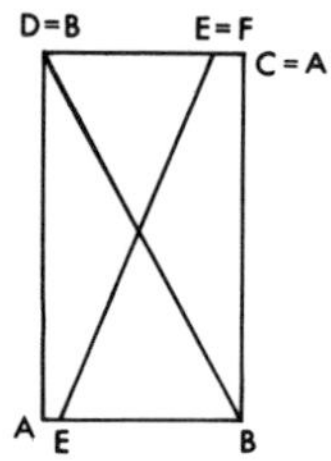

FIG. 105. Poincaréan polygon for a Möbius band

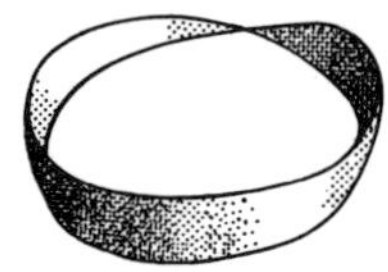

FIG. 106. Möbius band

Details of some models of the projective plane may be found in the following articles:
D. Dekker: 'A paper model of the projective plane', *Amer. Math. Monthly*, 61 (2) (1954) 113–115;
D. Gans: 'Models of projective and Euclidean space', *Amer. Math. Monthly*, 65 (10) (1958) 749–756.

§43 METRIC NON-EUCLIDEAN GEOMETRY

Earlier, mention was made of metrical notions in the non-Euclidean geometries with regard to the angle of parallelism and units of length. Here, we propose to determine the formulas for distance between two points, and angle between two lines, in both the elliptic and hyperbolic varieties. These may be compared with similar metrical expressions for the Euclidean plane.

Because elliptic trigonometry is often useful in physical theories such as Relativity, we find it convenient to examine this in a little detail. Corresponding results for hyperbolic trigonometry may then easily be derived by minor trigonometrical adjustments. In the ensuing treatment of the metrical properties of a triangle in Elliptic Geometry, we rely on O'Hara and Ward [76] pp. 239–240.

§ 43.1 Angular and Linear Metrics in Elliptic Geometry. Refer to Fig. 74 where $\mathscr{A}$ (the Absolute conic) is complex for Elliptic Geometry, i.e. the points O, U are complex. We may choose the equation of complex $\mathscr{A}$ to be

$x^2 + y^2 + z^2 = 0$, i.e. (21) with $\lambda = 1$. Taking P, Q real as previously and remembering that $\mathbf{O} = \mathbf{P} + \mu_1\mathbf{Q}$, $\mathbf{U} = \mathbf{P} + \mu_2\mathbf{Q}$, we see that O, U are complex only if μ_1, μ_2 are complex. This is so if, using the theory of § 35.3, we take $f_{12}^2 - f_{11}f_{22} < 0$, i.e. $\dfrac{f_{12}^2}{f_{11}f_{22}} < 1$. Write $\cos\theta = \dfrac{f_{12}}{\sqrt{f_{11}f_{22}}}\cdot$

Substitute in (20) and we obtain

$$\begin{aligned}(PQ) &= k \log\left\{\frac{\cos\theta + i\sin\theta}{\cos\theta - i\sin\theta}\right\}\\ &= k\log e^{2i(\theta + n\pi)}\\ &= k.2i(\theta + n\pi)\\ &= \theta + n\pi \text{ on putting } k = \frac{1}{2i} \text{ to make the distance}\end{aligned}$$

real. No loss of generality occurs in taking $n = 0$ so that $(PQ) = \theta$ and hence

$$\cos(PQ) = \frac{f_{12}}{\sqrt{f_{11}f_{22}}}\cdot$$

Dually, for angle we have

$$\cos(pq) = \frac{F_{12}}{\sqrt{F_{11}F_{22}}} \qquad (\alpha)$$

Now let ABC be a triangle whose vertices have co-ordinates $A(x_1, y_1, z_1)$, $B(x_2, y_2, z_2)$, $C(x_3, y_3, z_3)$, and whose sides BC, CA, AB have line co-ordinates (l_1, m_1, n_1), (l_2, m_2, n_2), (l_3, m_3, n_3) respectively. Denote by a, b, c the lengths of the sides BC, CA, AB respectively.

In the symmetrical determinant $\triangle = \begin{vmatrix} f_{11} & f_{12} & f_{13}\\ f_{21} & f_{22} & f_{23}\\ f_{31} & f_{32} & f_{33}\end{vmatrix}$, where $f_{ij} = x_ix_j + y_iy_j + z_iz_j = f_{ji}$ $(i, j = 1, 2, 3)$, let $F_{11}, F_{12}, \ldots, F_{33}$ be the cofactors of $f_{11}, f_{12}, \ldots, f_{33}$ respectively. Note that $\triangle = \begin{vmatrix} x_1 & y_1 & z_1\\ x_2 & y_2 & z_2\\ x_3 & y_3 & z_3\end{vmatrix}^2 > 0$. Bearing in mind that BAC is the angle between AB and AC (of which the line co-ordinates are $(-l_2, -m_2, -n_2)$), we have, from (α),

$\cos A = \dfrac{-F_{23}}{\sqrt{F_{22}F_{33}}}$ whence $\sin A = \sqrt{\dfrac{\triangle f_{11}}{F_{22}F_{33}}}$. (Check that $\cos^2 A + \sin^2 A = 1$.)

Also, $\cos a = \dfrac{f_{23}}{\sqrt{f_{22}f_{33}}}$, $\sin a = \sqrt{\dfrac{F_{11}}{f_{22}f_{33}}}$.

Similar results for the other angles and sides follow if we permute cyclically the letters and the suffixes.

Hence we obtain the elliptic trigonometrical formulas

$$\begin{cases} \cos a = \cos b \cos c + \sin b \sin c \cos A \\ \cos A = -\cos B \cos C + \sin B \sin C \cos a \\ \dfrac{\sin A}{\sin a} = \dfrac{\sin B}{\sin b} = \dfrac{\sin C}{\sin c} \left(= \sqrt{\dfrac{f_{11}f_{22}f_{33}}{F_{11}F_{22}F_{33}}} \right), \end{cases}$$

which coincide with the formulas of ordinary Spherical Trigonometry.

Further, $\cos A + \cos (B + C) = \sin B \sin C (\cos a - 1)$ on calculation,

$$\text{i.e. } \cos \left(\frac{A + B + C}{2}\right) \cos \left(\frac{B + C - A}{2}\right) = -\sin B \sin C \sin^2 \left(\frac{a}{2}\right)$$

$$= -\text{ve quantity.}$$

Consequently, the factors on the L.H.S. must have opposite signs.

Assume $A \geqslant B \geqslant C$, so that $-\pi < B + C - A < \pi$, whence

$$\cos \left(\frac{B + C - A}{2}\right) > 0.$$

$$\therefore \cos \left(\frac{A + B + C}{2}\right) < 0$$

i.e. $A + B + C > \pi$

$\therefore$ in Elliptic Geometry, the angle-sum is $> \pi$.

§ 43.2 Metrics in Hyperbolic Geometry. Here the Absolute conic $\mathscr{A}$ is real, so that O, U in Fig. 74 are real. Following the reasoning in the case of Elliptic Geometry, we have $f^2_{12} - f_{11}f_{22} > 0$.

Write $\cosh \theta = \dfrac{f_{12}}{\sqrt{f_{11}f_{22}}}$. Then (20) yields

$$(PQ) = k \log \left\{ \frac{\cosh \theta + \sinh \theta}{\cosh \theta - \sinh \theta} \right\}$$

$$= k \log e^{2\theta}$$

$$= 2k(\theta + n\,\pi i).$$

Choose $n = 0$ and $k = \frac{1}{2}$ and we have

$$(PQ) = \theta.$$

$$\text{i.e. } \cosh (PQ) = \frac{f_{12}}{\sqrt{f_{11}f_{22}}}.$$

$$\text{Dually, } \cosh (pq) = \frac{F_{12}}{\sqrt{F_{11}F_{22}}}.$$

Results in Hyperbolic Trigonometry corresponding to those in Elliptic Geometry may be deduced by substituting

$$\cosh a \text{ for } \cos a$$

etc. Write out a few of these for practice.

§ 44 APPLICATIONS OF NON-EUCLIDEAN GEOMETRY TO THE PHYSICAL UNIVERSE

§ 44.1 Lobatschewsky's Experiment. Lobatschewsky conceived his geometrical discoveries as not merely the end-product of a splendid intellectual exercise but as possibly applicable to the physical universe, e.g. in astronomical space.

Now we know (Exercises 9.9, 9.11) that in the immediate neighbourhood of a hyperbolic point Euclidean trigonometrical formulas hold. This prompted Lobatschewsky to enquire whether this approximation might not extend to the trigonometry of very great distances. As a relevant experiment to test his hypothesis, Lobatschewsky decided to fix a positive lower bound for the parallax of the star Sirius. Such an investigation involves an examination of the constant k, occurring in (31).

In Fig. 107, which explains the astronomical idea of parallax, we have, for the maximum parallax p of the star,

$$\Pi(x) > \frac{\pi}{2} - p$$

whence, following Bonola [14] p. 95, we obtain

$$\tan\left\{\frac{\Pi(x)}{2}\right\} > \tan\left(\frac{\pi}{4} - \frac{p}{2}\right) = \frac{1 - \tan(p/2)}{1 + \tan(p/2)} \qquad (\alpha)$$

But by (31), $\tan\left\{\frac{\Pi(x)}{2}\right\} = e^{-x/k}$.

From (α) and (31), it follows that

$$\frac{x}{k} < \log\left(\frac{1 + \tan p/2}{1 - \tan p/2}\right) \left(\text{assume } \frac{p}{2} < \frac{\pi}{4}\right)$$

$$= 2\left\{\tan\left(\frac{p}{2}\right) + \tfrac{1}{3}\tan^3\left(\frac{p}{2}\right) + \tfrac{1}{5}\tan^5\left(\frac{p}{2}\right) + \ldots\right\}$$

by the Maclaurin expansion for the logarithm function.

Moreover

$$\tan \mathrm{p} = \frac{2\tan(p/2)}{1 - \tan^2(p/2)} = 2\left[\tan\left(\frac{p}{2}\right) + \tan^3\left(\frac{p}{2}\right) + \tan^5\left(\frac{p}{2}\right) + \ldots\right]$$

whence $\frac{x}{k} < \tan p$.

Lobatschewsky took the parallax of Sirius as 1.24″ (now known to be 0.38″) whence $\frac{x}{k} < 0.000006012$.

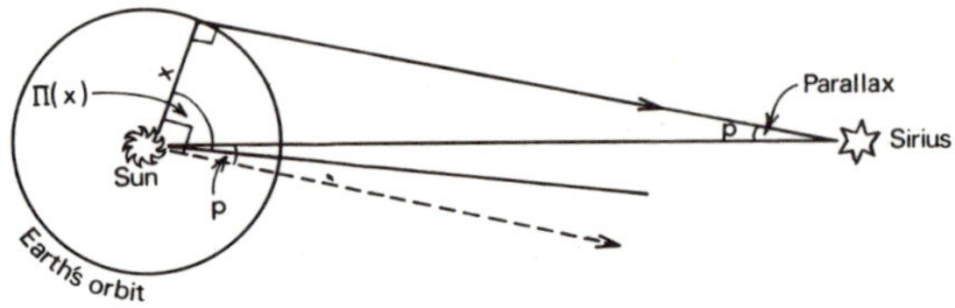

FIG. 107. Parallax of Sirius

This tells us that k is very great compared with the radius of the Earth's orbit though we do not know what k is. If we took a star whose parallax was 0.1″ we would find $k > 1{,}000{,}000 \times 2x$.

So if Euclidean Geometry and the Fifth Postulate are to hold in physical space $k \to \infty$, i.e. there must exist stars with parallaxes $\to 0$. Astronomical observations could never decide this with certainty. Yet considering the enormous value of k compared with actual measurable quantities, we must conclude (as Lobatschewsky did) that for all practical purposes Euclidean Geometry is valid.

Revert momentarily to the diagram for Lobatschewsky's parallels. Suppose $\theta = 0.000001''$. No instrument in the world is capable of distinguishing the infinite bundle of parallel lines in Lobatschewsky-Bolyai Geometry from the single parallel line of Euclidean Geometry. There is thus no way of telling whether the world in which we live is Euclidean or non-Euclidean. It is possible that θ, if it exists, is so small that we cannot detect it. It is abstractly conceivable, but not measurable.

Summing up, we may agree with Lobatschewsky and Bolyai that the geometry of small spaces is always approximately Euclidean, that no *a priori* grounds exist for a decision and that observation alone can give only an approximate answer, i.e. within the domain of experience Euclid's Parallel Postulate is justified.

§ **44.2 Gauss' Experiment.** Wishing to test the validity of the non-Euclidean and Euclidean geometries against the known structure of space, Gauss devised a famous experiment for the angle-sum of a triangle. Vertices of his triangle were three mountain peaks in the vicinity of Göttingen and the sides (straight lines) were paths of light rays (Fig. 108). Using the most delicate optical instruments, Gauss found that the deviation of the angle-sum from 180° was within the limits of observational and experimental errors. Consequently, his experiment was inconclusive. Even in the case of astronomical triangles whose sides are of the order of the distance of the Earth from the Sun, observations show that the angle-sum differs from 180° by a quantity less than 0.003″.

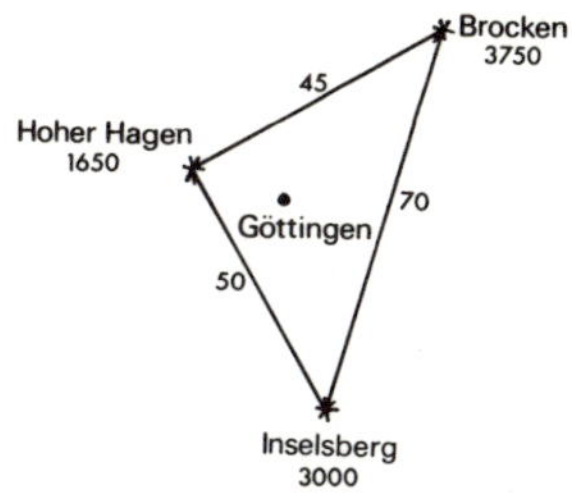

FIG. 108. Gauss' experiment (distances in miles, heights in feet)

It would, indeed, require just one convincing counter-example to disprove the complete validity of Euclid. But no such counter-example has ever been produced. To prove conclusively that Euclidean Geometry accurately represents reality it would be necessary to show that the angle-sum of every triangle is 180°, a task beyond human endeavour.

§ 44.3 Applications. Apart from the well-known use of the more general Riemannian Geometry in the Theory of Relativity, other applications of non-Euclidean Geometry occur in optics and the theory of wave propagation. In another sense, an application has been, not to the physical world, but to the world of the mind—our thinking has been freed from the bondage of the Kantian philosophy of space.

§ 44.4 Non-Euclidean Groups. Non-Euclidean geometries, whose metrical classification depends on the choice of the Absolute Conic (just as does Euclidean Geometry) fit into the ideas expounded by Klein in his *Erlanger Programm.* (Refer to the diagram.) Whereas the Euclidean group leaves invariant the circular points I, J, i.e. the degenerate Absolute, the non-Euclidean groups leave invariant the non-degenerate Absolute. More particularly, Hyperbolic Geometry is characterised by the group of transformations (*hyperbolic group*) leaving invariant a real conic, while Elliptic Geometry is specified by the group of transformations (*elliptic group*) which leave invariant a complex conic. Observe that Affine Geometry is not directly related to the non-Euclidean geometries. We have the hierarchical diagram:

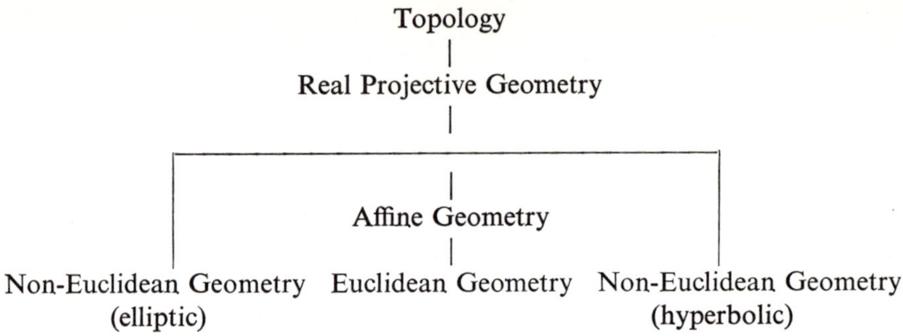

§45 PHILOSOPHICAL CONSIDERATIONS

Sometimes the question is asked: Which of the geometries is true? Meaning: Which most accurately measures the physical world of sense-perceptions? ('Is the Euclidean geometry true?' asks Poincaré. 'It [the question] is meaningless. As well ask whether the metric system is true. . . .') Mathematically, each geometry is equally valid since it follows logically from a given set of hypotheses. No assertion whatsoever is made of the physical properties of space. Historically, however, the subject is deeply embedded in empirical observations so that we may say that physical geometry is a practical interpretation of abstract mathematical geometry.

Space for the physicist is space revealed by measurements of material apparatus; whatever its geometry, his test of a geometry is the pragmatic one: Does it work? Does it describe his metrical space? Whether or not a given geometrical theory is factually correct when interpreted physically is not a mathematical matter but represents a problem for empirical science. Experiment and observations are the criteria for deciding the issue.

Gauss' crucial experiment on the mountain peaks was inconclusive since it did not disprove the validity of the Euclidean angle-sum. The deviation from 180° could have been due entirely to imperfections of human measurement, or it could have meant that this celestial triangle was really Lobatschewskian in character.

Actually, we are somewhat like the witness in the Graham Greene story who is asked by the lawyer: 'You are sure this is the man you saw?' 'Yes,' we say (thinking of Euclidean Geometry). 'Could you swear it was not his twin [Non-Euclidean Geometry] brother who is sitting at the back of the court?' Einstein has aptly summed up the dilemma in these words: 'As far as the propositions of mathematics refer to reality they are not certain, and insofar as they are certain they do not refer to reality.' (*Geometrie und Erfahrung*, p. 3.)

Since we cannot say which geometry most accurately describes the *local* properties of space (and we include objects distant many millions of miles in this category) the choice among the geometries is made solely on the basis of simplicity and convenience. But we should not necessarily expect Euclidean geometry to be a suitable means of describing the Universe as a whole, of which we occupy so tiny a portion. On this point, Einstein remarked: 'We know that the behaviour of measuring rods and clocks is influenced by gravitational fields, that is, by the distribution of matter. This in itself is sufficient to exclude the possibility of the exact validity of Euclidean geometry in our Universe.' In his Theory of Relativity, the geometry is Riemannian (i.e. space is finite but unbounded like the surface of a sphere), light travels along geodesics and the curvature of space is determined by the nature of the matter which fills it. Physical space may thus be warped, i.e. non-Euclidean. Of course, this is contrary to our common experience but our spatial experience is limited. Thus we might expect cosmological theories relating to astronomical distances, e.g. to extra-galactic space, to make use of non-Euclidean geometries in describing this space mathematically.

Clifford also has warned us against the danger of 'asserting dogmatically that an axiom based on the experience of a limited region holds universally'. The situation is analogous to that in Physics where the systems of Newton and Einstein give the same results for small distances and velocities, but diverge when very large magnitudes are involved. Interstellar space and the atomic world are beyond our spatial experience. May it not be reasonable to expect that in these fields our 'common-sense' Euclidean notions could play us false?

Further, we live on the surface of a sphere but since locally the surface of the Earth may approximate to a plane, Euclidean geometry here is a fairly reliable guide. Too rigid an adherence to Euclid may then lead to a belief in a flat Earth!

For all practical purposes, work based on any of the geometries of Euclid, Riemann, and Lobatschewsky and Bolyai will yield results far within the limits of accuracy of our measurements. But because of its simplicity and convenience, Euclidean geometry is the one we shall continue to use for engineering purposes (bridges, opera houses, tunnels, etc.) and for surveying, carpentry and navigation.

It is not our objective to plumb the depths of the philosophical problems raised by our geometrical ideas of space but merely to mention them. Two things, then, are certain:

(i) *The belief in the mathematical infallibility of Euclid's system as absolute truth*, as an immutable framework into which our empirical knowledge of physical reality must Procrustean-wise be fitted, *has been destroyed forever;*

(ii) *No proof is certain that Euclidean distance exactly measures the world of sight and touch.*

In the entire history of science there has seldom been a more revolutionary and sweeping intellectual development than that of non-Euclidean Geometry. For a comparable achievement, one must think of the astronomical revolution brought about by the heliocentric theory of Copernicus and perhaps of the biological theories of Darwin and of the Relativity theories of Einstein. The non-Euclidean movement shattered the entrenched belief that Euclid had revealed eternal truths. Competent and accurate as a measuring tool since Egyptian antiquity, intuitively appealing and full of common sense, sanctified and cherished as one of the richest intellectual legacies from Greece, the geometry of Euclid had stood for more than twenty centuries in lone, resplendent and irreproachable majesty. It was this that Lobatschewsky, Bolyai, Riemann and their disciples dethroned.

Says Mach: 'It is a sublime spectacle which these men offer: labouring for centuries from a sheer thirst of scientific elucidation, in quest of the hidden sources of a truth which no person of theory or of practice ever really doubted.' (Mach is referring to *all* who struggled with Postulate 5.)

Standard reference books (other than Bonola [14] and Wolfe [121] already mentioned) which might be consulted are, for instance: H. S. M. Coxeter [25] and S. Kulczŷcki [61].

Older texts such as J. L. Coolidge [23], D. M. Y. Sommerville [106], and Veblen and Young [113] 2, are worth looking at. Most of the above references deal with the historical aspects of non-Euclidean Geometry in varying degrees of detail. For the logical consistency and the empirical validity of non-Euclidean geometries, have a look at Prenowitz and Jordan [88].

Finally, two shorter articles might with benefit be studied, namely:

H. S. M. Coxeter: *Lectures in Modern Mathematics*, Vol. 3, ed. T. L. Saaty, Wiley (1965) ('Geometry' pp. 58–64).

H. Poincaré: 'Non-Euclidean Geometry', *Nature* 45, No. 1165 (February 25, 1892) 404–407.

Relativity ideas, particularly the relativistic cosmologies of Minkowski and de Sitter and the connection between Projective Geometry and Kinematics by way of relative velocity, are presented in the former. Poincaré's contribution summarises the achievements of the pioneers in Projective Geometry viewed against an historical background and concludes with a discussion of the relationship between Geometry and Astronomy.

It seems appropriate to conclude this Chapter on non-Euclidean Geometry with a quote from the book by the prominent Soviet geometer P. A. Shirokov, [101] p. 86, written on the hundredth anniversary of the death of Lobatschewsky:

'The significance of the various interpretations of non-Euclidean geometry is very great: on the one hand, they make possible applications of this geometric system (for instance, the Cayley-Klein interpretation can be used as the basis for the application of Lobachevskian geometry to the special theory of relativity, and the Poincaré interpretation is related to problems in the theory of analytic functions, . . .); on the other hand, they bind these two geometric systems—Euclidean and Lobachevskian—together, so that the problem of the consistency of the one reduces to the problem of the consistency of the other.'

Suggested topics for further study

(1) Saccheri's 'Euclid vindicated from every flaw'.
(2) Lobatschewsky's 'Theory of Parallels'.
(3) Bolyai's 'Science of Absolute Space'.
(4) Riemann's 'Hypotheses which lie at the foundations of Geometry'.
(5) Hyperbolic trigonometry (e.g. the circle, basic formulas, areas).
(6) Elliptic trigonometry.
(7) Horocycles.
(8) Work of Hjelmslev.
(9) Work of Nasîraddîn—refer to §§ 1.2, 1.3.
(10) Topology.
(11) Cosmology and Relativity.
(12) Poincaré model of the hyperbolic plane.

EXERCISES 9

In the hyperbolic plane, establish the results 1–3 for a Saccheri Quadrilateral:

1. The line joining the mid-points of the base and summit is perpendicular to both. (See Saccheri's Proposition I.)

2. Summit angles are equal. (See Saccheri's Proposition II.)

3. Base and summit are non-intersecting.

4. Following Wolfe [112] p. 72, let ABI be a triangle in which AI is parallel to BI and I is at infinity. Verify that Pasch's Axiom (§ 1) holds for this triangular figure in the hyperbolic plane.

5. Given a quadrilateral $ABCD$ in which $D\hat{A}B$ and $A\hat{B}C$ are right angles, prove that

$$B\hat{C}D \gtreqless C\hat{D}A \Rightarrow AD \gtreqless BC.$$

(When $B\hat{C}D = C\hat{D}A$, the figure is a Saccheri Quadrilateral.)

6. For the Lambert Quadrilateral in the hyperbolic plane, show that

(i) $AD > BC$,
(ii) $CD > AB$.

7. Which is greater in a Saccheri Quadrilateral in the hyperbolic plane, the base or the summit? Justify your answer. State the corresponding result in the elliptic plane.

8. Angle-sum of any quadrilateral in Elliptic Geometry is greater than four right angles. Reason?

9. A basic formula in the hyperbolic plane is

$$\cosh\left(\frac{c}{k}\right) = \cosh\left(\frac{a}{k}\right)\cosh\left(\frac{b}{k}\right)$$

where a, b, c are sides of a hyperbolic triangle, and k is a constant associated with the unit of length. Show that this is the equivalent of Pythagoras' Theorem in the Euclidean plane (i.e. for sufficiently small regions of the hyperbolic plane, Euclidean Geometry is applicable.) [Hint: Expand each cosh function as an infinite series by Maclaurin's Theorem, and then let $k \to \infty$.]

10. Corresponding to the fundamental formula for the hyperbolic plane in Exercise 9.9 is the formula for the elliptic plane obtained by replacing k by ik:

$$\cos\left(\frac{c}{k}\right) = \cos\left(\frac{a}{k}\right)\cos\left(\frac{b}{k}\right).$$

Deduce that when $k \to \infty$, this approximates to Pythagoras' Theorem in the Euclidean plane.

11. Other basic formulas in hyperbolic trigonometry are:

$$\begin{cases} \dfrac{\sinh\left(\frac{a}{k}\right)}{\sin\lambda} = \dfrac{\sinh\left(\frac{b}{k}\right)}{\sin\mu} = \dfrac{\sinh\left(\frac{c}{k}\right)}{\sin\nu} \\[2ex] \cosh\left(\frac{a}{k}\right) = \cosh\left(\frac{b}{k}\right)\cosh\left(\frac{c}{k}\right) - \sinh\left(\frac{b}{k}\right)\sinh\left(\frac{c}{k}\right)\cos\lambda \end{cases}$$

where λ, μ, ν are the angles of the hyperbolic triangle (sides a, b, c). Demonstrate, as in Exercise 9.9., that in the limiting form, these formulas become respectively, the Sine and, Cosine Formulas for the Euclidean plane.

12. State corresponding formulas to those in Exercise 9.11 for the elliptic plane. Hence deduce that in the limit, as $k \to \infty$, these formulas are the Sine and Cosine Formulas for the Euclidean plane.

Note: Exercise 9.9–9.12 reveal that Euclidean Geometry is a limiting form of both Hyperbolic Geometry and Elliptic Geometry, i.e. in going from one non-Euclidean Geometry to the other, we 'pass through' Euclidean Geometry (rather in the way in which a parabola is a critical limiting form of both hyperbola and ellipse—think of a rotating plane cutting a cone successively in a hyperbola and ellipse).

13. Angle of parallelism for any distance tends to $\frac{\pi}{2}$ as $k \to \infty$. Establish.

14. Deduce from the Lobatschewsky-Bolyai formula

$$\tan\left\{\frac{\Pi(p)}{2}\right\} = e^{-\frac{p}{k}}$$

the results

(i) $\cos \Pi(p) = \tanh\left(\frac{p}{k}\right)$; (ii) $\sin \Pi(p) = \operatorname{sech}\left(\frac{p}{k}\right)$.

15. In the hyperbolic plane, the area A of a circle of radius $r = a$ is known to be

$$A = 4k \int_0^a \int_0^{\frac{\pi}{2}} \sinh\left(\frac{r}{k}\right) d\theta\, dr.$$

Evaluate this double integral to derive $A = 4\pi k^2 \sinh^2\left(\frac{a}{2k}\right)$.

[Notice that $k \to \infty$ yields $A = \pi a^2$, as expected.]

16. Area of a Lambert Quadrilateral of acute angle α is known to be

$$L = \int_0^a \int_0^y \cosh\left(\frac{y}{k}\right) dy\, dx.$$

Establish that

$$L = k^2\left(\frac{\pi}{2} - \alpha\right).$$

[Use the formulas $\tanh\left(\frac{y}{k}\right) = \dfrac{\cosh\left(\frac{x}{k}\right)}{\cosh\left(\frac{m}{k}\right)}, \dfrac{\sinh\left(\frac{a}{k}\right)}{\sinh\left(\frac{m}{k}\right)} = \cos\alpha$, where a, m are constants associated with the Lambert Quadrilateral (Wolfe [121] p. 170).]

17. Gauss, in 1831, stated that the length of the circumference of a circle of radius r was (Bonola [14] p. 75)

$$\pi k\left(e^{\frac{r}{k}} - e^{-\frac{r}{k}}\right).$$

Verify that Gauss' expression becomes the usual one when $k \to \infty$.

18. Why is Euclid's theorem:
The exterior angle of a triangle is greater than an interior opposite angle
not necessarily true in Elliptic Geometry?

19. Cut the rectangular (Poincaréan) model of the Möbius Band (Fig. 105) lengthwise down the middle. What happens? Next, with a new model, cut lengthwise along a line GH parallel to AD where $AG = \frac{1}{3}AB$, $DH = \frac{1}{3}DC$. Now what happens?

20. Suppose ABC is a geodesic triangle on a surface whose area is given by $S = \iint_{ABC} dS$.

Assuming Gauss' Theorem for a surface of curvature K (constant or variable):

$$\iint_{ABC} K\,dS = A + B + C - \pi$$

establish that in the elliptic, Euclidean and hyperbolic planes, for which $K = \frac{1}{k^2} > 0, = 0, = -\frac{1}{k^2} < 0$ respectively, the angle-sum $A + B + C$ of the triangle ABC is $>, =, < \pi$ respectively.

[Thus, two-dimensional Elliptic, Euclidean or Hyperbolic Geometry may be described on surfaces of constant curvature which are positive, zero, or negative, respectively, e.g. a sphere, a plane, a pseudosphere. Bonola [14] p. 133 shows an illustration of a surface of constant negative curvature. This illustration is reproduced from a photograph of a surface constructed by Beltrami, the actual original model being in the Mathematical Institute of the University of Pavia.]

CHAPTER 10

FINITE GEOMETRY

'Not long ago, finite projective planes were regarded as a combinatorial curiosity. Today, they are basic in the foundations of geometry and in the analysis and design of experiments. Our new technology with its vital concern for the discrete has given the recreational mathematics of the past a new seriousness of purpose.'
(H. J. Ryser)

THE type of thinking to which we have been accustomed in the previous Chapters must now be adjusted to cope with a new range of ideas. Except for Axioms *PP*1–*PP*3 which they have in common, the Euclidean-based planes which we have been studying bear little resemblance to the ones we are about to investigate.

§ 46 FANO PLANES

Nothing in the Axioms *PS*1–*PS*5 prevents the projective space from having only a finite number of points, lines and planes. Such a geometry, naturally called *Finite Geometry*, was first studied in 1892, for *n* dimensions, by the Italian mathematician, G. Fano, in his article: '*Sui postulati fondamentali della geometria proiettiva in uno spazio a un numero qualunque di dimensioni*', *Giornale di Matematiche* 30 (1892) 106–132.

§ 46.1 Fano Planes. Initially, we restrict our thoughts to Finite Geometry in two dimensions. By employing only the minimum incidence properties allowed by *PP*1–*PP*3 and by Theorems 1–3, we can construct quite easily the simplest finite plane, namely one in which there are just three points on each line and three lines through each point. Such a finite plane (or *Fano plane*), consisting of just seven points and seven lines, is represented in Fig. 109(a). Aesthetic considerations prompt us to redraw the diagram as in Fig. 109(b). Observe that

(i) a circle (or similar visually pleasing curve) may be employed to represent one of the lines;

(ii) the lines, as drawn, are merely a skeletal framework to support the points; and

(iii) *the diagonal points of any complete quadrangle are collinear*—this is a characteristic feature of the seven-point plane. (Look back now to §5.5(4) and §5.6 (Fano's Axiom), also to Fig. 2(c).)

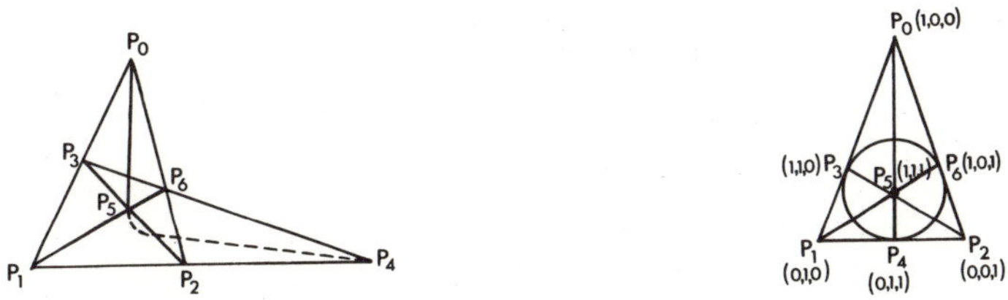

FIG. 109. (a) FIG. 109. (b)

7-points finite Fano plane

A special virtue of such a small finite geometry as this is that one can literally know everything about its incidence properties (even when the theory is extended to three dimensions).

When searching for a (finite) field whose elements may be used for homogeneous co-ordinates in this (finite) plane, we have to ask: What is the field whose total number of ordered triples (x, y, z), excluding $(0, 0, 0)$ is exactly seven?

It turns out to be GF(2), the Galois field of integers modulo 2. Details are given in Fig. 109(b), in which $P_0P_1P_2$ is the triangle of reference and P_5 is the unit point. (Notice, for example, that $\mathbf{P}_4 = \mathbf{P}_1 + \mathbf{P}_2$, $\mathbf{P}_2 = \mathbf{P}_1 + \mathbf{P}_4$, $\mathbf{P}_1 = \mathbf{P}_2 + \mathbf{P}_4$.)

A worthwhile exercise is the determination of the equations of the seven lines in the plane. For example, the equation of $P_3P_4P_6$ works out to be

$$x + y + z = 0$$

i.e. the unit line. (Remember that $-1 \equiv 1 \bmod 2$.)

Compare Figs. 109(a), (b) and Fig. 2(c) with Fig. 1, which relates to Axioms *PP*1–*PP*3, and with Fig. 13 (the construction of the fourth harmonic point). Observe in Fig. 109(b) that the diagonal triangle of Fig. 13 has disappeared, i.e. the diagonal points are collinear. This is readily understandable if we recall (§ 11, Example 5, Remark (4)) that, for GF(2), $D = C$ since $-\lambda \equiv \lambda$ mod 2. Actually Fano required that his diagonal triangle not collapse into a line, so there would be some justification for speaking of an 'anti-Fano plane' rather than a 'Fano plane' (§ 5.6). However, the later name is accepted phraseology. Incidentally, a nice theorem, due to A. M. Gleason ('Finite Fano planes', *Amer. J. Math.* 78 (1956) 797–807) asserts

Gleason's Theorem: Every finite Fano plane is Desarguesian.

Fano (Desarguesian) planes are planes over the finite Galois field GF(2^r). In passing, let us note that the finite plane can be fully represented by the following columnar incidence array (of lines) see the array in § 5.5:

l_0	l_6	l_5	l_4	l_3	l_2	l_1
P_0	P_1	P_2	P_3	P_4	P_5	P_6
P_1	P_2	P_3	P_4	P_5	P_6	P_0
P_3	P_4	P_5	P_6	P_0	P_1	P_2

Notice that this table of incidence is represented succinctly by $a + b \equiv 0, 1, 3 \bmod 7$ for the point P_a and the line l_b ($a, b = 0, 1, \ldots, 6$).

Question: Can (a) Desargues' Theorem (b) Pappus' Theorem exist in this finite plane?

§ 46.2 13-point plane. Next simplest plane after this seven-point plane occurs when there are just four points on each line and just four lines through each point. Following the procedure above, but omitting, for convenience, the label P and retaining only the numerical suffix for the point, we have the incidence array defined thus (and represented in Fig. 110):

l_0	l_{12}	l_{11}	l_{10}	l_9	l_8	l_7	l_6	l_5	l_4	l_3	l_2	l_1
0	1	2	3	4	5	6	7	8	9	10	11	12
1	2	3	4	5	6	7	8	9	10	11	12	0
3	4	5	6	7	8	9	10	11	12	0	1	2
9	10	11	12	0	1	2	3	4	5	6	7	8

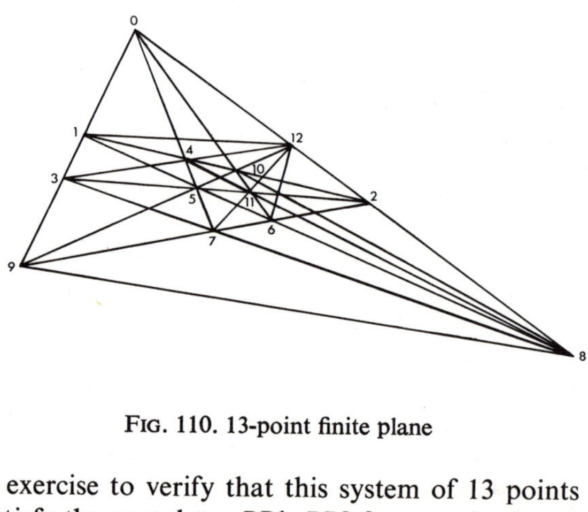

FIG. 110. 13-point finite plane

It is an easy exercise to verify that this system of 13 points and 13 lines does, in fact, satisfy the postulates $PP1$–$PP3$ for a projective plane.

Again observe that the point P_a and the line l_b are incident if $a + b \equiv 0, 1, 3, 9 \mod 13$.

Both the above arrays are special cases of the fact that in a finite projective plane having $n + 1$ points on every line $n = p^r$ with $q = p^{2r} + p^r + 1$ $(= n^2 + n + 1 = N)$, the points and lines of this plane, $PG(2, p^r)$—see §46.4 (iii) for the symbolism—can be exhibited as a rectangular array of q columns and $p^r + 1$ rows. Elements of the array are the points and elements in a column are the points of a line. Also notice that each row of the array is a cyclic permutation mod 7 (or mod 13, or, more generally, mod q) of the first row.

Reasoning as in the simpler case, we readily deduce that the co-ordinates of the 13 points belong to GF(3).

No seven points of a 13-point plane can form a seven-point plane, as may be verified. Consequently, diagonal points of any complete quadrangle in a 13-point plane are not collinear. (Check in Fig. 110.)

Example 47: in Fig. 110, show that

(i) (a) the two sets of triads P_0, P_2, P_4, and P_9, P_{11}, P_6, are in perspective and

(b) determine the four points of their Desargues line;

(ii) $P_2P_3P_5P_{11} \,\overline{\wedge}\, P_9P_3P_1P_0 \,\overline{\wedge}\, P_{11}P_3P_5P_2$ (find the centres of perspective).

Solution: (i) (a) Lines
$$\left.\begin{matrix} P_0 & P_9 & P_1 & P_3 \\ P_2 & P_{11} & P_5 & P_3 \\ P_4 & P_6 & P_{12} & P_3 \end{matrix}\right\} \text{are concurrent at } P_3$$

$\therefore$ triangles $P_0P_2P_4$, $P_9P_{11}P_6$ are in perspective.

(b) Now
$$\left.\begin{matrix} P_9 & P_6 & P_2 & P_7 \\ P_0 & P_4 & P_5 & P_7 \end{matrix}\right\} \text{meet in } P_7;$$
$$\left.\begin{matrix} P_9 & P_{11} & P_4 & P_8 \\ P_0 & P_2 & P_{12} & P_8 \end{matrix}\right\} \text{meet in } P_8;$$
$$\left.\begin{matrix} P_{11} & P_6 & P_0 & P_{10} \\ P_2 & P_4 & P_1 & P_{10} \end{matrix}\right\} \text{meet in } P_{10}.$$

But $P_7P_8P_{10}$ are collinear (with P_3)

$\therefore$ Desargues' Theorem is valid with Desargues line $P_3P_7P_8P_{10}$.

(ii) Further, $P_3P_5P_{11}P_2 \overset{P_6}{\overline{\wedge}} P_3P_1P_0P_9 \overset{P_8}{\overline{\wedge}} P_3P_5P_2P_{11}$.

§46.3 Affine Specialisations. Picking out a line in each finite plane and calling it the l_∞, we obtain the corresponding affine planes. Selecting, for example, the line $P_0P_1P_3$ in Fig. 109(a) and the line $P_0P_2P_8P_{12}$ in Fig. 110, we derive the affine specialisations shown in Figs. 111(a), (b).

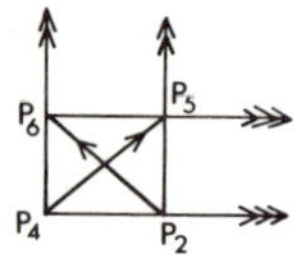

FIG. 111. (a)

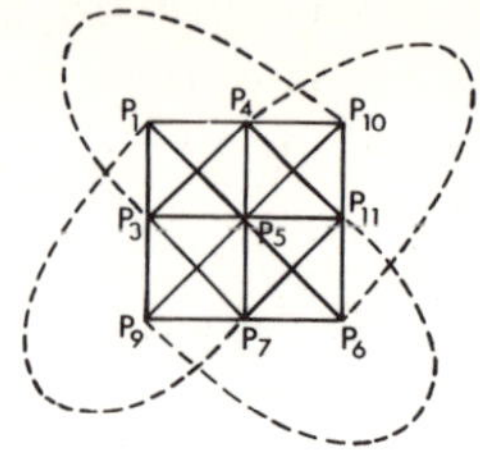

FIG. 111. (b)

Finite affine specialisations

Especially notice that in Fig. 111(a) the lines P_4P_6 and P_2P_5 are parallel. Fig. 111(a) contains four points and six lines, while Fig. 111(b) (sometimes known as the *Young plane*, after the American mathematician, J. W. Young) contains nine points and 12 lines.

§ **46.4 General Finite Plane of Order n.** Evidently, the Principle of Duality in relation to *PP*1 and *PP*2 implies that in a finite plane there must be the same number of points on a line as there are lines through a point. This fact is abundantly clear in the two cases which have been considered above.

In the general case, suppose that there are $n + 1$ points on each line. Then we say that the plane has *order n*, e.g. the planes in Figs. 109(a), 110 are of orders 2, 3 respectively. It follows that there are $n + 1$ lines through each point of the plane. We now ask: How many points and lines are there in projective and affine planes of order n?

Theorem 47: A projective finite plane of order n contains $N = n^2 + n + 1$ points and $N = n^2 + n + 1$ lines.

Proof: Consider a line l and a point P not lying on l.

Join P to each of the $n + 1$ points on l, thus producing $n + 1$ lines through P. On each of these lie n distinct points, in addition to P (draw a diagram) ∴ the number of points in the plane Pl is

$$N = n(n + 1) + 1 = n^2 + n + 1.$$

Dually, the plane contains N lines.

Try to invent some variations of this proof. For instance, consider the set of possible lines through a given point.

Theorem 48: An affine finite plane of order n contains n^2 points and $n^2 + n$ lines.

Proof: The truth of the Theorem follows immediately from Theorem 47 when we remove from the projective finite plane one line containing $n + 1$ points.

Clearly, the 7-point and 13-point projective planes (with their affine specialisation) occur when $n = 2, 3$ respectively.

Remarks:

(i) One may prove (Exercise 10.1) that the number of points for the general finite geometry in [3] is

$$n(n^2 + n + 1) + 1 = n^3 + n^2 + n + 1.$$

Dually, this is also the number of planes.

(ii) In $[k]$, the number of points (and the number of hyperplanes) is $\dfrac{n^{k+1} - 1}{n - 1} = n^k + n^{k-1} + \ldots + n + 1$ (Exercise 10.7).

(iii) O. Veblen and W. H. Bussey ('Finite Projective Geometries', *Transactions Amer. Math. Soc.* 7 (1906) 241–259) denoted the k-dimensional projective geometry of order n by the symbol $PG(k, n)$ where n is a prime or a power of a prime, $n = p^r$. Thus the finite plane geometry of order 2 is represented by the symbol $PG(2, 2)$—or by $P_2(GF(2))$ in the alternative symbolism of § 11.1—while in [3] its symbol is $PG(3, 2)$. Similarly, $PG(2, 3)$ signifies the plane 13-point geometry. For some details about $PG(2, 5)$, consult Coxeter [28] pp. 92–101, 126–129.

Since there is a Galois field of order n for every $n = p^r$, it follows that there is a $PG(k, p^r)$ for every pair of integers k, r and for every prime p. No finite planes other than those (of prime or prime-power orders) discovered by Veblen and Bussey are known to exist; however, there is no general proof that they are the only ones possible. The problem of the existence of planes of non-prime-power order is discussed in § 52. Because of the association of finite spaces with Galois fields, Finite Geometry is often nowadays called *Galois Geometry*.

§ 46.5 Subplanes of a Finite Plane. Mathematical systems possess subsystems which are usually interesting objects of study in themselves. Familiar examples are those relating to algebraic systems, e.g. subgroups of groups. With these thoughts in mind, we consider a similar property of the projective plane which, initially, need not be finite.

Given a projective plane π, consider a set of points $\{P_i\}$ and among them a subset of lines $\{l_j\}$, both sets satisfying Axioms $PP1$–$PP3$. Making all the joins and intersections possible, we can construct other points and lines from $\{P_i\}$ and $\{l_j\}$. In this way we form a projective plane called the *subplane* of π *generated* by $\{P_i\}$ and $\{l_j\}$.

Let π be a finite projective plane. The following result, due to the American mathematician R. H. Bruck, relates the order of π to the order of a subplane.

Bruck's Theorem: If a plane of order n has a subplane of order m, then

$$n = m^2$$
$$\text{or } n \geqslant m^2 + m.$$

For a proof of this theorem, the reader is referred to Hall [45] pp. 398–399, or Pedoe [85] p. 91.

According to Bruck's Theorem, the only possible subplanes of a plane of order nine are planes of orders two and three.

To give perhaps a little more of the feeling for subplanes, we append one further theorem relating to them. It is due to the Soviet mathematician, Lydia Kopejkina ('Decompositions of projective planes', *Izv. Akad. Nauk USSR*. Ser. Mat. 9 (1945) 495–526 (in Russian)), and concerns free planes, which have been referred to in § 5.7.

Kopejkina's Theorem: Every subplane of a free plane is a free plane.

Kopejkina's Theorem may be compared with an analogous theorem in group theory which asserts

Every subgroup of a free group is a free group.

Further information about subplanes, free planes and Kopejkina's Theorem may be found in, say,

R. Sandler: 'On subplanes of free planes', *Can. J. Math.*, 16 (1964) 379–385,

L. Siebenmann: 'A characterization of free projective planes', *Pacific J. Math.*, 15 (1) (1965) 293–298,

and the articles by Hall and Skornyakov referred to at the beginning of Chapter 11.

§ 46.6 Applications of Finite Geometry. Two of the main applications of Finite Geometry are in Statistics and in Theoretical Physics and Cosmology. Statistical applications relate mainly to experimental design, e.g. in the design of experiments in agricultural research.

Finite Geometry has also been suggested as a framework for the description of certain classical mechanics, quantum mechanics and cosmological theories. Particularly active in this respect is a group of Finnish scientists, notably Järnefelt and Kustaanheimo, at the Astronomical Observatory, Helsinki. See, for example, the paper, 'An Observation on Finite Geometries' by G. Järnefelt and P. Kustaanheimo, presented to the 11th Scandinavian Mathematical Congress held at Trondjheim, Norway, in 1949.

§ 47 INCIDENCE MATRICES

Associated with any finite projective plane of order n is a certain *incidence matrix* $\mathbf{A} = (a_{ij})$ $(i, j = 0, 1, \ldots, N-1 = n^2 + n)$ whose elements are defined thus:

If the N point and lines of the plane are labelled $P_0, P_1, \ldots, P_{N-1}$ and $l_0, l_1, \ldots, l_{N-1}$, then

$$a_{ij} = \begin{cases} 1 \text{ if } P_i \in l_j \\ 0 \text{ if } P_i \notin l_j. \end{cases}$$

With the labelling of points and lines in Fig. 109(a), we derive, for $n = 2$,

$$\mathbf{A} = \begin{bmatrix} 1 & 1 & . & 1 & . & . & . \\ 1 & . & 1 & . & . & . & 1 \\ . & 1 & . & . & . & 1 & 1 \\ 1 & . & . & . & 1 & 1 & . \\ . & . & . & 1 & 1 & . & 1 \\ . & . & 1 & 1 & . & 1 & . \\ . & 1 & 1 & . & 1 & . & . \end{bmatrix}$$

whence, after simple calculation

$$\mathbf{AA}^T = \mathbf{A}^T\,\mathbf{A} = 2\,\mathbf{I} + \mathbf{S} \qquad (\alpha)$$

Where $\mathbf{A}^T$ is the transpose of $\mathbf{A}$, $\mathbf{I}$ is the unit 7×7 matrix and $\mathbf{S}$ is the 7×7 matrix all of whose elements are unity ($\mathbf{S}^2 = 7\mathbf{S}$). Essential incidence features of the finite plane are exhibited by this matrix.

Remarks:

(i) For n general (integer) this result becomes

$$\mathbf{AA}^T = \mathbf{A}^T\mathbf{A} = n\,\mathbf{I} + \mathbf{S}$$

for appropriate sizes of $\mathbf{A}$, $\mathbf{I}$, $\mathbf{S}$. ($\mathbf{S}^2 = (n^2 + n + 1)\mathbf{S}$.)

(ii) Different labellings of $l_1, \ldots, l_N$ will produce different matrices $\mathbf{A}$ but the result (α) will be true in each case.

§ 48 SYMMETRIC BLOCK DESIGNS. THE (v, k, λ) CONFIGURATION

A (v, k, λ) *configuration* is an arrangement of v objects (or *varieties*) into v sets (or *blocks*) such that

(i) every block contains exactly k objects;
(ii) every object occurs in k blocks;
(iii) every pair of objects occurs in exactly λ blocks $(0 < \lambda < k < v)$.

In Statistics, the (v, k, λ) configurations arise naturally in the analysis and design of experiments where they are known as *symmetrical balanced incomplete block designs* (or *SBIBD* for short). A (v, k, λ) configuration is a special case of a wider class of experimental designs, generally non-symmetrical, called *balanced incomplete block designs* (abbreviated as *BIBD* in Statistics literature). For further information on this statistical aspect, the reader may consult, for example, R. C. Bose: 'On the construction of balanced incomplete block designs', *Annals of Eugenics* 9 (1939) 353–399.

Construction of a simple symmetrical block design is not difficult. Call the 'objects' points and the 'blocks' lines. Then it is evident that the seven-point plane in Fig. 109(a) represents a (7, 3, 1) configuration, i.e. a (v, k, λ)configuration for which $v = 7$, $k = 3$, $\lambda = 1$. Consequently, we assert

a finite plane represents a symmetrical block design.

Now, it may be verified that, for the (v, k, λ) configuration,

$$\lambda = \frac{k(k-1)}{v-1}$$

Putting $\lambda = 1$, $k = n + 1$ $(n \geqslant 2)$ in this formula, we deduce that $v = n^2 + n + 1 = N$. Thus the (v, k, λ) configuration reduces to the finite projective plane of order n *when* $\lambda = 1$ (in accordance with *PP*1). Consequently,

a (v, k, λ) *configuration generalises a finite projective plane.*

Associated with a (v, k, λ) configuration is an incidence matrix. Listing the objects x_i and the blocks B_i $(i = 0, 1, 2, \ldots, v - 1)$ in rows and columns respectively, we form the $v \times v$ incidence matrix $\mathbf{A} = (a_{ij})$ $(i, j = 0, 1, \ldots, v - 1)$ defined:

$$a_{ij} \begin{cases} = 1 \text{ if } x_i \in B_j \\ = 0 \text{ if } x_i \notin B_j. \end{cases}$$

This incidence matrix describes the essential features of the (v, k, λ) configuration. Choosing $v = 4$, $k = 3$, $\lambda = 2$, for example we have the $v \times v$ matrix

$$\mathbf{A} = \begin{bmatrix} . & 1 & 1 & 1 \\ 1 & . & 1 & 1 \\ 1 & 1 & . & 1 \\ 1 & 1 & 1 & . \end{bmatrix}$$

whence

$$\mathbf{AA}^T = \mathbf{A}^T\mathbf{A} = \begin{bmatrix} 3 & 2 & 2 & 2 \\ 2 & 3 & 2 & 2 \\ 2 & 2 & 3 & 2 \\ 2 & 2 & 2 & 3 \end{bmatrix}$$

Generally, for unspecified v, k, λ we may prove

$$\mathbf{AA}^T = \mathbf{A}^T\mathbf{A} = (k - \lambda)\,\mathbf{I} + \lambda\,\mathbf{S}$$

i.e. the right hand side is a matrix having k in the principal diagonal and λ in all other positions.

Closely connected with the (v, k, λ) configurations are certain matrices associated with the name of Jacques Hadamard (1865–1963, French).

§ 49 HADAMARD MATRICES

A *Hadamard matrix H* of order n is an $n \times n$ matrix ($n = 1, 2$, or a multiple of 4):

(i) elements are all ± 1;

(ii) $\mathbf{HH}^T = n\,\mathbf{I}$ (whence $|\mathbf{H}| = \pm n^{\frac{n}{2}}$).

For example

$$\begin{bmatrix} 1 & 1 & 1 & 1 \\ -1 & 1 & -1 & 1 \\ 1 & 1 & -1 & -1 \\ -1 & 1 & 1 & -1 \end{bmatrix}$$

is a Hadamard matrix.

Consider the $v \times v$ matrix $2\mathbf{A} - \mathbf{S}$. Write $k - \lambda = n = (n + 1) - 1$

$$\begin{aligned} \therefore\ (2\mathbf{A} - \mathbf{S})(2\mathbf{A}^T - \mathbf{S}) &= 4\mathbf{AA}^T - 2(\mathbf{AS} + \mathbf{SA}^T) + \mathbf{S}^2 \\ &= 4(n\mathbf{I} + \lambda\mathbf{S}) - 2.2k\mathbf{S} + v\mathbf{S} \because \begin{cases} \mathbf{SA} = \mathbf{AS} = k\mathbf{S}, \\ \mathbf{A}^T = \mathbf{A} \end{cases} \\ &= 4n\mathbf{I} + (v - 4n)\,\mathbf{S}. \end{aligned}$$

Then

(1) $v = 4n$ shows that $(2\mathbf{A} - \mathbf{S})(2\mathbf{A}^T - \mathbf{S}) = 4n\mathbf{I} = v\mathbf{I}$ $\qquad (\mathbf{S}^T = \mathbf{S})$

i.e. $2\mathbf{A} - \mathbf{S}$ is a Hadamard matrix of order v.

(2) $v = 2k + 1 = 4n - 1$ (i.e. $k = 2n - 1$, $\lambda = n - 1$) shows that the $4n \times 4n$ matrix obtained from $2\mathbf{A} - \mathbf{S}$ by adding a row and a column of units is a Hadamard matrix, i.e. the $(4n - 1, 2n - 1, n - 1)$ configuration is equivalent to a Hadamard matrix of order $4n$.

In other words

certain (v, k, λ) configurations (called Hadamard configurations) are described by Hadamard matrices.

In particular, the (7, 3, 1) configuration (Fig. 109(a)) is represented by a Hadamard matrix of order 8, i.e. *the* (7, 3, 1) *configuration is a Hadamard configuration.*

Construction methods for various infinite classes of Hadamard matrices are to be found in the article by R. E. A. C. Paley, 'On orthogonal matrices', *Journal of Mathematics and Physics*, 12 (1933) 311–320.

Hadamard matrices may be used in the construction of error correcting codes (Mann [129] p. 71). Recent research—carried out at the Jet Propulsion Laboratory of the California Institute of Technology—into the problem of constructing Hadamard matrices and their use in optimum codes for (radio) communication through space is outlined in 'The Search for Hadamard Matrices', *Amer. Math. Monthly* 70(1) (1963) 12–17, by S. W. Golomb and L. W. Baumert.

Hadamard matrices have been constructed for all orders (= multiples of 4) < 236.

§ 50 DIFFERENCE SETS

An excursion is now made briefly into number theory. At first sight this may seem to bear little relation to finite planes. However, this impression will soon be proved false.

Residues $a_0, a_1, \ldots, a_n \bmod v$ ($v = n^2 + n + 1$) are said to form a *perfect difference set* of order n if $d \not\equiv 0 \bmod v$ is expressible just once in the form $d \not\equiv (a_i - a_j) \bmod v$ $(i, j = 0, 1, \ldots, n)$. (There are $n^2 + n$ differences $a_i - a_j$; the a_i, a_j are integers.)

For instance, the sets 0, 1, 3 mod 7 and 0, 1, 3, 9 mod 13 form perfect difference sets ($n = 2, 3$ respectively) with tables of differences as follows:

	0	1	3
0	0	1	3
1	6	0	2
3	4	5	0

	0	1	3	9
0	0	1	3	9
1	12	0	2	8
3	10	11	0	6
9	4	5	7	0

If $k(k - 1) = \lambda(v - 1)$ and residues $a_1, \ldots, a_k \bmod v$ are such that for $d \not\equiv 0 \bmod v$ the congruence $a_i - a_j \equiv d \bmod v$ has exactly λ solutions, i.e. can be expressed thus in λ different ways, then $a_1, \ldots, a_k$ is called a *difference set* v, k, λ. The sets $a + i, \ldots, a_k + i \bmod v$ with $i = 0, \ldots, v - 1$ will form the blocks of a symmetric design. For example

$$0 + i,\quad 1 + i,\quad 3 + i \qquad \bmod 7$$

and

$$0 + i,\quad 1 + i,\quad 3 + i,\quad 9 + \mathrm{i} \bmod 13$$

give the lines of the finite planes of orders 2, 3 respectively ($\lambda = 1$). Refer to the incidence tables § 46 for verification. The integers of any column in the arrays of § 46 will obviously form perfect difference sets also. An example of a difference set for $\lambda = 2$ is the set 0, 2, 3, 4, 8 mod 11. A set for which $\lambda = 1$ is called a *planar difference set*.

It has been demonstrated by James Singer ('A Theorem in Finite Projective Geometry and Some Applications to Number Theory', *Transactions Amer.*

Math. Soc. 43 (1938) 377–385) that for a difference set to exist we must have as a sufficient condition that n is a prime or a power of a prime, $n = p^r$. Thus no difference set exists for, say, $n = 6$, 10, 12 or 14. Singer further proved that

the construction of difference sets is precisely equivalent to the construction of cyclic planes

and that

every finite Desarguesian plane is cyclic,

i.e. can be given by a difference set. (A plane π is *cyclic* w.r.t. a collineation if the cyclic group generated by the collineation is transitive on the points of π.)

Writing in *Sankhyā* (the Indian Journal of Statistics) 11 (3, 4) (1951) 357–364, T. A. Evans and H. B. Mann showed that

no difference set for $\lambda = 1$ *exists, unless n is a prime power, for* $n \leqslant 1600$,

i.e. for a cyclic plane, with $n \leqslant 1600$, $n = p^r$.

[*Sankhyā* is a Sanskrit word with several shades of meaning such as 'calculation', 'number'.]

The interested reader may care to consult M. Hall: 'A Survey of Difference Sets', *Proceedings Amer. Math. Soc.* 7 (1956) 975—998.

§ 51 ORTHOGONAL LATIN SQUARES AND GRAECO-LATIN SQUARES

A *Latin square* of order n ($n \geqslant 2$) is a square array of n^2 elements such that each element occurs exactly once in each row and in each column. Euler, in his 'Recherches sur une nouvelle espèce de quarrés magiques', *Verh. v. h. Zeeuwsch Genootsch. der Wetensch*, Vlissingen, 9 (1782) 85–239, used Latin letters to label the elements of these squares. Hence the name. A second such square might be then conveniently labelled with Greek letters. Clearly, the Cayley multiplication table for a group must appear as a Latin square. However, the converse is not necessarily true, i.e. a Latin square does not necessarily represent the multiplication table of a group. See Ledermann [62] p. 14 for a counter-example of a Latin square in which the associative law fails. (A 'non-associative group' is called a *loop*. More details about loops are given in § 54.6.)

Two Latin squares are mutually *orthogonal* if, when superimposed, the ordered pairs associated with the product elements give each of the n^2 elements just once. The resulting square array is called a *Graeco-Latin square* or *Euler square* (Euler's '*quarré magique complet*'), for, in them, Euler wrote the first symbol in Latin and the second in Greek.

An example of a pair of mutually orthogonal Latin squares is the pair of arrangements, written as matrices, with their corresponding Graeco-Latin square

$$\begin{bmatrix} 1 & 2 & 3 \\ 2 & 3 & 1 \\ 3 & 1 & 2 \end{bmatrix}, \quad \begin{bmatrix} 1 & 3 & 2 \\ 2 & 1 & 3 \\ 3 & 2 & 1 \end{bmatrix} \text{ or } \begin{bmatrix} a & b & c \\ b & c & a \\ c & a & b \end{bmatrix}, \quad \begin{bmatrix} \alpha & \gamma & \beta \\ \beta & \alpha & \gamma \\ \gamma & \beta & \alpha \end{bmatrix}$$

$$\begin{bmatrix} 11 & 23 & 32 \\ 22 & 31 & 13 \\ 33 & 12 & 21 \end{bmatrix} \qquad \begin{bmatrix} a\alpha & b\gamma & c\beta \\ b\beta & c\alpha & a\gamma \\ c\gamma & a\beta & b\alpha \end{bmatrix}.$$

Are there any other mutually orthogonal pairs when $n = 3$? Examine also the case $n = 2$. Construct a pair of mutually orthogonal 4×4 Latin squares. For the record, we have the following orthogonal set of Latin squares of order 4.

$$\begin{bmatrix} 1 & 2 & 3 & 4 \\ 2 & 1 & 4 & 3 \\ 3 & 4 & 1 & 2 \\ 4 & 3 & 2 & 1 \end{bmatrix}, \quad \begin{bmatrix} 1 & 2 & 3 & 4 \\ 3 & 4 & 1 & 2 \\ 4 & 3 & 2 & 1 \\ 2 & 1 & 4 & 3 \end{bmatrix}, \quad \begin{bmatrix} 1 & 2 & 3 & 4 \\ 4 & 3 & 2 & 1 \\ 2 & 1 & 4 & 3 \\ 3 & 4 & 1 & 2 \end{bmatrix}.$$

Are there any others? (No.)

Dorwart [32] shows in detail how to construct this set of orthogonal Latin squares of order 4 from the perfect difference set 0, 1, 4, 14, 16 mod 21 and its associated incidence table for $PG(2, 4)$. There is thus a strong connection between orthogonal Latin squares and finite planes, a fact which will be exploited in § 52. (For the finite plane of order 4, see W. L. Edge: 'Some implications of the geometry of the 21-point plane', *Math. Zeitschrift* 87(2) (1965) 348–362.)

For the case $n = 5$, one may consult Martin Gardner ('How three modern mathematicians disproved a celebrated conjecture of Leonhard Euler', *Scientific American*, 201(3) (November, 1959) 181–188) where four mutually orthogonal Latin squares of order 5 are given. Coxeter [28] p. 94 gives the incidence table based on the perfect difference set 0, 1, 3, 8, 12, 18 mod 31 for the associated finite plane $PG(2, 5)$. Further information on this 31-point plane geometry may be found in the article by W. L. Edge ('31-point Geometry', *Math. Gaz.*, 39 (1955) 113–121).

Pickert [86] p. 293 details the orthogonal Latin squares when $n = 9$.

§ 51.1 Application to Statistical Experiments. Suppose a scientist wishes to conduct an agricultural experiment on pasture or crop improvement. The research problem involves determining the effect of (say) three different chemical fertilisers on three different varieties of grass. Naturally, the agronomist experimenter in his field trial wishes to minimise the variation of soil fertility in the field. To get the most efficient estimates of the effects of the fertilisers he should organise the grass varieties so that each variety occurs just once in every row and just once in every column in a square arrangement of the grassland. This is represented by a Latin square. Similarly, another Latin square will represent the fertilisers. Superposition of one array upon the other will produce the Graeco-Latin square of order 3 given above which will characterise the experiment.

Not only agricultural research benefits from the use of Latin squares but the design of experiments in many other fields (Biology, Medicine, Sociology) is based on them. As Gardner (reference above) remarks, an element of the array '. . . need not, of course, be a piece of land. It may be a cow, a patient, a leaf, a cage of animals, the spot where an injection is made, a period of time or even an observer or group of observers. The Graeco-Latin square is simply the chart of the experiment.'

Apart from the systematic elimination of differences resulting from randomised arrangements, one of the special advantages of the use of orthogonal Latin squares in scientific experiments is that a considerable amount of information has been discovered about them. Rigorous theories, mathematically-based, can then be applied to the statistical experiments.

A text dealing with experimental analysis and design is Mann [137], while a useful article on the squares themselves is Mann, 'The Construction of Orthogonal Latin Squares', *Annals Math. Statistics*, 13 (1942) 418–423. Read also R. A. Fisher: 'The Mathematics of Experimentation', *Nature* 142 (September 3, 1938) 442–443.

§ **51.2 Application to Finite Planes.** A *complete set* of mutually orthogonal Latin squares of order $n \geqslant 3$ is a set of $n-1$ such squares. This is the maximum possible number. Thus for $n = 3, 4, 5$ the complete set consists of 2, 3, 4 mutually orthogonal Latin squares respectively. Pickert's list [86] of 8 for $n = 9$ is therefore complete.

Using the properties of Galois fields, Bose, then at the Statistical Laboratory, Calcutta, showed that

a complete set can always be constructed when n is the power of a prime.

That is, a complete set exists for $n = 3, 4 = 2^2, 5, 7, 8 = 2^3, 9 = 3^2, 11, 13$, but not, according to this theorem, for $n = 6, 10, 12, \ldots$ Full details of the Bose article are: 'On the Application of the Properties of Galois Fields to the Problem of Construction of Hyper-Graeco-Latin Squares', *Sankhyā* 3(4) (1938) 323–338. Independently, and at much the same time, Stevens (of the Galton Laboratory, London) obtained the same result ('The completely orthogonalized Latin square', *Annals of Eugenics* 9 (1939) 82–93). Bose showed that when $n-1$ mutually orthogonal Latin squares of order n are superimposed we obtain a *completely orthogonalized or Hyper-Graeco-Latin Square.*

Bose, in the article quoted, and Stevens (independently), also proved that *the existence of a finite projective plane of order n is equivalent to the existence of a complete set of mutually orthogonal Latin squares of order n.*

Since an affine plane is readily derived from a projective plane (and vice-versa) we may rephrase Bose's result as

a finite affine plane of order n produces a complete set of Latin squares of order n.

Conversely

the existence of a complete set of mutually orthogonal Latin squares of order n automatically ensures the construction of a finite affine plane of order n.

§ 52 EXISTENCE OF FINITE PLANES

Implied in the preceding work is some doubt about the existence of planes of certain orders. Ordinarily we would expect the automatic existence of a finite plane for each integral value of n. The disenchantment of this expectation makes one of the exciting episodes in modern mathematical research.

§ 52.1 Euler's Problem of the 36 Officers. Non-Existence of a Plane of Order 6. Genesis of our problem begins with the French Army Officer, Gaston Tarry, who, while serving at Kouba, Algeria, addressed himself to the famous problem posed by Euler in 1782 ('*Recherches sur une nouvelle espèce de quarrés magiques*'): Given that (i) an officer may have one of six distinct ranks, and (ii) there are six regiments, no two officers from the same regiment having the same rank, how can 36 officers be arranged on parade in 6×6 squares so that no row or column will contain two officers of the same rank or of the same regiment? Euler predicted that it was impossible to do this.

Tarry, justifying Euler, showed that the problem was insoluble ('*Le problème des 36 officiers*', *Compte Rendu de l'Association Française pour l'Avancement de Science Naturel*, 2 (1901) 170–203). This means that no pair of mutually orthogonal squares for $n = 6$ can exist and hence a complete set does not exist. By implication, therefore, Tarry proved that

a finite projective plane of order 6 *does not exist.*

(Had it existed, there would have been $6^2 + 6 + 1 = 43$ points, and lines, in it. The corresponding affine plane would have had $6^2 = 36$ points and $6^2 + 6 = 42$ lines.)

Now that the Euler problem and its solution by Tarry have been indicated, let us savour something of the spirit of Euler when posing this problem. (Euler was by then blind and relied on dictating his thoughts to others.) Tarry begins his article thus:

'Le troisième mémoire d'Euler, publié dans les Comptes rendus de la Société des sciences de Flessingue, a pour titre: Recherches sur une nouvelle espèce de carrés magiques. Il commence ainsi:

Une question fort curieuse, qui a exercé pendant quelque temps la sagacité de bien du monde, m'a engagé à faire les recherches suivantes, qui semblent avoir ouvert une nouvelle carrière dans l'analyse, et en particulier sur la doctrine des combinaisons.

Cette question roulait sur une assemblée de 36 officiers de 6 différents grades et tirés de 6 régiments différents qu'il s'agissait de ranger dans un carré, de manière que sur chaque ligne, tant horizontale que verticale, il se trouvât 6 officiers tant de différents grades que de régiments différents. Or, après toutes les peines qu'on s'est données pour résoudre ce problème, on a été obligé de reconnaître qu'un tel arrangement est absolument impossible, QUOI QU'ON NE PUISSE PAS EN DONNER UNE DÉMONSTRATION RIGOUREUSE.

Nous allons donner la demonstration de cette impossibilité.'

§ **52.2 Euler's Conjecture** (1782, '*Recherches . . .*'):
There do not exist mutually orthogonal Latin squares of order n for $n \equiv 2$ *mod* 4.

Tarry's investigation bore this out for $n = 6$. Consult also R. A. Fisher and F. Yates: 'The 6 × 6 Latin squares', *Proc. Camb. Phil. Soc.*, 30(4) (1934) 492–507.

The falsity of Euler's conjecture is discussed in § 52.4.

§ **52.3 Bruck-Ryser Theorem.** No real progress was made until 1949 when a breakthrough was made by R. H. Bruck and H. J. Ryser ('The Non-existence of Certain Finite Projective Planes', *Can. J. Math.* 1(1) (1949) 88–93) who proved, using some deep number theory

> *Bruck-Ryser Theorem* (*Theorem 49*)*:* If $n \equiv 1, 2 \bmod 4$ there cannot be a plane of order n unless n is expressible as a sum of two integral squares, $n = a^2 + b^2$.

Actually Bruck and Ryser expressed their Theorem in a rather different, though equivalent, form. (Check with the original article by them and look at Hardy and Wright [46] p. 299.)

This theorem asserts the non-existence of infinitely many planes of orders $n = 6, 14, 22, \ldots$, but it says nothing about the case $n = 10\ (= 3^2 + 1^2)$ or $n = 18\ (= 3^2 + 3^2)$.

§ **52.4 Bose-Shrikhande-Parker Theorem.** One of the major breakthroughs in Mathematics in our time occurred only a few years ago when Euler's conjecture was decisively refuted by Bose and Shrikhande (then at the University of North Carolina). In their joint paper, 'On the Falsity of Euler's conjecture about the non-existence of two orthogonal Latin squares of order $4t + 2$', *Proceedings National Academy of Science, U.S.A.*, 45 (1959) 734–737, they constructed a Graeco-Latin square of order 22 ($\equiv 2 \bmod 4$; $\neq a^2 + b^2$). In fact, they showed the falsity of Euler's conjecture for an infinity of values of v $(= 4t + 2) \geqslant 22$. The method of constructing this square utilised the solution of the famous *Kirkman Schoolgirl Problem* (Kirkman, 'Schoolgirl problem', *The Lady's and Gentleman's Diary* (1850)): How can 15 girls go walking every day for seven days, in five rows of three each, so that any two girls walk in the same row exactly once during the seven days?

Soon afterwards E. T. Parker was able to construct a Graeco-Latin square of order 10. Because of its significance we reproduce it here

$$\begin{bmatrix} 00 & 47 & 18 & 76 & 29 & 93 & 85 & 34 & 61 & 52 \\ 86 & 11 & 57 & 28 & 70 & 39 & 94 & 45 & 02 & 63 \\ 95 & 80 & 22 & 67 & 38 & 71 & 49 & 56 & 13 & 04 \\ 59 & 96 & 81 & 33 & 07 & 48 & 72 & 60 & 24 & 15 \\ 73 & 69 & 90 & 82 & 44 & 17 & 58 & 01 & 35 & 26 \\ 68 & 74 & 09 & 91 & 83 & 55 & 27 & 12 & 46 & 30 \\ 37 & 08 & 75 & 19 & 92 & 84 & 66 & 23 & 50 & 41 \\ 14 & 25 & 36 & 40 & 51 & 62 & 03 & 77 & 88 & 99 \\ 21 & 32 & 43 & 54 & 65 & 06 & 10 & 89 & 97 & 78 \\ 42 & 53 & 64 & 05 & 16 & 20 & 31 & 98 & 79 & 87 \end{bmatrix}$$

Note the Graeco-Latin square of order 3 in the right hand bottom corner. It is not known whether this is a characteristic feature of 10×10 Graeco-Latin squares.

Not long afterwards, Bose, Shrikhande and Parker ('Further Results on the Construction of Mutually Orthogonal Latin Squares and the Falsity of Euler's Conjecture', *Can. J. Math.* 12(2) (1960) 189–203) proved the following important result:

Bose-Shrikhande-Parker Theorem (*Theorem 50*):
There exist at least two orthogonal Latin squares of any order $v > 6$.

Readers of this article will notice (p. 200) a Graeco-Latin square different from the one printed above.

In order for a finite plane of order 10 to exist, it would be necessary to find $9(= 10 - 1)$ mutually orthogonal Latin squares. This has not yet been done. Thus, the smallest finite plane whose existence is in doubt is that of order $n = 10$ ($n = 12$ is also uncertain).

§ 52.5 Computers and the Existence of Finite Planes. Now we know that a Galois field of prime or prime-power order determines a finite Desarguesian plane. Accordingly, there exist finite planes of orders

2, 3, 4, 5, , 7, 8, 9, , 11, , . . .

containing

7, 13, 21, 31, , 57, 73, 91, , 133, , . . .

points and lines respectively.

All these orders have been scrutinised to determine whether all planes of each order are Desarguesian, equivalently: whether there is only one plane of each order to within isomorphisms. For $n \leqslant 5$, it has been known for some time that there is essentially only one plane of each of these orders.

For $n = 7$, the breakthrough came with the discovery by W. A. Pierce (applying the classical Kirkman Schoolgirl Problem, a 'schoolgirl' being regarded as a point), that the diagonals of a complete quadrilateral are never concurrent when $n = 7$ ('The Impossibility of Fano's Configuration in a Projective Plane with Eight Points Per Line', *Proceedings Amer. Math. Soc.*, 4 (1953) 908–912). Using Pierce's result, Marshall Hall gave a proof of the theorem that any projective plane with eight points per line is Desarguesian ('Uniqueness of the Projective Plane with 57 Points', *Proceedings Amer. Math. Soc.* 4 (1953) 912–916).

In the case $n = 8$, Hall and his colleagues used the SWAC electronic digital computer at the Numerical Analysis Research Center in the University of Southern California, Los Angeles, to establish that there is only one plane of that order; hence it must be the Desarguesian one (M. Hall, J. D. Swift and R. J. Walker: 'Uniqueness of the Projective Plane of Order Eight', *Mathematical Tables and Other Aids to Computation*, 10 (1956) 186–194).

There are four known projective planes of order 9. They are

(i) the unique Desarguesian plane corresponding to the Galois field $GF(3^2)$;

(ii) three non-Desarguesian planes, viz:

- (a) the *Veblen-Wedderburn plane* (1907) co-ordinatisable by a Veblen-Wedderburn system, which is not a field—see Chapter 11 (§ 56) for definitions and references;
- (b) the *dual of the Veblen-Wedderburn plane*; and
- (c) the *Hughes plane* of order 9 (after D. R. Hughes, 1957)—see the reference at the end of this Chapter, Topic (9).(Hughes planes exist for orders p^{2r} where p is an odd prime and r is a positive integer. They cannot be co-ordinatised by Veblen-Wedderburn systems.)

It is natural to ask whether or not any other planes of order 9 exist. In 1959 Hall and colleagues carried out this search partly by hand and partly by machine (the SWAC computer), but could not discover any such planes (M. Hall, J. D. Swift and R. Killgrove: 'On Projective Planes of Order Nine', *Mathematical Tables and Other Aids to Computation*, 13 (1959) 233–246). The authors concluded: 'The failure to find, in this extensive search, any planes beyond those known for many years strengthens the conjecture that the four planes are a complete set for order 9. This number is the smallest order still in doubt as to the list of possible planes and also is the smallest order with more than a single plane.'

Computers have been employed unsuccessfully in attempts to discover the existence of more complicated planes. Hall, quoted in Gardner, reported with regard to Latin squares: 'Extensive searches on SWAC, . . . have failed to produce an orthogonal 10-by-10 pair. But even with more than 100 hours of high-speed search, the part of the possible cases tried is so microscopic that no conclusion may be drawn.' And Ryser ('Geometries and Incidence Matrices', *Contributions to Geometry*, *Slaught Memorial Paper No.* 4, Math. Assoc. of America (1955) 25–31) stated: 'For $n = 10$, there are something like $2^{10,000}$ possible candidates for an incidence matrix. Machines cannot cope with numbers of this magnitude.' Viewed against this knowledge, the achievement of Bose, Shrikhande and Parker is all the more astonishing.

For further information on the problem of the existence of the plane of order 10, the reader could consult E. T. Parker: 'Computer Investigation of Orthogonal Latin Squares of Order 10', *Proceedings of Symposia in Applied Mathematics* (American Mathematical Society) 15 (1963) 73–81. The electronic machine referred to in this article is the UNIVAC 1206 Military Computer.

The situation relating to the existence and types of projective planes of orders $n = 2, 3, \ldots, 16$ is set out in the accompanying Table:

Order ($= n$)	Does it exist?	Classification	Number of points (lines) $= n^2+n+1$
2	Yes	Desarguesian	7
3	Yes	Desarguesian	13
4	Yes	Desarguesian	21
5	Yes	Desarguesian	31
6	No (Tarry 1901)		
7	Yes	Desarguesian	57
8	Yes	Desarguesian	73
9	Yes	Desarguesian (1), non-Desarguesian (3)	91
10	Unknown		
11	Yes	Desarguesian	133
12	Unknown		
13	Yes	Desarguesian	183
14	No (Bruck, Ryser 1949)		
15	Unknown		
16	Yes	Desarguesian, non-Desarguesian	273

Table: Finite Planes of Orders $n = 2, 3, \ldots, 16$.

§ 53 OTHER CURRENT RESEARCH IN FINITE GEOMETRY

Besides developments in the areas already mentioned, there are many other currently fruitful fields for investigation. Just a few of these are briefly stated.

§ 53.1 Non-Euclidean Finite Planes. Several articles on finite Bolyai-Lobatschewsky planes have recently appeared. One which may be consulted is L. Graves: 'A finite Bolyai-Lobatschevsky plane', *Amer. Math. Monthly* 69 (1962) 130–132 which contains appropriate references. The chief distinguishing axiom for this type of finite plane is the Lobatschewsky one, namely

Through each point P not on a line l there pass at least *two* lines not meeting l.

Another article which contains further references and relates the non-Euclidean plane to Statistics and to *partial planes* (What are they? Refer to § 5.7) is
Esther Seiden: 'On a Method of Construction of Partial Geometries and Partial Bolyai-Lobachevsky Planes', *Amer. Math. Monthly* 73 (2) (1966) 158–161.

§ 53.2 Conics and Ovals. An *oval* in a projective plane of order n is a set of $n + 1$ points, of which no three are collinear. A non-singular *conic* is a set of $n + 1$ points (n odd) which can be represented by an equation of the second degree over an appropriate co-ordinate field. Some relevant articles on ovals and conics are
Bertil Qvist: 'Some remarks concerning curves of the second degree in a finite plane', *Annales Acad. Scient. Fenn.* (A, 1) 134 (1952) 27 pages;
Beniamino Segre: 'Ovals in a finite projective plane', *Can. J. Math.* 7 (1955) 414–416;
W. L. Edge: 'Conics and projectivities in a finite plane', *Can. J. Math.* 8 (1956) 362–382;
F. Buekenhout: '*Plans projectifs à ovoïdes Pascaliens*', *Archiv der Mathematik* 17 (1966) 89–93.

Ovals, and the previous topic (§ 53.1), are covered in the article by T. G. Ostrom: 'Ovals and finite Bolyai-Lobachevsky planes', *Amer. Math. Monthly* 69 (9) (1962) 899–901.

Two important theorems on ovals are those due to Segre (1955) and Buekenhout (1966):

Segre's Theorem: In a finite Desarguesian plane (of odd order) every oval is a conic.

Buekenhout's Theorem: If $\mathcal{O}$ is a Pascalian oval in a finite plane, then the plane is Pappian and $\mathcal{O}$ is a conic.

By a *Pascalian oval* is meant an oval for which Pascal's Theorem (Theorem 24) for an inscribed hexagon is valid.

It is worth noting that Segre's Theorem had been surmised as early as 1949 by Järnefelt and Kustaanheimo. Concerning the value of his Theorem, Segre wrote: 'This result fills up a gap in the finite congruence axiomatics set up by Kustaanheimo [1950], and has important implications if we accept the idea, advanced by Järnefelt [1951], of a possible connection between the physical world and the geometry of a finite linear space.'

Elementary properties of ovals were determined in 1952 by Qvist.

A *k-arc* is a set of $k > 3$ points of which no three are collinear. A line containing 2, 1, 0 points of a k-arc or oval is called a *secant*, *tangent* or *exterior line* respectively. Many interesting combinatorial results have been discovered concerning points and lines related to k-arcs and ovals. For example, we have the following result of Qvist:

Qvist's Theorem: In a finite plane of even order n, all the tangents of an $(n + 1)$-arc are concurrent.

Bose proved in 1947 that the k-arcs in a plane of order n contain at most

$$\begin{cases} n + 1 \text{ points } (n \text{ odd}) \\ n + 2 \text{ points } (n \text{ even}). \end{cases}$$

(These maximal k-arcs are called *ovals* by some authors, such as Segre.)

Research work on ovals and k-arcs is of course not restricted to Desarguesian planes, but relates also to non-Desarguesian planes, e.g. Hughes planes.

The study of k-arcs and ovals extends naturally to non-collinear sets of points in spaces of dimension $\geqslant 3$ leading to the concept of *k-caps* and *ovaloids*. Interested readers could consult the contributions by Segre in the Miscellaneous Research Articles and in the *Proceedings* of the Edinburgh Congress mentioned at the end of this Chapter.

§ 53.3 Galois Fields and Finite Planes. Associated with every prime power p^r there is, as we know, a finite Galois field $GF(p^r)$ with p^r elements ('marks') where p is a prime, r an integer. With these elements as co-ordinates, we are assured of the existence of finite planes of order p^r.

Next we use an elegant theorem of J. H. M. Wedderburn (1882–1948, Scottish-American) ('A theorem on finite algebras', *Transactions Amer. Math. Soc.* 6 (1905) 349–352):

Theorem 51 (*Wedderburn's Theorem*): A finite division ring is necessarily commutative and so is a finite field $GF(p^r)$.

[Another proof of Wedderburn's Theorem, due to E. Witt, is reproduced in Hall [45] pp. 375–376.]

Remembering (§§ 11, 13.3) that a division ring co-ordinatises a Desarguesian plane while a field implies the validity of Pappus' Theorem, we are led to the conclusion:

Planes of order p^r are both Desarguesian and Pappian.

It is a matter of numerical enumeration to verify that the number of possible triads from the field $GF(p^r)$, always excluding (0, 0, 0), is exactly $p^{2r} + p^r + 1$. Check in the cases (say) $p^r = 2, 3, 4, 5$.

§ **53.4 Collineations.** A study of collineations (Chapter 4) tells us much about the nature of a plane. In Chapter 11 it will be seen that there are fewer collineations in a non-Desarguesian plane than in a Desarguesian plane.

To appreciate something of the flavour of collineations in a finite plane, note the following two numerical results:

(1) A Desarguesian plane of order $n = p^r$ has a collineation group of order

$$rn^3(n^2 + n + 1)(n + 1)(n - 1)^2.$$

(2) (Singer): A Desarguesian plane of order n has a collineation group of order $n^2 + n + 1$ which is cyclic on the points and also on the lines.

A simple example of the first result is that, for $n = 2$, the 7-point Fano plane has a collineation group of order 168 (which, incidentally, is the second smallest group of non-prime order which is *simple*, i.e. possesses no invariant subgroups—the smallest non-prime order simple group (of order 60) is associated with the symmetrical rotations of a regular icosahedron and with the solution of a quintic equation). Coxeter [28] calculates by geometrical methods that the order of the collineation group when $n = 5$ is 372,000. Using his methods, we could without difficulty verify the answer given for the case $n = 2$.

For a summary of the contributions to the theory of collineations in a finite plane by Baer, Bruck, Gleason, Hughes, Ostrom, Parker, Singer and Wagner, the reader is referred to Hall [45]. In particular, Hughes has proved a theorem, which places restrictions on possible collineations in finite planes, similar to the restrictions of the Bruck-Ryser Theorem.

Information about the general theory of collineations in finite geometry may be found in Carmichael [19].

§ **53.5 Octaves.** A good deal of recent research into *Octave Geometry* has been carried out by a group of Dutch mathematicians (H. Freudenthal and others—see below). Octave Geometry is an algebraic extension of *Quaternion Geometry*, i.e. Projective Geometry over the quaternions, that is to say, in which each co-ordinate is a quaternion. (This, itself, is an extension of Complex Projective Geometry.)

An *octave* (or *Cayley number*) is an expression of the form $\mathbf{p} + \mathbf{q}\mathbf{e}$ where $\mathbf{p}, \mathbf{q}$ are quaternions and the symbol $\mathbf{e}$ is such that

$$\mathbf{p} + 0\mathbf{e} = \mathbf{p},\ 0 + \mathbf{q}\mathbf{e} = \mathbf{q}\mathbf{e}.$$

Addition and multiplication (based on quaternion operations) are defined, along with other details, in Artzy [4]. The *Cayley* (or *octave*) *algebra* so formed is both non-commutative and non-associative (Cayley [20] 1,p. 127).

For the allied geometry, consult

T. Springer: 'The Projective Octave Plane', *Indagationes Mathematicae*, 22 (1960) 74–101,

F. van der Blij: 'History of the Octaves', *Simon Stevin* (1961) 106–125.

§ **53.6 Group Theory and Block Design.** According to Marshall Hall (as conjectured by him at the International Group Theory Conference, in Canberra, 1965), there appears to be a relation between block designs (in particular, Steiner triple systems) and groups similar to that between projective planes and co-ordinate systems. For details of the connection of group theory with block designs consult the article by Hall in Kovács and Neumann [135].

§ **53.7 Summary.** Apart from its intrinsic geometric interest, the finite plane has been shown to be the focal point of many different areas of Mathematics, a situation which may be schematically represented below:

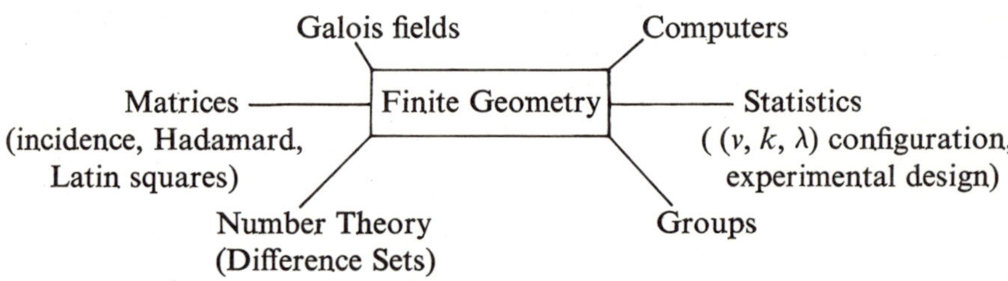

Ramifications of Finite Geometry

As a simple example of the interrelationships of many of the concepts introduced in this Chapter, we note that

the (7, 3, 1) configuration is simultaneously equivalent to

(i) a finite projective plane;
(ii) a Hadamard configuration;
(iii) a Steiner triple system;
(iv) a planar difference set.

Before leaving this fascinating topic, we mention a few general basic references on Finite Geometry, not already given: F. W. Levi [63]; L. Lombardo-Radice [67]; H. J. Ryser [94]; Beniamino Segre [97]. Additionally, consult the articles: M. Hall: 'A Survey of Combinatorial Analysis', *Surveys in Applied Mathematics IV*, Wiley (1958) 35–104; M. Hall: 'Block Designs', *Applied Combinatorial Mathematics*, Wiley (1964) 369–405; B. Segre: 'Galois Geometries', *Proceedings of the International Congress of Mathematicians, Edinburgh*, 1958, C.U.P. (1960) 488–499, and M. Hall: *Combinational Theory*, Blaisdell (1967); *Proceedings of the Projective Geometry Conference*, 1967, University of Illinois, Chicago.

Reluctantly, we have to leave the absorbing study of finite planes with the observation that they are a fruitful and rewarding area of research activity. Perhaps, as Ryser remarked in 1963 of Combinatorial Mathematics generally, the 'greatest truths are still to be revealed'.

Suggested topics for study

(1) Curves in Finite Geometry: Ovals, k-arcs.
(2) Bolyai-Lobatschewsky planes.
(3) Kirkman's Schoolgirl Problem.
(4) Special finite planes, e.g. $n = 5, 7, 8, 9$.
(5) Finite Geometry in three dimensions (E. J. F. Primrose: 'Quadrics in Finite Geometries', *Proceedings Cambridge Philosophical Soc.* 47(2) (1951) 299–304).
(6) Pascal's Theorem in finite planes (C. E. Rickart: 'The Pascal configuration in a finite projective plane', *Amer. Math. Monthly* 47 (1940) 89–96).
(7) Statistical aspects of Finite Geometry (Esther Seiden: 'A Theorem in Finite Projective Geometry and an Application to Statistics', *Proc. Amer. Math. Soc.* 1 (1950) 282–286; also the reference in (5) above).
(8) Hadamard matrices.
(9) Hughes planes (D. R. Hughes: 'A class of non-Desarguesian projective planes', *Can. J. Math.*, 9 (1957) 378–388; D. R. Hughes: 'Review of some results in collineation groups', *Proceedings of a Symposium in Pure Mathematics*, Amer. Math. Soc., 1 (1959) 42–55; Lombardo-Radice [67]).
(10) Kustaanheimo planes (B. Segre [97] pp. 305–306; P. Kustaanheimo: 'On the relation of congruence in finite geometries' *Mathematica Scandinavica* 5 (1957) 197–201).

(11) Inversive (= Möbius) planes (P. Dembowski and D. R. Hughes: 'On Finite Inversive Planes' *J. Lond. Math. Soc.* 40 (1965) 171–182).
(12) Steiner triple systems.

EXERCISES 10

1. Show that the number of planes of order n, and of points, in $PG(3, n)$ is $n^3 + n^2 + n + 1$.

2. Establish that the number of triangles in a plane of order n is

$$\frac{n^3(n^2 + n + 1)(n + 1)}{6}.$$

In $PG(3, 2)$,

3. How many lines (or planes) pass through each point?

4. How many lines are there in the whole of the space?

5. How many planes are there through each line?

6. Answer the questions of Exercises 10.3, 10.4 for $PG(3, n)$.

7. Show that $PG(k, n)$ contains $n^k + n^{k-1} + \ldots + n + 1 = \dfrac{n^{k+1} - 1}{n - 1}$ points.

8. Verify Theorem 4 in $PG(3, 2)$.

9. In $PG(2, n)$ how many points are left invariant by
(i) an elation;
(ii) a homology?

10. Verify that 0, 1, 4, 14, 16 mod 21 form a perfect difference set.

11. Establish that 0, 1, 3, 8, 12, 18 mod 31 form a perfect difference set.

12. An 18th century card puzzle, quoted by Gardner is as follows: Aces, kings, queens and jacks are taken from a deck and arranged in a square so that every row and column will contain all four values and all four suits. Can you solve this problem?

[Card problems of this nature appear to pre-date Euler (1782) and are probably traceable to Claude Gasper Bachet de Méziriac (1581–1638, French) in 1624. Have a look at: 'Who discovered the Latin Square?' *The American Statistician*, 11 (4) (1948) 13, by M. G. Kendall (then Statistician to the United Kingdom Chamber of Shipping), who surmised that Bachet's problem may have originated in the early 14th century.]

13. Show that the 13-point plane, in its affine specialisation, contains four sets of three parallel lines.

14. Prove the incidence matrix result: $\mathbf{AA}^T = \mathbf{A}^T\mathbf{A} = 2\mathbf{I} + \mathbf{S}$ (Consult Hall [45] p. 395, or Pedoe [85] p. 90 if in difficulty.)

15. Is $\begin{bmatrix} 1 & 1 & 1 & 1 \\ 1 & 1 & -1 & -1 \\ 1 & -1 & 1 & -1 \\ 1 & -1 & -1 & 1 \end{bmatrix}$ a Hadamard matrix? Try to construct some other Hadamard matrices.

CHAPTER 11

TERNARY RINGS AND PROJECTIVE PLANES

'This ternary ring . . . reduces to the usual co-ordinate ring in an analytic plane, but, without conditions of Desarguesian type, it will have very few of the ordinary ring properties of a skew-field.'
(Marshall Hall, Jr.)

HOMOGENEOUS co-ordinates provide us with an entirely adequate description of the projective plane. Pappus' Theorem, the proof of which involves the commutative law of multiplication, suggests that algebraic structures relate directly to geometrical figures, and conversely. Underlying this concept is the possibility that the projective plane may be greatly richer in geometrical wealth than we have hitherto imagined.

This, in fact, turns out to be true for, by a suitable algebraic treatment, various projective planes closely related to the general projective plane are shown to exist. The algebraic apparatus which so powerfully reveals the latent geometry of the plane is based on the concept of a ternary ring, an idea originating with the American mathematician, Marshall Hall, ('Projective planes', *Transactions Amer. Math. Soc.* 54 (1943) 229–277. In conjunction with this article consult also M. F. Smiley: 'A remark on a theorem of Marshall Hall', *Proceedings. Amer. Math. Soc.* 1 (1950) 342–343; M. Hall: 'Correction to "Projective planes"', *Transactions Amer. Math. Soc.* 65 (1949) 473–474. Another valuable approach is that of Emil Artin by way of homomorphisms (References: Artin [3], Pedoe [85]).

Basic material on recent research work done on ternary rings may be found in several sources, e.g. Artzy [4], Blumenthal [12], Hall [45], Heyting [49], Pedoe [85], Segre [97] and Skornyakov (*'Proektivnye ploskosti'*, *Uspehi Matematičeskih Nauk* 6(6) (1951) 112–154 (in Russian); [English translation (by K. A. Hirsch) 'Projective planes', *Amer. Math. Soc.*, *Translations* Series 1, Vol. 1. 'Algebra' (1962) 51–107]. Two other papers which ought to be consulted are those by R. H. Bruck ('Recent Advances in the Foundations of Euclidean Geometry') and M. Hall ('Finite Projective Planes'), both of which appeared in *Contributions to Geometry*, *Slaught Memorial Paper No.* 4, Math. Assoc. of America (1955) · 2–17, 18–24, respectively. Consult also the pamphlet by H. G. Forder, 'Co-ordinates in Geometry', Auckland University College, Math. Series No. 1 (1953).

Rather than attempt to reproduce material which is already well presented in these texts, I propose to limit my scope by proceeding in the following way:

(1) initially, the concept and basic properties of a ternary ring will be introduced;
(2) a major piece of theory illustrating the application of previous knowledge (the minor Desargues' Theorem) to ternary rings will follow;
(3) finally, an outline, with appropriate references, of the main development of ternary rings in relation to projective planes will be given.

It is hoped that this modest glimpse of an exciting development, one of the significant advances in recent Mathematics, will inspire the reader to study this work in depth.

§ 54 TERNARY RINGS

§ 54.1 Introduction of Co-ordinates. As the raw material for the construction of a co-ordinate system in the projective plane π defined by Axioms *PP*1–*PP*3, we consider

(i) a set S of elements $S = \{0, 1, a, b, \ldots, m, \ldots, x, y, \ldots\}$;
(ii) four points O, X, Y, U in the plane such that no three are collinear.

No assumptions are made about the elements (numbers) of the set S from which co-ordinates are chosen.

Call XY the *line at infinity*, l_∞.

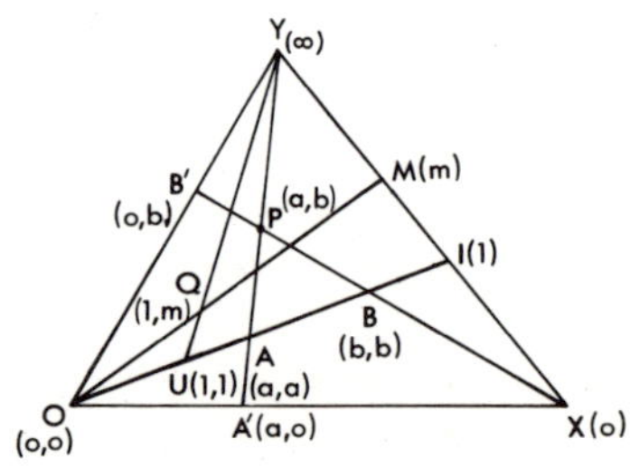

FIG. 112. Ternary ring co-ordinates

Assign co-ordinates as follows (Fig. 112):

(a) points *not* on l_∞: $O(0, 0)$
$U(1, 1)$
$A(a, a)$
$B(b, b)$

	$P(a, b)$	$(= YA \cap XB)$
	$A'(a, 0)$	$(= OX \cap YA)$
	$B'(0, b)$	$(= OY \cap XB)$
	$Q(1, m)$	$(= YU \cap OM)$
(b) points on l_∞:	$X(0)$	
	$I(1)$	$(= OU \cap XY)$
	$M(m)$	$(= OQ \cap XY)$
	$Y(\infty)$.	

Thus points not on l_∞ are represented by means of an ordered pair of numbers, say (x, y) in general where $x, y \in S$. Points on l_∞, on the other hand, are (except for Y) represented by a single number $\in S$, and Y is assigned the symbol ∞. A single co-ordinate such as (m) may be thought of intuitively as a slope.

§ 54.2 Equations of Particular Lines. Certain important lines in the plane are given equations as below

unit line OU	$: y = x$	
any line through Y	$: x = c$	(l_∞ excepted)
any line through X	$: y = d$	(l_∞ excepted)
OY	$: x = 0$	
OX	$: y = 0$.	

§ 54.3 Introduction of Addition and Multiplication. Algebraic operations, called addition and multiplication, involving the co-ordinates will now be defined in terms of the lines of the plane.

If (x, y) is any point other than I on the line $B'I$ joining $(0, b)$ and (1), we *define* the binary operation of *addition* by writing

$$y = x + b$$

which we take as the equation of the line. Addition is thus determined by the system of lines parallel to $y = x$, i.e. by the pencil of lines through (1).

If (x,y) is a point on the line OM joining $(0, 0)$ and (m) (but not at M itself), we *define* a binary operation of *multiplication* by putting

$$y = xm$$

and take this to be the equation of the line. Multiplication is thus determined by the pencil of lines through (0, 0). There is nothing special about the choice here of the order xm (rather than mx) though it is the usual one occurring in the literature. Equally well, we could have taken $y = mx$ with appropriate subsequent modifications.

Elements 0, 1 have the usual properties

$$0 + a = a + 0 = a$$
$$0m = m0 = 0$$
$$1m = m1 = m$$

(which the reader may easily verify. (Exercises 11.1–11.3.)).

§ 54.4 Introduction of a Ternary Operation. Any line not through Y must intersect l_∞ in some point (m) and OY in some point $(0, b)$. Let (x, y) be a point of this line.

Define a *ternary operation*

$$y = x{\cdot}m * b$$

and take this as the equation of the (general) line joining (m) and $(0, b)$.

Variations of the notation $\cdot$ and $*$ suggest themselves. Hall, for instance, uses $\cdot$ and $\circ$ as his notation. A useful alternative notation for the ternary operation is $T(x, m, b)$, so that the equation of the general line is $y = T(x, m, b)$.

Both addition and multiplication, as defined above, are special cases of the ternary operation. In fact

$$x + b = x{\cdot}1 * b \equiv T(x, 1, b)$$

while

$$xm \quad = x{\cdot}m * 0 \equiv T(x, m, 0).$$

Under this operation, the set S becomes a (planar) *ternary ring* R, sometimes called *Hall's ternary ring*.
Thus it appears that the plane can be represented by a ternary ring R whose ternary operation possesses certain properties. Conversely, a ternary ring having these properties determines a plane uniquely.

Note:

(i) Every line of the plane except l_∞ and lines $x =$ const., has an equation expressible in terms of the operation of R. No attempt has been made to assign an equation to l_∞.

(ii) If $a{\cdot}b * c = ab + c$, then the ring is said to be *linear*.

(iii) $y = x{\cdot}m * b$ generalises the equation $y = xm + b$ in the real Euclidean plane w.r.t. an orthonormal co-ordinate system. (Refer to Artzy [4] for a diagram.)

§ 54.5 Ternary Ring.

Theorem 52: For every choice of four points X, Y, O, U in a plane, no three collinear, there is determined a ternary ring R whose elements include a zero 0 and a unit $1 \neq 0$. The ternary operation $x \cdot m * b$ satisfies the following laws:

$$\begin{cases} TR\ 1. & 0 \cdot m * b = a \cdot 0 * b = b \\ TR\ 2. & 1 \cdot m * 0 = m \cdot 1 * 0 = m \\ TR\ 3. & \text{Given } a, m, b, \exists \text{ exactly one } z \in S\colon a \cdot m * z = b. \\ TR\ 4. & \text{If } b_1, b_2, m_1 \neq m_2, \text{ are given}, \exists \text{ a unique } x \in S\colon x \cdot m_1 * b_1 \\ & = x \cdot m_2 * b_2. \\ TR\ 5. & \text{If } c_1, c_2, a_1 \neq a_2 \text{ are given}, \exists \text{ a unique pair } m, b \in S\colon \\ & a_1 \cdot m * b = c_1,\ a_2 \cdot m * b = c_2. \end{cases}$$

(The symbol $\exists$ means 'there is (are)'.)

Proof: Construct a ternary ring based on X, Y, O, U with an operation $x \cdot m * b$.

*TR*1, *TR*2 are then consequences of the definition of a ternary ring.

*TR*3 says that the line joining (m) and (a, b) intersects OY in a unique point $(0, z)$. (*PP*1, *PP*2 have been invoked here.)

*TR*4 says that two lines $y = x \cdot m_1 * b_1$ and $y = x \cdot m_2 * b_2$ with different slopes intersect in a unique point (x, y) $(\notin l_\infty)$. (*PP*2 has been used.)

*TR*5 says that if (a_1, c_1) and (a_2, c_2) $(a_1 \neq a_2)$ are two points $\notin l_\infty$ then there is a unique line of the form $y = x \cdot m * b$ passing through them. (*PP*1 has been used.)

Note:

(i) The meaning of this Theorem is that a projective plane, defined by the Axioms *PP*1–*PP*3, can be represented by a ternary ring R whose ternary operation obeys certain rules.

(ii) Conversely, a ternary ring R satisfying *TR*1–*TR*5 uniquely defines a projective plane, i.e. satisfies Axioms *PP*1–*PP*3 (Exercise 11.4).

(iii) Properties of a field which may be lacking in a ternary ring, as defined, are the commutative laws of addition and multiplication, the associative laws of addition and multiplication, the distributive laws and the existence of multiplicative inverses. Nor has the ternary ring, as defined, all the properties which are usually embodied in the structure of a ring. For this reason, the name *ternary* is a useful alternative to 'ternary ring'.

It is our intention to impose successively certain geometrical restrictions of Desarguesian type and to observe their effects on the structure of the ternary ring. To facilitate subsequent reference, we assemble here the symbolism for those Desarguesian theorems which are of significance for the ternary ring:

(i) in the projective plane

Δ = the (general) Desargues' Theorem (Fig. 5(a) in [2])

δ = the minor Desargues' Theorem (Fig. 5(b) in [2]);

(ii) in the affine plane

Δ_a = the affine (general) Desargues' Theorem (Fig. 113(a), Exercise 8.53(ii))

δ_a = the affine minor Desargues' Theorem (Fig. 113(b), Exercise 8.53(iii)).

When Δ_a, δ_a are valid, we may say that Δ, δ are satisfied *affinely*.

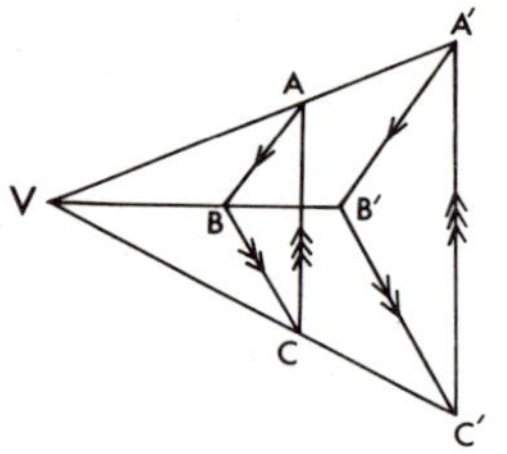

FIG. 113. (a) Affine Desargues' theorem Δ_a

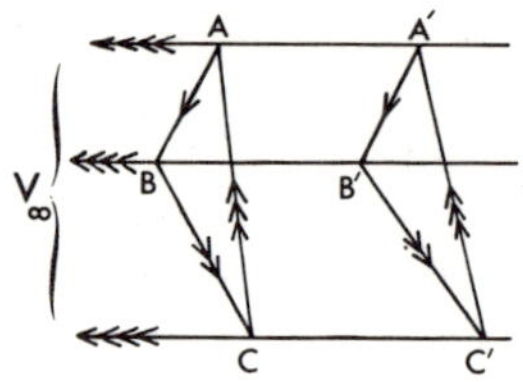

FIG. 113. (b) Affine minor Desargues' theorem δ_a

§ **54.6 Loops.** At this stage, we introduce the concept of a loop, an algebraic system having a looser structure than a group. Loops will be required when we impose more restrictions on the projective plane.

Definition: A *loop* is a set of elements $S = \{a, b, c, \ldots\}$ and a binary operation denoted by $\circ$ with the properties

$$\begin{cases} L1: & S \text{ is closed under } \circ, \text{ i.e. } a \circ b \in S \text{ for all } a, b \in S. \\ L2: & \text{In the equation } a \circ b = c, \text{ any two of } a, b, c\ (\in S) \text{ determine the third uniquely.} \\ L3: & \text{For all } a \in S, \exists \text{ an identity } e: e \circ a = a \circ e = a. \end{cases}$$

With the help of § 54.3, we may verify that $L1$–$L3$ are valid for

(i) S under addition ($e = 0$) and

(ii) S (excluding 0) under multiplication ($e = 1$),

i.e. $\begin{cases} S \text{ is an additive loop} \\ S \text{ (excluding 0) is a multiplicative loop,} \end{cases}$

or, as we may say, following Hall [45],

$\begin{cases} \text{addition is a loop} \\ \text{multiplication (excluding 0) is a loop.} \end{cases}$

We shall need these results in §§ 55, 56.

An associative loop is a group.

§ 55 LINEARITY AND THE MINOR DESARGUES' THEOREM

Naturally, we want lines to be represented by linear equations, i.e. we desire to establish the linearity of the ternary operation $x.m*b$, i.e. we wish to prove that $x.m*b = xm + b$, which expresses the ternary operation as a combination of two binary operations. In the process, it transpires that the minor Desargues' Theorem is involved, i.e. that the centre of perspective of two triangles lies on the axis of perspective. This means that we are investigating the effect of elations (§ 16) in terms of the ternary ring which co-ordinatises the plane.

A careful analysis of the method is undertaken below. For the original, consult Hall [45] pp. 360–361, or Hall, 'Projective planes', pp. 264–265.

Assume now that π is (Y, l_∞) transitive (§ 17). Recall (Exercise 4.11) that this implies that δ_a is valid for vertex Y and axis l_∞.

Theorem 53 (Linearity Theorem): A plane π is (Y, l_∞) transitive if, and only if, the ternary ring R which co-ordinatises it has the properties
(i) $a{\cdot}m*b = am + b$ (i.e. the ring is linear);
(ii) addition is a group.

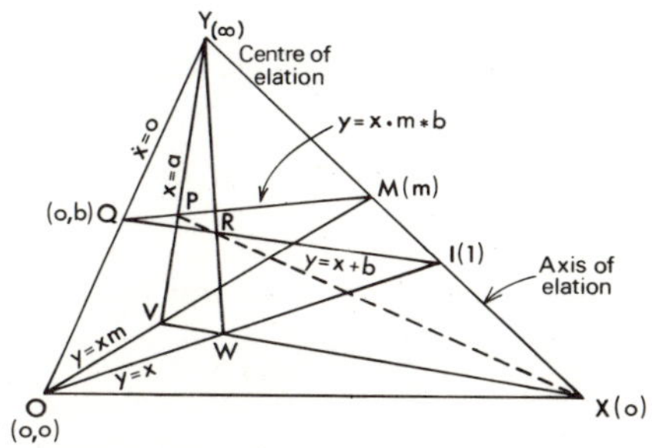

FIG. 114. Linearity theorem

Proof:
(i) Consider the diagram (Fig. 114) in which the points O, Q, X, Y, I, M are given, and the lines whose equations are shown are also given.
On MQ, take $P \equiv (a, a{\cdot}m*b)$.
Now $V \equiv (a, am)$ $(= OM \cap YP)$
$\therefore$ VX has equation $y = am$
$\therefore$ $W \equiv (am, am)$ $(= OI \cap VX)$
$\therefore$ YW has equation $x = am$

$\therefore\ R \equiv (am, am + b)\ (= YW \cap QI)$
$\therefore\ RX$ has equation $y = am + b$

We wish to show that P, R, X are collinear. From this, (i) will follow immediately.

Let α be the (Y, l_∞) collineation (= elation) which leaves invariant all points on l_∞ and all lines through Y.

Thus, in particular, α leaves YPV $(x = a)$ and YRW $(x = am)$ invariant. Also, let $O\alpha = Q$, i.e. $(0, 0)\ \alpha = (0, b)$.

Under the elation α then, we have

$$\begin{array}{lrcll} & OM & \xrightarrow{\alpha} & QM & \\ & OI & \longrightarrow & QI & \\ \therefore & V & \longrightarrow & P & \\ & W & \longrightarrow & R & (\therefore\ OWI \to QRI) \\ \text{and} & X & \longrightarrow & X & \\ \therefore & VWX & \longrightarrow & PRX & \end{array}$$

(i.e. P, R, X are collinear $\therefore$ elations take lines into lines)

$$\begin{array}{rl} \text{i.e. } y = am \longrightarrow & y = am + b \\ \therefore \qquad & P = (a, am + b) \\ & \quad = (a, a{\cdot}m * b) \text{ given} \end{array}$$

$$\therefore\ a{\cdot}m * b = am + b$$

$$\therefore\ \delta_a \Rightarrow \textit{linearity of the ternary operation.}$$

Geometrically, we may prove P, R, X collinear as follows:

$$PQR \overset{Y}{\overline{\wedge}} VOW$$

$\therefore\ PQ \cap VO$, $QR \cap OW$, $PR \cap VW$ are collinear
i.e. M, I, $PR \cap VW$ are collinear, by Desargues' Theorem. This implies that PRX is a line.

(ii) We may tabulate the effects of α on various points and lines thus:

$$\begin{array}{lrcll} & y = x & \xrightarrow{\alpha} & y = x + b & \text{(proved above)} \\ & x = c & \longrightarrow & x = c & ((Y, l_\infty) \text{ transitivity}) \\ \therefore & (c, c) & \longrightarrow & (c, c + b) & \\ \therefore & y = c & \longrightarrow & y = c + b & \\ & x = a & \longrightarrow & x = a & ((Y, l_\infty) \text{ transitivity}) \\ \therefore & (a, c) & \longrightarrow & (a, c + b) & \end{array}$$

$$\text{Similarly } (u, v) \xrightarrow{\beta} (u, v + d)$$

where β is the (Y, l_∞) collineation (elation) determined by

$$(0, 0)\,\beta = (0, d).$$

$$\begin{aligned}
\text{Now } [\,(0, 0)\,]\,\alpha\beta &= [\,(0, 0)\,\alpha]\,\beta \\
&= (0, b)\,\beta \\
&= (0, b + d) \\
\therefore\ [\,(a, c)\,]\,\alpha\beta &= [a, c + (b + d)\,] \\
\text{But also } [\,(a, c)\,]\,\alpha\beta &= [a, c + b]\,\beta \\
&= [a, (c + b) + d] \\
\therefore\ c + (b + d) &= (c + b) + d
\end{aligned}$$

i.e. the associative law of addition is satisfied,
i.e. addition is a group (i.e. an associative loop—refer to § 54.6).

Note:

(i) The diagram representing the associativity of addition is called the *Reidemeister figure* in honour of the modern German mathematician, Kurt Reidemeister.

(ii) The Converse of Theorem 53 is left as an exercise (Exercise 11.6).

Our immediate task has been completed. By means of a ternary ring, the basis has been laid for the co-ordinate representation of π. As the story of the subsequent development of the interaction between the algebraic structure of the ternary ring and the geometric structure of the plane has been fully told elsewhere—see the references at the beginning of this Chapter—we reluctantly content ourselves with a rough outline of this fascinating story.

§ 56 VEBLEN-WEDDERBURN PLANES

Next we investigate the conditions imposed on the ternary ring by the existence of a translation plane w.r.t. l_∞, i.e. (§ 17) a plane π which is (C, l_∞) transitive for all $C \in l_\infty$ and not transitive just for the case $C = Y$ (as in the Linearity Theorem).

Plainly this is a more restrictive condition than any previously imposed. Consequently we might expect the structure of R to be tightened.

Definition: A *Veblen-Wedderburn system* (or *Veblen-Wedderburn ring*) is an algebraic system such that

- $VW1$. Addition is an Abelian group.
- $VW2$. Multiplication, excluding 0, is a loop (§ 54.6).
- $VW3$. $(a + b)\,m = am + bm$.
- $VW4$. $xr = xs + t$ has a unique solution x in S $(r \neq s)$.
- $VW5$. $a{\cdot}m * b = am + b$.

This definition tells us that a Veblen-Wedderburn system is a special type of linear ternary ring. Such a system was first described by O. Veblen and J. H. M. Wedderburn ('Non-Desarguesian and non-Pascalian geometries', *Transactions Amer. Math. Soc.* 8 (1907) 379–388).

Theorem 54: A plane π is a translation plane (= *Veblen-Wedderburn plane*) w.r.t. l_∞ if, and only if, the corresponding ternary ring is a Veblen-Wedderburn system.

Proof: See Hall [45] pp. 362–364, Pedoe [85] pp. 127–130.

Note:

(i) Theorem 54 says that the co-ordinates of points in a Veblen-Wedderburn plane (translation plane) belong to a Veblen-Wedderburn system.

(ii) Notice that $VW1$ requires the commutative law of addition. This depends on a theorem due to R. Baer ('Homogeneity of projective planes' *Amer. J. Math.* 64 (1942) 137–152):

Baer's Theorem: The translation group of elations is Abelian.

Refer to Exercise 4.18 for the definition of a translation group.

(iii) The diagram for the commutativity of addition is sometimes called the *Thomsen figure*, in honour of the modern German mathematician, Gerhard Thomsen.

(iv) A Veblen-Wedderburn plane, by its definition, increases the number of collineations possible in a plane. It has more structure than the projective plane of Theorem 53, just as a Veblen-Wedderburn system has more structure than the ternary ring of Theorem 53.

Our ternary ring may still lack the following properties of a field: the commutative law of multiplication, the associative law of multiplication, left distributivity, and multiplicative inverses (i.e. 'division').

§57 MOUFANG PLANES

Of these remaining deficiencies, the left distributive law and the existence of multiplicative inverses are the next ones remedied in the natural development of the theory.

Using a pencil of lines through Y, we may show (e.g. Hall [45] pp. 367–369) that the following new properties of the co-ordinates (in addition to those in Theorems 53, 54) are true:

(i) $a(s + t) = as + at$;

(ii) inverses a^{-1} exist: $aa^{-1} = 1 = a^{-1}a\ (a \neq 0)$;
(iii) $a^{-1}(ab) = b$.

Expressed in other words, these properties require the plane to be a translation plane for every line passing through Y including l_∞, i.e. they require more than the restrictive form δ_a, but they do not require δ yet.

Working through the properties of the co-ordinates in a ternary ring, one may establish that

$$[y(zy)]x = y[z(yx)]$$

for any three numbers $x, y, z \in S$. This is called the *Moufang identity*.

Putting $z = 1$, we have immediately

$$(yy)\,x = y\,(yx)$$

which is called the *left alternative law*. Observe that this is in actuality a weakened form of the associative law of multiplication which is a goal we still seek.

An algebraic structure possessing all the previous properties and a right alternative law is called an *alternative division ring*, i.e. an alternative division ring is a special ternary ring such that

(i) addition is an Abelian group;
(ii) $(a + b)m = am + bm$;
(iii) $a(s + t) = as + at$;
(iv) $aa^{-1} = a^{-1}a = 1\ (a \neq 0)$;
(v) $a^{-1}(ab) = b$;
(vi) $(ba)a^{-1} = b$;
(vii) $a(ab) = (aa)b\ [= a^2b]$ (left alternative law);
(viii) $(ba)a = b(aa)\ [= ba^2]$ (right alternative law.)

That is, an alternative division ring is a special non-associative skew-field (i.e. commutative and associative laws of multiplication are lacking) which obeys the two alternative laws. An alternative division ring is, of course, a special Veblen-Wedderburn system. Properties (vi) and (viii) are the only new properties; (viii) is deducible from (vi) just as (vii) is deducible from (v).

Definition: A *Moufang plane* is a translation plane for *every* line in the plane. (Some authors refer to a Moufang plane as a *microdesarguesian plane*.)

This requires the validity of the minor Desargues Theorem δ for every line of the plane.

Theorem 55: A plane is a Moufang plane if, and only if, every co-ordinatising ternary ring is (1) linear;
(2) an alternative division ring.

Proof: Consult Hall [45] pp. 370–371, Pedoe [85] pp. 138–139.

Note:

(i) Moufang planes are named after the contemporary German mathematician, Ruth Moufang (1906–), who first studied them in 1933 (Ruth Moufang: '*Alternativkörper und der Satz von vollständigen Vierseit*', *Abhandlungen Math. Seminar, Hamburg* 9 (1933) 207–222). She established the Moufang identity in 1934.

(ii) Because of the relationship between a Moufang plane and the alternative laws, a Moufang plane is also called an *alternative plane*. From its definition, a Moufang plane is a more restricted type of Veblen-Wedderburn plane and is much richer in collineations. For a specific example of a Veblen-Wedderburn plane which is not an alternative (Moufang) plane, we borrow from Skornyakov (p. 85 of his article quoted in the remarks preceding § 54) to whom the illustration was communicated by G. Kleĭnerman.

Consider the set of all expressions $a + be_1 + ce_2 + de_3$ where a, b, c, d belong to the field of real numbers, and e_1, e_2, e_3 are 'units' with the following multiplication table

1	e_1	e_2	e_3
e_1	-1	e_3	$-2e_2$
e_2	$-e_3$	-1	$3e_1$
e_3	$2e_2$	$-3e_1$	-1 .

Details of the calculations which verify that this is indeed a Veblen-Wedderburn system may be found in Blumenthal [12] pp. 137–139. However

$$\begin{aligned} & (e_1e_1)e_3 = -e_3 \\ \text{while}\quad & e_1(e_1e_3) = e_1(-2e_2) \\ & \qquad\quad = -2e_1e_2 \\ & \qquad\quad = -2e_3 \\ \text{i.e.}\quad & (e_1e_1)e_3 \neq e_1(e_1e_3) \end{aligned}$$

i.e. the left alternative law does not hold.

Thus the plane of the Kleĭnerman example is a Veblen-Wedderburn plane but not a Moufang plane. Try to establish other properties of the Kleĭnerman plane.

Without going into the matter at this stage, we might mention that a good deal of important research into rings of characteristic 2 and characteristic different from 2 has recently been carried out by the Americans, Erwin Kleinfeld and R. L. San Soucie, and the Russian, Leo Skornyakov.

By the *characteristic of a ring* we mean the smallest integer $p \geqslant 2$ such that, if a is any element of the ring, $pa = 0$. If no such integer exists then the ring has *characteristic* 0. Similar remarks apply to the characteristic of a field, e.g. $GF(2)$, which co-ordinatises the finite Fano plane, has *characteristic* 2.

One of the most significant results relating to alternative rings was obtained by R. H. Bruck and Erwin Kleinfeld ('The structure of alternative division rings', *Proceedings Amer. Math. Soc.* 2 (1951) 878–890), namely

Bruck-Kleinfeld Theorem (1951) *Theorem 56:* Either an alternative division ring is associative or it is a Cayley-Dickson Algebra over its centre.

[The *centre* of a Cayley-Dickson Algebra is the set of elements c such that $cx = xc$ for all elements x of the algebra.]

Proof: Read the Bruck-Kleinfeld article.

Here we meet again the Cayley-Dickson Algebra (or just Cayley Algebra) of octaves (see § 53.5). An interesting result proved in Octave Geometry is: The affine plane in the projective octave plane is isomorphic with the Moufang plane over the Cayley-Dickson (octave) division algebra, i.e. the projective octave plane is the projective closure of this Moufang plane. (Refer to the Springer article quoted in § 53.5.)

§ 58 DESARGUESIAN PLANES

In order to give the alternative division ring further properties of a field, we need to strengthen the alternative laws to make them associative. To do this we vary the restrictions previously placed on the Desargues' Theorem, namely δ_a and δ. It is now the general Desargues' Theorem Δ (of which δ, δ_a are special cases) with which we are concerned, i.e. we are interested in homologies (§ 16.2) rather than elations. We recall that a plane in which Desargues' Theorem is always valid is called a *Desarguesian plane*, so that all the planes discussed earlier in this Chapter are non-Desarguesian. The following important theorem may be proved, though it is not easy to do so.

Theorem 57: A Desarguesian plane is co-ordinatised by a skew-field (= associative division ring), and conversely.

That is,

Desargues' Theorem is valid if the associative law of multiplication holds, and conversely.

Proof: See Hall [45] pp. 372–374, Hall [44] pp. 25–27, Pedoe [85] pp. 140–144.

It is also proved (Hall [45] p. 374) that
the collineation group for a Desarguesian plane π is transitive on quadrilaterals, and that

every ternary ring for π is the same associative division ring.

Naturally, from their construction, Desarguesian planes constitute an essentially more restricted class than do Moufang planes. Desarguesian planes are richer in collineations than Moufang planes.

Summary of main facts relating to Desargues' Theorem Δ

(1) Δ is valid in $[n]$ for $n \geqslant 3$.
(2) Δ is not necessarily valid in [2] (thus giving rise to non-Desarguesian planes).
(3) Δ valid in [2] $\Leftrightarrow$ co-ordinates belong to an associative division (ternary) ring (= 'non-commutative field' = skew-field = (ordinary) division ring).
(4) A plane co-ordinatised by an associative division ring is Desarguesian and may be embedded in an arbitrary $[n]$.

It is hoped that these few comments on Desargues' Theorem will encourage the reader to seek fuller details.

§ 59 PAPPIAN PLANES

The remaining algebraic law which must be satisfied if the ternary ring is to become a field is the commutative law of multiplication. When this is so, Pappus' Theorem Π is valid and the now highly-structured plane is *Pappian*.

Theorem 58: A Pappian plane is co-ordinatised by a field, and conversely.

That is,
Pappus' Theorem is valid if the commutative law of multiplication holds, and conversely.
Proof: Refer to Artzy [4] pp. 234–235, Hall [44] p. 45, and Pedoe [85] pp. 146–148.

From Hessenberg's Theorem we know (for finite and non-finite planes) that

$$\Pi \Rightarrow \Delta.$$

But

$$\Delta \nRightarrow \Pi.$$

This completes the story of the projective plane. Certain interesting features of Pappus' Theorem in the affine plane (e.g. the *Thomsen property* Θ—nothing to do with the previous use of this symbol in Chapter 7) and the various relationships among Π_a (*affine Pappus Theorem*), Θ, Δ_a, δ_a, Π, Δ are to be found in Artzy [4].

Although the texts of Hall, Pedoe and Artzy in particular have been constantly referred to in this brief sketch (and in Exercises 11), the reader will lose much if he does not consult some of the other major references, namely Blumenthal, Segre and Skornyakov and, especially, the original research articles.

§ 59.1 Finite Pappian Planes. Reverting to Wedderburn's Theorem (Theorem 51), we re-emphasise the result that

every finite Desarguesian plane is Pappian.

Another important theorem in finite fields is as follows:

Theorem 59 (*Artin-Levi-Zorn Theorem*)*:* Every finite alternative division ring is a field.

Immediately we deduce that

every finite Moufang plane is Pappian.

Combining Theorems 51 and 59, we have the following identification of planes

finite Moufang = *finite Desarguesian* = *finite Pappian*.

A finite Veblen-Wedderburn plane, if not corresponding to a field, is non-Desarguesian. (Finite Hughes planes, though non-Desarguesian, are not translation planes.)
References for Theorem 59 are:
M. Zorn: '*Theorie der alternativen Ringe*', *Abh. Math. Seminar Univ. Hamburg* 8 (1931) 123–147;
F. W. Levi [63].

A research article on finite Pappian planes which the reader may wish to consult is that of Chung-Tao Yang: 'Certain chains in finite projective geometry' *Duke Math. Jour.* 15 (1948) 37–47, written in Chekiang, China.

§ 59.2 Summary of the Chapter. Thus we have shown how the 'rudimentary' projective plane (axiomatised by *PP*1–*PP*3) is given more and more geometrical structure as the ternary ring is given more and more algebraic structure. All this is summarised in the following table (suggesting the structural hierarchy of projective planes '*rudimentary*'→*Veblen-Wedderburn*→*Moufang*→*Desarguesian*→*Pappian*):

Theorem	Transitivity	Type of plane	Restriction on ternary ring	
δ_a	(Y, l_∞)		linearity additive group	Veblen- Wedderburn system
	(C, l_∞)	Veblen- Wedderburn (= trans- lation) plane	right distribution additive commutativity	
	translation plane for a pencil	division ring plane	left distribution inverses	division ring
δ	translation plane for every line	Moufang (= alternative) plane	alternative laws	alternative division ring
Δ	(X, OY)	Desarguesian plane	multiplicative associativity	associative division ring
Π		Pappian plane	multiplicative commutativity	field

§ 60 COMPARISON OF TERNARY RING CO-ORDINATES AND HOMOGENEOUS CO-ORDINATES

By means of the equivalence stated below we can move from one set of co-ordinates to the other

Ternary Ring Co-ordinates	Homogeneous Co-ordinates
(x, y)	$\lambda(x, y, 1)$
(m)	$\lambda(1, m, 0)$
(∞)	$\lambda(0, 1, 0)$

where $\lambda \neq 0$.

Which system of co-ordinates is to be preferred? Both have their advantages and disadvantages. While the homogeneous co-ordinates account for all points in the plane in a uniform manner, they do not reveal entirely the hidden treasures in the plane and the deep relationship between the geometric structure of a plane and the algebraic structure of its co-ordinate system. Ternary ring co-ordinates do illuminate these matters though they cannot cope uniformly with the problem of infinity. On balance, though, it does appear that this is a price worth paying because of the power and insight they make possible, despite their lack of symmetry.

§ 60.1 Applications of Projective Planes. In his systematic treatment of projective planes, Hall has this to say regarding their usefulness: 'Projective planes are the logical basis for the investigation of combinatorial analysis, such topics as the Kirkman schoolgirl problem and the Steiner triple systems (= block designs) being interpretable directly as plane configurations, various problems in non-associative or non-distributive algebras, and the problems of universal identities in groups and skew-fields, as well as certain problems on structure (lattice) theory and the theory of partially ordered sets.'

By extension of some of these topics, the theory has application to, say, the theory of configurations in Combinatorial Topology and to Quantum Theory (related to non-associative algebras). Geometric algebras connected with Quantum Theory were investigated by Pascual Jordan, J. von Neumann and E. Wigner: 'On an algebraic generalisation of the quantum mechanical formalism', *Ann. Math.* (2) 35 (1934) 29–64.

Previously, in Chapter 10, we mentioned Steiner triple systems. These are special balanced incomplete block designs (BIBD), which though studied in 1853 by Steiner, had been earlier investigated by Kirkman: 'On a problem in combinations', *Cambridge and Dublin Math. J.* 2 (1847) 191–204. For Steiner triple systems and the related Kirkman triple systems, consult Ryser [94].

The applications of projective planes specified by Hall should be noted in conjunction with earlier comments on applicability in §§ 3.7, 6.4.

§ 60.2 Concluding Survey of Projective Planes. Starting from a small number of axioms, we established the basic nature of the projective plane as far as the theory of conics. Many important plane geometries (models) satisfying these axioms were discussed in some detail: analytical number planes, the augmented affine and Euclidean planes, non-Desarguesian planes and finite planes. Analytical methods enabled us to co-ordinatise different planes by means of different fields. Working with an algebraic system (ternary ring) having a looser structure than a field, and imposing successive restrictions on the system, we saw how the projective plane was increasingly enriched in structure. Introduction of measurement into the projective plane led to the metric geometries, Euclidean and non-Euclidean.

Behind this development of the theory lay the unifying power of groups, which by Klein's *Erlanger Programm* assured us of an intimate relationship between the hierarchy of the non-finite projective planes and the structure of certain infinite groups.

Suggested topics for further study

(1) Axioms of order and continuity in a projective plane (Artzy [4] pp. 239–247; Coxeter [27]; O. Wyler: 'Order in projective and descriptive geometry', *Composito Mathematica*, 11 (1) (1953) 60–70).

(2) Non-Archimedean Geometry (a geometry in which the Archimedean axiom of continuity is denied).

(3) Semi-translation planes (T. G. Ostrom: 'Semi-translation planes' *Trans. Amer. Math. Soc.* 111 (1964) 1–18).

(4) Near-fields (= Veblen-Wedderburn systems with associative multiplication, used for co-ordinatising Hughes planes and other planes); (Hall [45] pp. 388–392; Hall [44] pp. 50–52; H. Zassenhaus: '*Über endliche Fastkörper*', *Abh. Math. Sem. Hamburg*, 11 (1936) 187–220). (Unlike finite division rings which are fields, by Theorem 59, finite near-fields are not fields.)

(5) Loops (R. H. Bruck [16]; D. R. Hughes: 'Planar division neo-rings' *Trans. Amer. Math. Soc.* 80 (1955) 502–527; J. M. Osborn: 'New Loops from Old Geometries', *Amer. Math. Monthly* 68(2) (1961) 103–107; L. Paige: 'Neofields', *Duke Math. J.* 16 (1949) 39–60).

(6) Lattices (Hall [44] pp. 7–15; Skornyakov [102]).

(7) Moufang planes.

(8) Veblen-Wedderburn planes. André planes. Hall planes.

Some additional topics and problems are suggested in Lombardo-Radice [67]. Alert readers wishing to find out about new developments and trends in Projective Geometry should keep in touch with the mathematical journals which have been mentioned in this book. For example, two articles worth looking at are 'On Veblen-Wedderburn Systems', *Amer. Math. Monthly* 64(9) (1957) 631–635, and 'The Construction of Projective Planes from Generalised Ternary Rings', *Amer. Math. Monthly* 73(1) (1966) 36–40, both by J. R. Wesson. Another one is by M. J. Kallaher, entitled 'Moufang Veblen-Wedderburn Systems', *Math. Zeitschr.* 105 (1968) 114–127.

EXERCISES 11

Arguing only from the geometrical definitions for addition and multiplication, show that elements 0, 1 have the properties 1–3 where $a, m \in S$ (Pedoe [85] pp. 121–122):

1. $0 + a = a + 0 = a$

2. $0m = m0 = 0$

3. $1m = m1 = 1$.

4. A ternary ring satisfying *TR*1–*TR*5 uniquely defines a projective plane. Prove (Hall [45] p. 356, Pedoe [85] pp. 123–124).

5. Draw the Reidemeister figure for the affine plane.

6. If a ternary ring R of π satisfies:

(i) $a \cdot m * b = am + b$,

(ii) addition is a group

prove that π is (Y, l_∞) transitive (Hall [45] p. 362, Pedoe [85] pp. 126–127).

7. Prove Theorem 54 (including its Converse) for a Veblen-Wedderburn plane.

8. Establish the result: A ternary ring which is both linear and an alternative division ring co-ordinatises a Moufang plane.

9. Show that a reflection is an involution. (Also see Hall [45] p. 371, Pedoe [85] p. 140.)

10. Prove that if a ternary ring of a projective plane is linear and a division ring, then the plane is Desarguesian.

In the theory of Alternative Rings, Bruck and Kleinfeld define the *commutator* (x, y) and the *associator* (x, y, z) for elements x, y, z belonging to a ring, as follows:

$$(x, y) = xy - yx$$
$$(x, y, z) = (xy)z - x(yz).$$

[Thus, the associator vanishes identically in an associative ring and the commutator vanishes identically in a commutative ring.]

Establish the properties 11–19 (Hall [45] p. 377, Bruck and Kleinfeld (article quoted w.r.t. Theorem 56):

11. $(x, x, y) = 0$

12. $(y, x, x) = 0$

13. $(x, y, z) + (x, z, y) = 0$ [Hint: Consider $(x, y + z, y + z)$.]

14. $(x, y, z) + (y, x, z) = 0$.

15. Under the permutations of the symmetric group on x, y, z, the associator (x, y, z) is unchanged by the alternating group, but is changed in sign by the odd permutations.

16. $(xy)x = x(yx)$ (*reflexive law*)

17. $((x, y, z), x, y) = (x, y)(x, y, z) = -(x, y, z)(x, y)$

18. $((x, y, z)^2, x, y) = 0$

19. $(v^2, r, s) = 0$ ($v = (x, y)$; r, s arbitrary).

Defining the function $f(w, x, y, z)$ in an alternative ring by $f(w, x, y, z) = (wx, y, z) - x(w, y, z) - (x, y, z)w$, establish 20–22:

20. $f(w, x, y, z)$ is skew-symmetric (Hall [45] p. 378–379)

21. $3f(w, x, y, z) = (w, (x, y, z)) - (x, (y, z, w)) + (y, (z, w, x)) - (z, (w, x, y))$

22. $f(w, x, y, z) = ((w, x), y, z) + ((y, z), w, x)$.

For x, y, z elements of an alternative ring, prove 23–28:

23. $(x^2, y, z) = x(x, y, z) + (x, y, z)x$

24. $(x, xy, z) = (x, y, xz) = (x, y, z)x$

25. $(x, yx, z) = (x, y, zx) = x(x, y, z)$

26. $(xy)(zx) = x((yz)x) = (x(yz))x$

27. $x(y(xz)) = ((xy)x)z$

28. $((zx)y)x = z(x(yx))$.

(Results 26–28 are known as *Moufang identities*.)

29. If in a field $ab = ba$ holds for all pairs of elements a, b, prove that the plane co-ordinatised by the field is Pappian.

30. Taking the Pappus line to be l_∞, state Pappus' Theorem in the affine plane. Illustrate with a diagram.

MISCELLANEOUS RESEARCH ARTICLES

(mostly relating to Chapters 10 and 11)

Besides the mathematicians already mentioned in the text, either as the originators of research articles or as the authors of reference books, there are many others with whose discoveries and expository contributions the reader may wish to be familiar. The following list is by no means exhaustive. A great range of mathematical depth is encompassed in this selection, arranged alphabetically:

R. ARTZY: 'Non-Euclidean Incidence Planes', *Israel J. Math.* 4 (1966) 43–53.

R. BAER: 'Polarities in finite projective planes', *Bull. Amer. Math. Soc.* 52 (1946) 77–93.

F. BUEKENHOUT: '*Étude intrinsèque des ovales*', *Rend. Mat.* (3–4) 25 (1966) 1–61.

F. BUEKENHOUT: '*Ovales et Ovales projectifs*', *Acc. Naz. Lincei.* Serie 8 40(1) (1966) 46–49.

L. CARLITZ: 'A Theorem on Permutations in a Finite Field', *Proc. Amer. Math. Soc.* 11 (1960) 456–459.

G. M. CONWELL: 'The 3-space PG (3, 2) and its group', *Ann. Math.* 11 (1910) 60–76.

A. C. COULSON: 'Mathematics and the Real World', *Bull. Malayan Math. Soc.* 6(50) (1959) 81–86.

R. FEINMAN: 'Geometrical transformations of Veblen-Wedderburn Systems', *Amer. Math. Monthly* 73(5) (1966) 484–486.

K. D. FRYER and I. HALPERIN: 'Co-ordinates in Geometry', *Trans. Roy. Canadian Soc.* 48 Series 3 (1954) Section 3,11–26.

V. HAVEL: 'Harmonic quadruplet in a Moufang plane', *Czechoslovakian Math. J.* 5(80) 1 (1955) 76–81.

D. E. KNUTH: 'A Class of Projective Planes', *Trans. Amer. Math. Soc.* 115(3) (1965) 541–549.

D. E. KNUTH: 'Finite semi-fields and Projective Planes', *J. Algebra* 2(2) (1965) 182–217.

F. KOSIER: 'A generalisation of Alternative Rings', *Trans. Amer. Math. Soc.* 112(1) (1964) 32–42.

N. H. KUIPER: 'A Real Analytic non-Desarguesian plane', *Nieuw Archief voor Wiskunde* 5(3) (1957) 19–24.

Sahib Ram MANDAN: 'Pascal's theorem in *n*-space', *J. Australian Math. Soc.* 5(4) (1965) 401–408.

C. R. MACINNES: 'Finite Planes with less than 8 points on a line', *Amer. Math. Monthly* 14 (1907) 171–174.

H. F. MACNEISH: 'Euler Squares', *Ann. Math.* 23 (1921–22) 221–227.

H. F. MACNEISH: 'Four Finite geometries', *Amer. Math. Monthly* 49 (1942) 15–23.

G. E. MARTIN: 'On arcs in a finite projective plane', *Can. J. Math.* 19(2) (1967) 376–393.

N. S. MENDELSOHN: 'Non-Desarguesian Projective Plane geometries which satisfy the Harmonic Point Axiom', *Can. J. Math.* 8(4) (1956) 532–562.

D. L. MORGAN and T. G. OSTROM: 'Co-ordinate Systems of some semi-translation planes', *Trans. Amer. Math. Soc.* 111 (1964) 19–32.

HANNA NEUMANN: 'On some Finite non-Desarguesian planes', *Archiv der Math.* 6 (1953) 36–40.

T. G. OSTROM: 'Ovals, dualities and Desargues' Theorem', *Can. J. Math.* 7 (1955) 417–431.

E. R. OTT: 'Finite Projective Geometries $PG(k, p^n)$', *Amer. Math. Monthly* 44 (1937) 86–91.

L. PAIGE and C. WEXLER: 'A Canonical Form for Incidence Matrices of Finite Projective Planes and their associated Latin Squares', *Portugaliae Mathematica* 12(3) (1953) 105–112.

M. RICHARDSON: 'On Finite Projective Games', *Proc. Amer. Math. Soc.* 7 (1956) 458–464.

L. ROSATI: '*Piani desarguesiani non ciclici*', *Convegno Internazionale Reticoli e Geometrie Proiettive*, Palermo and Messina, Sicily (1957) 87–99.
H. SCHWERDTFEGER: 'Projective geometry in the one-dimensional affine group', *Can. J. Math.* 16 (1964) 683–700.
H. SCHWERDTFEGER: 'Group-theoretical interpretation of projective incidence theorems', [135] 307–314.
B. SEGRE: 'Introduction to Galois Geometries', *Atti Acc. Naz. Lincei* 8 (1967) *Memoire*.
B. SEGRE: 'Le Geometrie di Galois', *Ann. Mat. Pura Appl.* (4) 47–48 (1959) 1–97.
M. F. SMILEY: 'Von Staudt Projectivities of Moufang Planes', *Algebraical and Topological Foundations of Geometry*, Utrecht Colloquium (1959) 165–166.
JILL C. D. SPENCER (Mrs. Yaqub): 'On the Lenz-Barlotti classification of projective planes', *Quart. J. Math.* 11(44) (1960) 241–247.
T. A. SPRINGER and F. D. VELDKAMP: 'Elliptic and Hyperbolic Octave Planes I–III', *Koninklijke Nederlandse Akademie van Wetenschappen Proceedings*, Series A, Mathematical Sciences, 56 (1963) 413–451.
M. SUSEELA (Miss): 'Non-Desarguesian Geometry', *Math. Student* 14 (1, 2) (1946) 1–13.
G. THOMSEN: 'The treatment of elementary geometry by a group-calculus', *Math. Gaz.* 17(225) (1933) 230–242.
J. TITS: '*Ovoïdes et groups de Suzuki*', *Archiv der Math.* 13 (1962) 187–198.
J. TITS: '*Ovoïdes à translations*', *Rendiconti di Matematica* 21 (1, 2) (1962) 37–59.
B. J. TOPEL: 'Bolyai-Lobatschevsky Planes with Finite Lines', *Reports of a Mathematical Colloquium*, University of Notre Dame, Indiana (1944) 40–42.
A. G. WALKER: 'Finite Projective Planes', *Edinburgh Math. Notes* 36 (1947) 12–17.
G. ZAPPA: 'Sui gruppi di Collineazoni dei Piani di Hughes', *Boll. U.M.I.* (Italian Mathematical Union) 12(3) (1957) 507–516.

Comments:

(1) Conwell's paper gives a complete solution to Kirkman's Schoolgirl Problem.

(2) MacNeish's four finite geometries are (i) Fano's 7-point plane; (ii) the affine 13-point plane, consisting of 12 lines and nine points; (iii) Pappus' configuration; (iv) Desargues' configuration. See also Eves [36] 1 pp. 425–433.

(3) Fryer and Halperin deal with the converse of von Neumann's generalisation to complemented modular lattices of the theorem (Theorem 57) that every projective Desarguesian plane can be co-ordinatised with co-ordinates in a suitable division ring.

(4) Thomsen's article shows how geometrical properties of points and lines are expressible as relations among the corresponding half-turns and reflections. Also see Bachmann [5].

(5) Some researchers whose work is relevant to our themes but is not specifically mentioned by title include, say, J. André, W. S. Connor, A. J. Hoffman, W. Klingenberg, H. Lenz, F. Maisano, H. Naumann, P. Rachevsky, D. K. Ray-Chaudhuri, M. Sce, G. Tallini, N. Tschetweruchin, A. Wagner and A. Zaddach, along with many others.

CHAPTER 12

THROUGH THE LOOKING GLASS

'The best is yet to be.'
(Robert Browning)

RETROSPECT AND PROSPECT

FROM the simple decorative patterns of the pre-scientific Neolithic craftsmen, from the clay tablets of the Babylonians and the Rhind papyrus of Ahmes (the ancient Egyptian scribe), through Euclid to Lobatschewsky and Einstein, the cavalcade of geometric thought sweeps by in the mind's eye.

In the relatively brief space spanned by this book, an attempt has been made to outline some of the intellectual achievements (and frustrations) in the long, unrelenting search for a logical, coherent description of the space (particularly two-dimensional) around us. Resulting from this process there have been such major developments as the introduction of analytical methods, the establishment of the abstract approach and the discovery of different geometries (non-Euclidean, non-Desarguesian, finite), each with its own special significance for Mathematics and the Sciences. Indeed, it might be said that we have been observing one aspect of the creative imagination of man, and we have had dramatic expression of the tremendous potential of this imagination.

It is a sobering thought that more mathematics has been published in the last two or three decades than in all the rest of recorded history. Likewise, it is a chastening thought that a new field of study—*Continuous Geometry*, invented in 1935 by John von Neumann (1903–1957, German-American)—is based on the notion that 'there can be no minimal element—that is, no atomic element or *point*', to quote from a description of Continuous Geometry.

Thoughtful readers might well ponder on the directions of study which follow on naturally from our exposition of theory for the plane. Some of these would include

Plane Curves;
Higher-dimensional Geometry;
Topology;
Abstract Algebraic Geometry,

where, by the last term, we mean that area of study of geometrical entities in which the emphasis is on high-powered contemporary algebra, namely, the algebra of groups, rings, ideals, modules, fields, matrices and (say) valuation theory (Hodge and Pedoe [51]).

Each of these possesses a beauty and a fascination of its own, each has its applications in the world of modern technology and commerce. There is something, we feel, for each temperament. It is sincerely to be hoped that the reader who has found some pleasure in this book, will use it as the beginning of a deep and abiding interest in Geometry.

> 'Alas, . . .
> That Youth's sweet-scented manuscript should close.'
>
> (Omar Khayyám)

REFERENCES

1. ADAMS, L. J. and WHITE, P. A.: 'Analytical Geometry and Calculus', C.U.P. (1961).
2. ADLER, Claire F.: 'Modern Geometry', McGraw-Hill (1967).
3. ARTIN, E.: 'Geometric Algebra', Interscience (1957).
4. ARTZY, R.: 'Linear Algebra', Addison-Wesley (1965).
5. BACHMANN, F.: '*Aufbau der Geometrie aus dem Spiegelungsbegriff*', Springer-Verlag (1959).
6. BAER, R.: 'Linear Algebra and Projective Geometry', Academic (1952).
7. BAKER, H. F.: 'Principles of Geometry', Vols. 1, 2, C.U.P. (1930).
8. BAKER, H. F.: 'Introduction to Plane Geometry', C.U.P. (1943).
9. BELL, E. T.: 'The Development of Mathematics', McGraw-Hill (1945).
10. BELL, E. T.: 'Men of Mathematics', Simon and Schuster (1937).
11. BIRKHOFF, G. and MACLANE, S.: 'A Survey of Modern Algebra', Macmillan (1965).
12. BLUMENTHAL, L. M.: 'A Modern View of Geometry', Freeman (1961).
13. BOLYAI, J.: 'The Science of Absolute Space' (translated by G. B. Halsted) in Bonola [14].
14. BONOLA, R. 'Non-Euclidean Geometry', Dover (1955).
15. BORZUK, K. and SZMIELEW, Wanda: 'Foundations of Geometry', North-Holland (1960).
16. BRUÇK, R. H.: 'A Survey of Binary Systems', Springer-Verlag (1958).
17. BUSEMANN, H. and KELLY, P. J.: 'Projective Geometry and Projective Metrics', Academic (1953).
18. CAJORI, F.: 'A History of Mathematics', Macmillan (1919).
19. CARMICHAEL: 'Introduction to the Theory of Groups of Finite Order', Dover (1956).
20. CAYLEY, A.: 'Collected Mathematical Papers', Vols. 1, 2, C.U.P. (1889) ('A Sixth Memoir on Quantics', 561–592).
21. CHASLES, M.: '*Traité des Sections Coniques*', Gauthier-Villars (1865).
22. CHASLES, M.: '*Traité de Géométrie Supérieure*', Gauthier-Villars (1880).
23. COOLIDGE, J. L.: 'The Elements of non-Euclidean Geometry', O.U.P. (1927).
24. COOLIDGE, J. L.: 'A History of Geometrical Methods', O.U.P. (1947).
25. COXETER, H. S. M.: 'Non-Euclidean Geometry', University of Toronto Press (1942).
26. COXETER, H. S. M.: 'Introduction to Geometry', Wiley (1961).
27. COXETER, H. S. M.: 'The Real Projective Plane', McGraw-Hill (1949).
28. COXETER, H. S. M.: 'Projective Geometry', Blaisdell (1964).
29. CREMONA, L.: 'Elements of Projective Geometry', O.U.P. (1885).
30. DESARGUES, G.: '*Oeuvres de Desargues*', Vols. 1, 2 (ed. M. Poudra), Paris (1864).
31. DESCARTES, R.: 'The Geometry of René Descartes' (translated from the French and Latin by D. E. Smith and Marcia L. Latham), Dover (1954).
32. DORWART, H.: 'The Geometry of Incidence', Prentice-Hall (1966).
33. DURELL, C. V.: 'Projective Geometry', Macmillan (1948).
34. DURELL, C. V.: 'Algebraic Geometry', Bell (1955).
35. EUCLID: 'The Thirteen Books of Euclid's "Elements" ' (translated by T. Heath), Vol. 1, Dover (1956).
36. EVES, H.: 'A Survey of Geometry', Vols. 1, 2, Allyn and Bacon (1963–1965).
37. EVES, H. and NEWSOM, C. V.: 'An Introduction to the Foundations and Fundamental Concepts of Mathematics', Holt, Rinehart and Winston (1958).
38. FAULKNER, T. E.: 'Projective Geometry', Oliver & Boyd (1949).
39. FILON, L. N. G.: 'An Introduction to Projective Geometry', Arnold (1935).
40. FISHBACK, W. T.: 'Projective and Euclidean Geometry', Wiley (1962).

41. FORDER, H. G.: 'The Foundations of Euclidean Geometry', Dover (1958).
42. GOODSTEIN, R. L. and PRIMROSE, E. J. F.: 'Axiomatic Projective Geometry', Leicester (1953).
43. GRAUSTEIN, W. C.: 'Introduction to Higher Geometry', Macmillan (1948).
44. HALL, M.: 'Projective Planes and Related Topics', California Institute of Technology (1954).
45. HALL, M.: 'The Theory of Groups', Macmillan (1964).
46. HARDY, G. H. and WRIGHT, E. M.: 'An Introduction to the Theory of Numbers', O.U.P. (1956).
47. HATTON, J. L. S.: 'The Principles of Projective Geometry', C.U.P. (1913).
48. HEADING, J.: 'The Methods of Pure Projective Geometry', Macmillan (1958).
49. HEYTING, A.: 'Axiomatic Projective Geometry', Noordhoff (1963).
50. HILBERT, D.: 'Foundations of Geometry', Open Court (1950).
51. HODGE, W. V. D. and PEDOE, D.: 'Methods of Algebraic Geometry', C.U.P. (1947).
52. HOPKINS, E. J. and HAILS, J. S.: 'Plane Projective Geometry', O.U.P. (1953).
53. IVINS, W. M.: 'Art and Geometry: A Study of Space Intuitions', Dover (1964).
54. KENDALL, M. G.: 'A Course in the Geometry of *n* Dimensions', Griffin (1961).
55. KEPLER, J.: '*Paralipomena in Vitellionem*' (in Band 2, '*Gesammelte Werke*', Munich (1939)).
56. KLEIN, F.: '*Gesammelte Mathematische Abhandlungen*', Vol. 1, Springer (1921).
57. KLEIN, F.: 'Elementary Mathematics from an Advanced Standpoint', Vol. 2 ('Geometry') (translated by E. R. Hedrick and C. A. Noble), Dover (1939).
58. KLINE, M.: 'Mathematics in Western Culture', Allen and Unwin (1954).
59. KOSTOVSKII, A. N.: 'Geometrical Constructions using Compasses only', Pergamon (1961).
60. KUIPER, N. H.: 'Linear Algebra and Geometry', North-Holland (1962).
61. KULCZŶCKI, S.: 'Non-Euclidean Geometry', Pergamon (1956).
62. LEDERMANN, W.: 'Introduction to the Theory of Finite Groups', Oliver and Boyd (1964).
63. LEVI, F. W.: 'Finite Geometrical Systems', Calcutta (1942).
64. LEVY, H.: 'Projective and Related Geometries', Macmillan (1964).
65. LIMING, R. A.: 'Practical Analytic Geometry with Applications to Aircraft', Macmillan (1944).
66. LOBATSCHEWSKY, N. I.: 'Geometrical Researches on The Theory of Parallels', Open Court (1914); also in Bonola [14].
67. LOMBARDO-RADICE, L.: '*Piani grafici finiti non Desarguesiani*', Palermo (1959), translated into English in the Appendix to Segre [97] pp. 365–455.
68. MASCHERONI, L.: '*La Geometria del Compasso*', Pavia (1797), Palermo (1901).
69. MAXWELL, E. A.: 'The Methods of Plane Projective Geometry based on the Use of General Homogeneous Co-ordinates', C.U.P. (1948).
70. MAXWELL, E. A.: 'General Homogeneous Co-ordinates in Space of Three Dimensions', C.U.P. (1951).
71. MAXWELL, E. A.: 'Fallacies in Mathematics', C.U.P. (1959).
72. MESCHKOWSKI, H.: 'Evolution of Mathematical Thought', Holden-Day (1965).
73. MESERVE, B.: 'Fundamental Concepts of Geometry', Addison-Wesley (1955).
74. MÖBIUS, A.: '*Der Barycentrische Calcül*', (1827).
75. MOHR, G.: '*Euclides Danicus*', Amsterdam (1672), Copenhagen (1928).
76. MORITZ, R. E.: 'On Mathematics and mathematicians', *Memorabilia Mathematica*, Dover (1958).
77. MOISE, E.: 'Elementary Geometry from an Advanced Standpoint', Addison-Wesley (1963).
78. NARAYAN, S. and KOCHHAR, M. L.: 'A Text-book of Modern Pure Geometry', Chand (1959).
79. O'HARA, C. W. and WARD, D. R.: 'An Introduction to Projective Geometry', O.U.P. (1946).
80. PAPPUS of ALEXANDRIA: '*La Collection Mathématique*', Vols. 1, 2 (translated by ver Eecke), Blanchard (1933).

81. PASCAL, B.: '*Essay pour les coniques*', Paris (1640); facsimile, Sanford [96], translated by Frances Clarke in Smith [103], pp. 326–330.
82. PASCH, M.: '*Vorlesungen über neuere Geometrie*', Leipzig (1882).
83. PATTERSON, B. C.: 'Projective Geometry', Wiley (1948).
84. PEDOE, D.: 'Circles', Pergamon (1957).
85. PEDOE, D.: 'An Introduction to Projective Geometry', Pergamon (1963).
86. PICKERT, G.: '*Projektive Ebenen*', Springer-Verlag (1955).
87. PONCELET, J. V.: '*Traité des propriétés projectives*', Vols. 1, 2, Gauthier-Villars (1866).
88. PRENOWITZ, W. and JORDAN, M.: 'Basic Concepts of Geometry', Blaisdell (1965).
89. PRIMROSE, E. J. F.: 'Plane Algebraic Curves', Macmillan (1955).
90. RICHARDSON, G. and RAMSEY, A. S.: 'Modern Plane Geometry', Macmillan (1940).
91. RIEMANN, G. F. B.: '*Über die Hypothesen welche der Geometrie zu Grunde liegen*' in '*Gesammelte mathematische Werke*', Dover (1953) pp. 272–287 (translated in full by H. S. White in Smith [103] pp. 411–425).
92. ROBINSON, G. de B.: 'The Foundations of Geometry', Toronto (1946).
93. ROSENBAUM, R. A.: 'Introduction to Projective Geometry and Modern Algebra', Addison-Wesley (1963).
94. RYSER, H. J.: 'Combinatorial Mathematics', Carus Monograph No. 14,Mathematical Association of America (1963).
95. SACCHERI, G.: '*Euclides ab omni naevo vindicatus*' (translated by G. B. Halsted), Open Court (1920).
96. SANFORD, Vera: 'A Short History of Mathematics', Harrap (1930).
97. SEGRE, B.: 'Lectures on Modern Geometry', Cremonese (1961).
98. SEIDENBERG, A.: 'Lectures in Projective Geometry', Van Nostrand (1963).
99. SEMPLE, J. G. and KNEEBONE, G. T.: 'Algebraic Projective Geometry', O.U.P. (1952).
100. SEMPLE, J. G. and ROTH, L.: 'Introduction to Algebraic Geometry', O.U.P. (1949).
101. SHIROKOV, P. A.: 'A Sketch of the Fundamentals of Lobachevskian Geometry', Noordhoff (1963).
102. SKORNYAKOV, L. A.: 'Complemented Modular Lattices and Regular Rings', Oliver and Boyd (1964).
103. SMITH, D. E.: 'A Source Book in Mathematics', Vol. 2, Dover (1959).
104. SMOGORZHEVSKII, A. S.: 'The Ruler in Geometrical Constructions', Pergamon (1961).
105. SOMMERVILLE, D. M. Y.: 'Analytical Conics', Bell (1946).
106. SOMMERVILLE, D. M. Y.: 'The Elements of Non-Euclidean Geometry', Bell (1914).
107. STEINER, J.: '*Systematische Entwicklung der Abhängigheit geometrischer Gestalten von einander*', Leipzig (1896).
108. STEINER, J.: 'Geometrical Constructions with a Ruler', *Scripta Mathematica* (1950).
109. STRUIK, D. J.: 'Lectures on Analytical and Projective Geometry', Addison-Wesley (1953).
110. STRUIK, D. J.: 'A Concise History of Mathematics', Vols. 1, 2, Dover (1948).
111. TODD, J. A.: 'Projective and Analytical Geometry', Pitman (1947).
112. VAN DER WAERDEN, B. L.: 'Science Awakening', Noordhoff.
113. VEBLEN, O. and YOUNG, J. W.: 'Projective Geometry', Vols. 1, 2, Ginn (1938).
114. VON STAUDT, K. G. C.: '*Geometrie der Lage*', Nürnberg (1847).
115. VON STAUDT, K. G. C.: '*Beiträge zur Geometrie der Lage*', Nürnberg (1856).
116. WALKER, R. J.: 'Cartesian and Projective Geometry', Arnold (1953).
117. WELCHMAN, W. G.: 'Introduction to Algebraic Geometry', C.U.P. (1950).
118. WHITEHEAD, A. N.: 'The Axioms of Projective Geometry' (Cambridge Tracts in Mathematics and Mathematical Physics, No. 4), Hafner (1906).
119. WHITTAKER, E. T.: 'From Euclid to Eddington: a Study of Conceptions of the External World', Dover (1958).
120. WINGER, R. M.: 'An Introduction to Projective Geometry', Dover (1962).
121. WOLFE, H. E.: 'Introduction to non-Euclidean Geometry', Holt-Dryden (1945).
122. WOODS, F. S.: 'Higher Geometry', Ginn (1922).
123. YAGLOM, I. M.: 'Geometric Transformations' (Part I: translated from the Russian by Allen Shields), Random House (1962).
124. YOUNG, J. W.: 'Projective Geometry', Open Court (1930).

ADDENDUM

125. Albert, A. A. and Sandler, R.: 'An Introduction to Finite Projective Planes', Holt, Rinehart and Winston (1968).
126. Ayres, F.: 'Theory and Problems of Projective Geometry', Schaum (1967).
127. Barlotti, A.: 'Some Topics in Finite Geometrical Structures', Institute of Statistics Mimeo Series No. 439, University of North Carolina (1965).
128. Dembowski, P.: 'Finite Geometries', Springer (1968).
129. Dickson, L. E.: 'Linear Groups with an Exposition of the Galois Field Theory', Dover (1958).
130. Fraleigh, J. B.: 'A First Course in Abstract Algebra', Addison-Wesley (1968).
131. Gruenberg, K. W. and Weir, A. J.: 'Linear Geometry', van Nostrand (1967).
132. Hartshorne, R.: 'Foundations of Projective Geometry', Benjamin (1967).
133. Hardy, G. H.: 'A Course of Pure Mathematics', C.U.P. (1946).
134. Jones, B. W.: 'The Arithmetic Theory of Quadratic Forms', Carus Mathematical Monographs Number 10, Wiley (1950).
135. Kovács, L. G. and Neumann, B. H. (ed.): 'Proceedings of the International Conference on The Theory of Groups held at the Australian National University, Canberra, 10–20 August, 1965', Gordon and Breach (1967).
136. Loney, S. L.: 'The Elements of Co-ordinate Geometry. Part II: Trilinear Coordinates, etc.', Macmillan (1946).
137. Mann, H. B.: 'Analysis and Design of Experiments', Dover (1949).
138. Mann, H. B.: 'The Addition Theorems of Group Theory and Number Theory' (Chapter 6: 'Difference Sets in Abelian Groups'), Wiley (1965).
139. Maxwell, E. A.: 'Deductive Geometry', Pergamon (1962).
140. Modenov, P. S. and Parkhomenko, A. S.: 'Geometric Transformations' (2 volumes, translated from the Russian by Michael B. P. Slater), Academic (1965).
141. Newton, I.: *Principia* (in modern English, translated by F. Cajori), Univ. of California Press (1946).
142. Perfect, Hazel: 'Topics in Geometry', Pergamon (1963).
143. Plücker, J.: '*Neue Geometrie des Raumes*', Teubner (1868).
144. Postnikov, M. M.: 'Fundamentals of Galois Theory', Hindustan (1961).
145. Pogorelov, A. F.: 'Lectures on the Foundations of Geometry' (translated from the Russian by Leo F. Boron with the collaboration of Ward D. Bouwsma), Noordhoff (1966).
146. Room, T. G.: 'A Background (Natural, Synthetic and Algebraic) to Geometry', C.U.P. (1967).
147. Springer, C. E.: 'Geometry and Analysis of Projective Spaces', Freeman (1964).
148. Tuller, Annita: 'A Modern Introduction to Geometries', van Nostrand (1967).
149. Vajda, A.: 'The Mathematics of Experimental Design', Griffin (1968).
150. Vajda, S.: 'Patterns and Configurations in Finite Spaces', Griffin (1968).

ANSWERS TO SELECTED EXERCISES

EXERCISES 1

1. Let a, b, c, d be the four lines defined by Theorem 1 (dual of *PP*3). Consider the three possible cases (based on the line a):

(i) a point not on the line a;
(ii) a point on the line a but not $a \cap b$;
(iii) $a \cap b$.

Then a point of (i), by *PP*1, lies on the three distinct lines formed by its join with $a \cap b$, $a \cap c$ and $a \cap d$;

a point of (ii), by *PP*1, lies on a and on the two distinct lines formed by its join with $b \cap c$ and $b \cap d$;

$a \cap b$ is on the three lines a, b, $(a \cap b)(c \cap d)$, by *PP*1.

2. Hall's list:

(i) the empty plane,
(ii) a plane containing a single point and no lines,
(iii) a plane containing a single line and no points,
(iv) a plane containing a single point and one or more lines through the point,
(v) a plane containing a single line and one or more points on the line,
(vi) a plane containing a single line and a single point not on the line,
(vii) a plane containing a point P on a line l with one or more additional lines through P and one or more additional points on l.
(viii) a plane containing a line l and one or more points on l, a point P not on l and one or more lines through P joining P to the points of l.

11.

Ax.0	←— duals —→	*Ax*.0
1		*Ax*.3
2		*Ex*.1.5
3		*Ax*.1
4		*Ex*.1.9
5		*Ex*.1.10.

24. Let the given skew lines be a, b, c. Take the other given lines as l, m, assumed coplanar, i.e. non-skew. But a, b, c all meet l and m. Thus, a, b, c are coplanar, which is contrary to the hypothesis that they are skew. Hence l and m are skew.

25. Assume the existence of another distinct line d' satisfying the conditions of Theorem 4. Thus $d' = \alpha \cap \beta \,(= d)$. But two distinct planes are on exactly one line (Deduction IV, § 6.12) $\therefore\ d' = d$.

26. Two concurrent lines lie in a plane.

27. Two points lie on a line.

28. A line and a point (not on the line) lie in a plane.

40. For n even or odd, we have

[0] is the dual of $[n-1]$
[1] .... $[n-2]$
[2] .... $[n-3]$
....
$[r]$ .... $[n-r-1]$ $\quad (r \leqslant n-1)$.

When n is even, there is no self-dual space. However, when n is odd, the self-dual space is of dimension $r = n - r - 1$, i.e. $r = \frac{n-1}{2}$. (Thus, in [3], a line (= [1]) is self-dual; in [5], a plane is self-dual.)

41. Three points lie in a plane.

42. Two $[n-3]$'s having an $[n-4]$ in common lie in an $[n-2]$. Since the two planes intersect in a line, they must lie in a [3].)

46. (i) 3, 4

(ii) 1.

EXERCISES 2

1. *PQR* is the Desargues line of the perspective triangles *ABC*, *DEF*. (The centre of perspective is the point of concurrency.)

2. Triangles *BCD*, *ZYX* are in perspective from *A*.

3. Triangles *ADC*, *PRM* are in perspective from *B*.

4. Desargues' Theorem in [2]: If *ABC*, $A'B'C'$ are two triangles, such that AA', BB', CC' are concurrent at a point *V*, then the points $AB \cap A'B'$, $BC \cap B'C'$, $CA \cap C'A'$ are collinear on the line *d*.

Dual of Desargues' Theorem in [2]: If *abc*, $a'b'c'$ are two triangles such that $a \cap a'$, $b \cap b'$, $c \cap c'$ are collinear on a line *v*, then the lines $(a \cap b)\,(a' \cap b')$, $(b \cap c)\,(b' \cap c')$, $(c \cap a)\,(c' \cap a')$ are concurrent at a point *D*. The Converse is the same as the Dual.

5. He employs the diagram for Desargues' Theorem in [2].

6. Apply the Converse of Desargues' Theorem to the triangles *AFH*, *UCG*.

9. Let $ABC \overset{U}{\underset{\wedge}{=}} A'B'C'$ and $ABC \overset{V}{\underset{\wedge}{=}} B'C'A'$.

Now the Pappus line of the Pappus triads $AA'U$, $BC'V$ contains the points $AC' \cap BA'$, $A'V \cap UC' = C$, $AV \cap BU = B'$

i.e. AC', BA', CB' are concurrent

$\therefore\ ABC \overset{W}{\underset{\wedge}{=}} C'A'B'$ (letting the centre of perspective be *W*).

10. *HTY* ($Y = US \cap KV$) is the Pappus line of the Pappus triads *U*, *V*, *P*; *K*, *S*, P'.

14. Triangles *WZA*, *MNR* are in perspective from *V*.

16. Use the Pappus triads *PXR*, *QYS*.

Use the Desargues triangles *SAP*, *RBQ*.

17. (i) Use $AB'C' \overset{O}{\underset{\wedge}{=}} A'BC$.

(ii) Use the Converse of Desargues' Theorem applied to triangles *AQR*, *PBC*.

EXERCISES 3

1. $(1, -1, 0)$.

2. $L \equiv (0, n, -m)$; $M \equiv (n, 0, -l)$; $N \equiv (m, -l, 0)$.

$LX: my + nz = 0;$

$MY: lx + nz = 0;$

$NZ: lx + my = 0.$

3. Dual: Equations of the 6 vertices of a complete quadrilateral may be written as $l_i \pm i_j = 0\ (i < j;\ i, j = 1, 2, 3)$ where the homogeneous line co-ordinates are l_1, l_2, l_3.

4. $A \equiv (0, 1, 1)$; $B \equiv (1, 0, 1)$; $C \equiv (1, 1, 0)$.

6. (x_1, y_1, z_1).

7. $XA: y = z$; $YB: z = x$; $ZC: x = y$. $(0, 1\ -1)$; $(1, 0\ -1)$; $(1, -1, 0)$ resp.

8. (i) $G = AB \cap ZU$
(ii) $GX \cap YZ$
(iii) $BC \cap YZ$.
9. $y + z = 0$.
10. $x - y = 0$.
11. $(1, -1, 0)$.
17. $\frac{x}{a - a'} + \frac{y}{b - b'} + \frac{z}{c - c'} = 0$.
20. Yes.

EXERCISES 4

4. $\begin{bmatrix} \cdot & \cdot & 1 \\ 1 & \cdot & \cdot \\ \cdot & 1 & \cdot \end{bmatrix}$.
5. (i) (5, 3, 2)
(ii) $x - 5y - 4z = 0$
(iii) $(1, 1, -3)$
(iv) $3x - 3y - z = 0$.

EXERCISES 5

2. $xx' + 3x - 2x' - 2 = 0$. Double points are given by $x = 1, -2$ $\therefore$ the projectivity is hyperbolic.
4. $-\frac{q}{p}$.
7. $\frac{4}{5}$ or $\frac{5}{4}$.
10. $(A, B, C, P) \overset{A'}{\overline{\overline{\wedge}}} (Q, N, M, R) \overset{A}{\overline{\overline{\wedge}}} (A', B', C', P'')$ where $N = AB' \cap A'B$, $M = AC' \cap A'C$, m is the line determined by N, M, $R = m \cap A'P$, and $AR \cap l' = P''$, $Q = m \cap AA'$.
i.e. $(A, B, C, P) \overline{\wedge} (A', B', C', P'')$.
But $(A, B, C, P) \overline{\wedge} (A', B', C', P')$, given.
Since a projectivity is completely determined by three pairs of corresponding points, it follows that $P'' = P'$, i.e.
$AP' \cap A'P = R \in m$.
11. $AZ \cap A'O = O \in m$
$A'Z \cap AO' = O' \in m$
i.e. $O, O' \in m$, i.e. $OO' = m$.
$O' = O = Z \in m$. It follows, by repeated use of the Converse of Desargues' Theorem, that $AA', BB', CC', \ldots, PP'$ are concurrent. (See Exercise 2.11.)
12. True. If there are more than two transversals, then there is an infinite number.
13. (i) $\frac{b - d}{b - c}$
(ii) k
(iii) -1

(iv) -1

(v) $\dfrac{a-2b}{b}$

(vi) -1.

15. $\lambda = \dfrac{1 \pm i\sqrt{3}}{2} = -\epsilon, -\epsilon^2$ (where ϵ is a cube root of unity).

16. $x = \dfrac{5}{2}$.

22. $k = 1$.

EXERCISES 6

2. Let AT cut the conic again in R. Hence, P, P'; Q, Q'; R, A are pairs of points in involution on the conic, i.e. AP, AP'; AQ, AQ'; AR, AA are pairs in an involution pencil on the conic. Evidently, AR may be replaced by AT, and AA by AS.

3. Pairs of tangents from a given point to a system of conics touching four fixed lines form an involution pencil.

4. This is the dual of Notes (ii), § 23.1. In Sturm's Theorem the involution pencil has two double lines through the vertex L of the pencil. To each of these there corresponds a conic of the system which passes through L.

6. Let the conics S_1, S_2, S_3 all meet in U, V.

Let $S_1 \cap S_2 = A, B$, $S_2 \cap S_3 = H, K$, $S_3 \cap S_1 = P, Q$, $UV \cap PQ = F$, $PQ \cap AB = G$, $PQ \cap HK = G'$, $PQ \cap S_2 = X, Y$.

Consider conics S_1, S_2 through A, B, U, V.

$\therefore$ P, Q; X, Y; F, G are pairs of points in involution (Theorem 23).

Consider conics S_2, S_3 through H, K, U, V.

$\therefore$ P, Q; X, Y; F, G' are pairs of points in involution (Theorem 23)

$\therefore$ $G' = G$.

$\therefore$ AB, HK, PQ are concurrent (at G).

7. Consider the degenerate quadrangle $IIJJ$ inscribed, separately, in both the conics.

8. Apply Brianchon's Theorem to the degenerate hexagon $ABCDEF$.

9. Let the vertices of the quadrangle be A, B, C, D. Apply Pascal's Theorem to (say) the degenerate hexagon $AABCCD$.

10. Firstly, prove AA' is the polar of P (Fig. 49).

16. Let the four points A, B, C, D be collinear on l. Let O be the pole of l w.r.t. the conic. Let the polars of A, B, C, D w.r.t. the conic be a, b, c, d (intersecting l in A', B', C', D' respectively). Show that a, b, c, d are concurrent at O. Then show that A, A'; B, B'; C, C'; D, D' are pairs in involution on l.

21. See O'Hara and Ward [79] pp. 130–131.

22. Consider the quadrangle based on $A, B, C, BB' \cap CC'$. The result follows from Hesse's Theorem (= Exercise 6.21), of which it is really a restatement.

24. Let AH meet the tangent at B in A'.

Let BH meet the tangent at A in B'.

$\therefore$ $(A') \barwedge (B')$ since there is clearly a projectivity between (A'), the set of points A', and (B'), the set of points B' (why?)

$\therefore$ $A(A') \barwedge B(B')$.

But $H = AA' \cap BB'$, i.e. H is the intersection of corresponding lines of two related pencils, with vertices A, B

$\therefore$ by the definition of a conic, the locus of H is a conic S' which must pass through A, B.

It is reasonably easy to show that S, S' have common tangents at A, B, i.e. have double contact at A, B.

EXERCISES 7

1. $m^2 - 4nl = 0$.
2. $(x^{\frac{1}{2}} \pm y^{\frac{1}{2}})^2 = z \to (x + y - z)^2 = 4\,xy$ which simplifies to
$x^2 + y^2 + z^2 - 2yz - 2zx - 2xy = 0$ (the required equation).
4. Secant: $x - (\theta + \theta')y + \theta\theta' z = 0$.
Tangent: $x - 2\theta y + \theta^2 z = 0$.
5. $(s, -q, p)$ if the involution has equation $pxx' + q(x + x') + s = 0$.
17. The conic S.
19. Desargues line $\equiv$ unit line $x + y + z = 0$.

EXERCISES 8

1. Pairs a, b; b, c; c, d meet l_∞ in *two* points harmonically separating I, J. (These two points must be the same for each pair.) Thus, d, a meet l_∞ in precisely these two points $\therefore d \perp a$.

2. Let F, E be the mid-points of the sides AB, AC respectively of the triangle ABC. Let $AB \cap l_\infty = F'$, $AC \cap l_\infty = E'$, $BC \cap l_\infty = X$, $FE \cap l_\infty = Y$. Then $(AB, FF') = (AC, EE') = -1$ (since F, E are mid-points). This must mean that BC, FE, $F'E'$ are concurrent, i.e. $X = Y$, i.e. $FE // BC$.

10. Consider l_∞. It cuts the conics of the pencil in pairs of points in involution. The double points of the involution clearly correspond to parabolas of the pencil, i.e. there are just two parabolas in the pencil. One pair of points of the involution harmonically separates I, J (Exercise 5.25). That is, there is just one rectangular hyperbola of the pencil.

12. Focus is the centre; directrix is l_∞.

23. By considering (AB, CS), establish that $O(AB, CL) = O(AB, CM) = O(AB, CN)$ where O is the circumcentre.

29. $(PQ) = (pq) = \frac{\pi}{2} + n\pi$ (n an integer) if $k = \frac{1}{2i}$.

32. $m' = m$.

33. Note that the equation of the director circle of the given ellipse is $x^2 + y^2 = a^2 + b^2$. Use Example 34.

34. Use Example 34, replacing ellipse by hyperbola.

35. Use Example 35 (i).

37. Use Example 35 (i).

38. The circle and the rectangular hyperbola cut l_∞ in the pairs of points I, J; P, Q respectively for which $(PQ, IJ) = -1$ (by the definition of a rect. hyperbola) $\therefore l_\infty$ belongs to the envelope Φ (by the definition of Φ) $\therefore \Phi$ is a parabola.

39. The F-conic is the director circle of S.

40. $\Theta = ab - h^2$, $\Theta' = a + b$.
(i) $\Theta' = 0 \Rightarrow S$ is a rectangular hyperbola;
(ii) $\Theta = 0 \Rightarrow S$ is a parabola.

43. $\Theta^3 - 4\Delta\,\Theta\,\Theta' + 8\Delta^2\Delta' = 0$.

44. $p = \dfrac{|ax_1 + by_1 + c|}{\sqrt{a^2 + b^2}}$ for point $P(x_1, y_1)$ and line $ax + by + c = 0$.

For the isotropic line OI: $y = ix$ (say) we have

$$p = \frac{|y_1 - ix_1|}{\sqrt{1^2 + i^2}} = \frac{|y_1 - ix_1|}{0} = \infty \text{ if } p \notin OI$$

$$= \frac{0}{0} = \text{indeterminate if } P \in OI, \text{ i.e. } y_1 = ix_1.$$

48. Ceva's Theorem: $\frac{BD}{CD} \cdot \frac{CE}{AE} \cdot \frac{AF}{BF} = -1.$

51. Equiangular (= equilateral).

53. (i) Two triangles in parallel perspective.
(ii) Two similar triangles in perspective.
(iii) Two congruent triangles in parallel perspective.

55. Castelnuovo's Theorem: The radical axes of three circles, in pairs, are concurrent. If the point of concurrency lies on l_∞, then the three radical axes are parallel, i.e. the centres of the three circles are collinear.

59. O, I, J.

EXERCISES 10

3. Seven lines (seven planes) through each point.

4. 35.

5. Three planes through each line.

6. (i) $n^2 + n + 1$;
(ii) $(n^2 + 1)(n^2 + n + 1)$.

9. (i) $n + 1$;
(ii) $n + 2$.

15. Yes.

INDEX

Absolute the 219–26, 272–4
Alternative laws 319
Angle
 Euclidean formula for 223–4
 of parallelism 259, 265
 -sum 8, 224, 262–3, 274
 summit- 261–2, 267
Apolarity 200–1
Art and perspective 16–17
Associator 327
Asymptote 213
Axiom(s)
 Archimedes' 11
 Congruence 11, 13
 connection (= incidence) 11–12, 47–71
 continuity 11, 116, 326
 Euclid's 11–12
 Fano's 59–60
 Greek notion of 3
 Hall's incidence 49
 Heyting's 62
 Hilbert's 11, 12
 homogeneous co-ordinate 91, 93
 Huntington's 13, 14
 nature of 36, 37
 order 11, 116
 other systems of 12–13, 49–50
 parallels 11, 12
 partial plane 60
 Pasch's 4, 281
 Playfair's 5
 Posidonius' 41
 Proclus' 41–42
 qualities of 37–39
 Seidenberg's 69
 value of 39
 Veblen and Young's 49–50
Axis
 of conic 215
 radical 250
 rectangular Cartesian 227

Babylon (ians) 1
Bachet's card puzzle 308
Base
 of range 140
 of quadrilateral 255–6
Betweenness 23–24, 116, 209
Bisection 208, 236
 Angle- 228
Block design 291–3, 306
 balanced incomplete 292
 symmetrical 291–3
Braikenridge-Maclaurin construction
 181-2

Cardioid 206
Centroid 250
Cevian 165
Characteristic equation (polynomial)
 128–9, 132–3
 of a ring (field) 320–1
Chromosomes 3
Circle(s) 211–3, 229
 centre diameter of 211
 coaxal 250
 concentric 213
 degenerate 211, 233
 director 251
 great 9–10, 270–1
Codes 293–4
Cofactor 120, 191
Collineation 32, 119–39, 305
 anti- 138
 axis of 133
 canonical equation of 132–3
 central 33–34, 133–9
 centre of 133
 elation 132–9, 315–20
 homology 132–9, 321–2
 invariant points of 128–39
 involutory 135
Commutator 327
Computers 300–2
Cone of lines, planes 171
Configuration
 definition 61
  of Desargues 61, 73–6
   Fano 58, 60, 285
   Hadamard 293
   Pappus 61, 79–81
   (v, k, λ) 291–5, 306–7
Congruence 226, 249, 314, 341
Congruent by translation 210

Conic(s)
Absolute (gauge) 219–26, 272–4
Apollonius and 166
Brackenridge-Maclaurin 186–7
central 213
centre of 213
circum- 173, 193–4
confocal 251
degenerate 168–9, 222
diameter of 213
conjugate 251
directrix of 214–5
discriminant of 190–1
11-point 188
-envelope (line-conic) 170–1
equation of 191
focus of 214–5
Kepler's 'blind' 21, 215
harmonic 201–4
hexagon inscribed in 179
in- 173, 194
in finite geometry 303–4
point- 167–70
equation of 189–90
polar reciprocal 204
projective invariance of 191
self-polar 194, 198
special forms of 193–6
Steiner-Chasles 166–71, 185–6
uniqueness of 169–70
vertex of 215
Von Staudt's 185–6
Contact (of conics) 177
Co-ordinates
Artin's homorphism system 117
Beltrami 264
Hall's ternary ring 117, 310–27
homogeneous 27–30, 225–33, 238–48
Cartesian general 30, 91–139, 144–6, 189-205, 220–5, 324–5
hyperplane (prime) 106
line (axial, ray) 106, 118
(= tangential) 94-5
Lobatschewskian 264
non-homogeneous 28, 92, 227
rectangular Cartesian 27, 226–7
transformation of 119–22, 144
Weierstrassian 264–5
Correlation 124
Coset 125–7
Cosmology 280–1, 290
Cross-ratio 24–25, 150–65, 171–88, 208-9, 212–20
invariance of 25, 153
metrical 227–8
Möbius 152, 155, 164
property of involution 160

Curve
cubic 190, 197
of order n 190

Difference set 294–5
Digon (= biangle) 267
Dimension(s) 61–62, 199
Discriminating cubic 197
Doge's Palace 22–23
Drosophila 13
Duality
of pole and polar 185
principle of, in [2] 52–55, 95
in [3] 64

Eigenvalue (eigenvector) 128–30, 132–3
Eigenspace 139
Einstein and Relativity 8, 68, 248
Electronics 69
Ellipse 22, 213
Erlanger Programm 31–33, 127, 248–9, 277–8
Euclidean
angle 223–5
distance 225–6
Euler's conjecture 299
falsity of 299–300
Experiment
Gauss' 276–8
Lobatschewsky's 275–6
statistical 296–7

Field 114, 322
Galois 91, 285–9, 304–5
of points 114
near- 326
Figure
Reidemeister 317
Thomsen 318
Formula
cosine (sine) 282
Laguerre's 224–5
Lobatschewsky–Bolyai 265

Gas molecules 69
Generalife Gardens 23
Geodesic 271
Geometry
affine 22, 25–26, 32, 207, 278
compass 14–15
complex 104
continuous 330
Desarguesian 75
elliptic 8–10, 221, 266–8, 270–4, 278, 282

Geometry *(cont.)*
empirical 2
Euclidean 2–6, 222–53, 278, 282
finite (= Galois) 68, 284–308
Hilbert 250
hyperbolic 7–9, 36, 221, 257–65, 268–70, 274, 278, 282
inversion 40
Line 106, 118
Minkowski 207, 248
Neolithic 1
non-Archimedean 326
non-Desarguesian 75, 85–8
non-Euclidean 6–11, 221–2, 254–83
octave 306, 321
origins of 1–2
parabolic 222
pseudo-Euclidean 253
projective 14–26, 39–40, 49, 127
in [3] 62–65
quaternion 306
real 104, 278
Saccheri's 6–7, 254–7
straight-edge 15
subordinate 98
Gödel 39, 41
Group(s)
affine 33, 207, 238
and block design 306
cross-ratio 155
elliptic 277
Euclidean (= rigid motion) 33, 241, 246–9
factor 125, 145
full linear 125, 145
hyperbolic 277
isometry 244
Klein 163
linear fractional (projective) 146
non-Euclidean 277
projective (= collineation) 32–33, 123–8, 145, 305
simple 305
translation 139, 241–2, 318
triply transitive 143
Gudermannian relation 155

Hadamard matrix 293–4
Harmonic
conjugate 99–102, 108
net 115
property of complete quadrangle 100, 158, 236
property invariance of 141, 142
range 153–4

Harmonic *(cont.)*
sequence 110
re-entrant 110
set 55–56
Homogeneous products 205
Homomorphism 127
Horocycle 281
Hydrodynamics 69
Hyperbola 22, 213
of Apollonius 251
rectangular 213–4
Hyperplane (prime) 66, 105

Image 17
Incidence
relation 94–95
matrix 291
Inclusion 13
Infinity
circle at 31
circular points at 30–31, 51, 207, 222–36, 246–9
line at 19–21, 206, 310
equation of 29, 222
plane at 21, 62
point at 17–21, 51, 104, 206, 259, 311
co-ordinates of 28, 310
Inpolarity 200
Inside 167, 211
Intersection 49, 66
Invariant subgroup 125–7, 145, 241–2
Invariants
affine 24, 209, 238–9
dimension of 199
Euclidean 241, 249
fundamental numerical 246
of a pair of conics 196–204
projective 22–23, 209
Involution 158–65
absolute 208, 247–8
conic 173–8
elliptic 159, 247–8
fundamental equation of 159–60
hyperbolic 159–248
metrical 229
orthogonality 208, 247–8
-pencil 159
Isometries 244

Joachimsthal's ratio equation 192
Join 49, 66

k-arc 304
k-cap 304
Kantian philosophers 3, 8

Kirkman Schoolgirl problem 300
Kleinermann example 320
Krypton 86 266

Lanyon quoit 2
Latin squares 295–300
Lattices 326
Length 226
 unit of 265–6
Line(s)
 affine 207
  augmented 207
 augmented Euclidean 18
 complex 104–5
 concurrent 48
 conjugate 184–5
 coplanar 49
 cutting (non-cutting) 258
 definitions of 4, 11, 41, 93
 Desargues 73, 99
  equation of 118, 205
 double 147
 elliptic 9–10, 271
 equation of 94–5
 exterior 304
 Hessian 182
  equation of 195
 hyperbolic 269–70
 ideal ($= l_\infty$) 206
 intermediary 141
 isotropic 208, 232–3
 Moulton 86
 of centres 250
 orthogonal 252
 Pappus 79, 103, 180
 parallel 5–10, 11, 12, 27–29, 206–239
 Pascal 180
  equation of 205
 perpendicular 208
 Plücker 180
 projective 18
  ordered 116
  real 104–5, 115–6
 Simson (= Wallace) 251
 skew 62–65
 undefined 12, 47, 62
 unit 96
 vanishing 18–19
Loop 295, 314

Map 17
Mathematical
 science 2
 space 3

Midpoint 208, 212
Möbius band 271–2, 283
Model(s) 51, 56–59, 62–63, 70, 92–95
 Beltrami 268
 (Cayley-) Klein 269
 isomorphism of 59
 non-Euclidean 9–10, 268–72
 Poincaré 269–70
 spherical 9–10, 270-71
Moufang identity 319, 327
Mustang fighter 183
Mystic hexagram 179

Net of rationality 115
Number 90

Occam's razor 37
Octaves 306, 321
Origin 227
Orthocentre 214
Outside 167
Outpolarity 200
Ovaloids 304
Ovals 303
 Pascalian 304

Pappus triad 79–80
Parabola 22, 213, 230
Parallax of Sirius 275–6
Parallel(s)
 Lobatschewskian 258, 269
 left (right) 259
Parallelogram 236
Parameter 98, 146–52, 189
 for conic 195–6
Pascal pentagon 181–2
Peaucellier's linkage 40
Pencil 95, 141
 axial 141
 vertex of 140
Perspective 16–17, 72, 140–65
 axial 77
 axis of 73
 central 72
 centre of 72
 doubly 89
 plane 89
 triply 89
Perspectivity 33
 axial 141
 axis of 141
 central 140

Perspectivity *(cont.)*
centre of 140
self corresponding point of 140
Pharaoh Sesostris 2
Plane(s)
affine 26, 207
augmented 26, 56, 207
real 209–10
Analytical 56, 95
André 326
Bolyai-Lobatschewsky (finite) 303
complex 104–5
cyclic 295
degenerate 50
Desarguesian 75, 113, 321–2
finite 301–2, 305, 323
double 147
dual 54
elliptic 221–2, 271
Euclidean 11, 20, 26
augmented 20, 26, 51, 56
finite
affine 287–8
existence of 56–59, 300–2
Fano 57–60, 284–6, 288
of order *n* 288–9
13-point 286–8
free 61, 290
freely generated 61
Hall 326
Hughes 301, 323
hyperbolic 221, 269–70
inversive (= Möbius) 307
Kustaanheimo 307
Moufang (= alternative) 318–21
finite 323
Moulton 85–88
augmented 86–87, 89
generalised 88
non-Desarguesian 56, 75, 84, 310–21
non-Euclidean (= elliptic, hyperbolic)
non-Pappian 79, 84, 310–22
number 94
octave 306, 321
Pappian 79, 114, 322
finite 323
partial 60–61, 303
Pickert-Moulton 87–88
projective 20, 26, 49, 63, 65, 324–6
real 104–5, 208–10
semi-translation 326
undefined 12, 62
Veblen-Wedderburn (= translation) 138, 301, 317–8
dual 301
finite 323
skew 71
Plane(s) *(cont.)*
sub- 289–90
Young 288
Platonic School 2
Poincaréan polygon 271–2, 283
Point(s)
antipodal 9–10, 270–1
Brianchon 181
collinear 48
complex 104–5
conjugate 184–5
coplanar 49
definitions of 3, 4, 11, 41
distinct 48
double 147
elliptic 9, 221–2, 270–1
equation of 94–95
fourth harmonic 99–102
Frégier 252
Hessian 182
hyperbolic 221, 269–70
ideal (= P_∞) 206, 259, 269
ultra- 259, 269
independent 71
limiting 250
Miquel 212
Moulton 86
of contact 170–71
orthogonal 252
Pappus 80
product (quotient) of 111–3
metrical 231
real 104–5
Steiner 180
sum (difference) of 107–10
metrical 231
undefined 12, 47, 62
unit 29, 96
Polarities 138, 184
Pole and polar 183–5, 192–3
Poncelet's principle of continuity 177
Porism 199
Postulate
5 (Euclid) 5, 12
Greek notion of 3
Lobateschewsky-Bolyai 258, 259, 303
Nasîraddîn 6, 41
Ptolemy 41
Riemann 266–7
Saccheri 6–7
Pre-dynastic Egyptian pottery 1–2
Primal 205
Projecting to infinity 17–22, 26
Projection 17–26, 33–34, 140
axis of 17
conical 17–22, 33–34
cylindrical 21–22, 34

Projection *(cont.)*
orthogonal 22
vertex (= centre) of 17
Projective
angle 218–21
distance 218–21
property 22–23
Projectivity 141–65
anti- 147
cross-axis of 144, 181
cyclic 149–50
fundamental equation of 145–6
inverse 142
period of 150
product of 142
unit 142
Pseudosphere 268
Pythagorean Brotherhood 2

Quadrangle
complete 54–55, 236
co-ordinates of vertices 103–4
degenerate 177–8
property of involution 160
Quadrangular set 55
Quadrialteral
complete 54–55
Lambert 257, 262–3, 281–3
Nasîraddîn 6, 257
Saccheri 6, 254–7, 261–3,281
Quadrillaterally inscribed 252

Rabatment 34
Railway lines 10, 271
Range 95, 140
equianharmonic 164
related 142
Ray 141
of light 4, 86, 276–7
Rectangle 250
Result of
Argunov-Skornyakov 82–83
Bose–Stevens 297–8
Gauss 283
Pedoe 84
Plücker-Gaskin 251
Singer 295, 305
Steiner 188
Tarry 298–9
von Staudt 188
Rigid bodies, dynamics of 106
Ring
division 113, 305
alternative 319
associative 321

Ring *(cont.)*
linear 312
ternary 310–27
operations 312

Saccheri's Propositions 6–7, 254–6
Samos tunnel 2
Scale 231
-constant 218
Secant 168, 304
Segment 5, 209–10, 226
division 23–24, 239
Separation 19, 116, 209
Similar
matrices 122
triangles 263, 268
Simplex of reference 106
Solid 61
Space(s) 61–71, 105-6
Spectrum 139
Sphere 13
inch- 14
Standard metre 266
Statistics 296–7
Steiner triple system 307
Summit 255–6
Symbol ∞ 90–91

Tangent 168, 193, 304
Tetrahedron 70, 89
Theorem
Artin-Levi-Zorn 323
Baer 318
Bose-Shrikhande-Parker 299–300
Brianchon 181
Bruck 290
Bruck-Kleinfeld 321
Bruck-Ryser 299
Buekenhout 304
Cayley-Hamilton 139
Castelnuovo 187
Ceva 165, 252
Chasles 171–2, 236
converse 172
dual 172
Cronheim 84
Desargues (perspective triangles) 16
17, 73–89, 99, 137 157, 286–7, 313,
321–2 W
affine 314
converse 75, 77, 99, 321
dual in [2] 77
minor 137, 313, 319
affine 314, 315–8
Desargues (involution) 175–6
dual (= Sturm) 176

Theorem *(cont.)*
8 associated points 250
Enriques 101–2
Frégier 252
fundamental 144
Gaskin 252
Gauss 283
Gleason 286
Hesse 188
Hessenberg 83–84
Hilbert 113–4
converse 114
Hjelmslev 244–5
Jacobi 192
Kopejkina 290
Lame 187
dual 187
linearity 315–7
Menelaus 252–3
Miquel 212
Pappus 16, 79–89, 102–3, 158, 180
affine 323
dual 89
Pascal 179–80
converse 181
dual (= Brianchon) 181
Qvist 304
Segre 304
Seydewitz 186, 236
Steiner 172–3, 235
Sturm 176
Thales 212
Wedderburn 305
Thomsen property 322
Topology 33, 278
Trace 130
Transformation(s)
affine 238–9
alias 120–1
alibi 120–1
biological 35–36
equiareal 250
Euclidean 239–243, 246–8
Transformation(s) *(cont.)*
glide-reflection 243–4
half-turn 253
Möbius 250
projective 32, 119–39
reflection
line 242–3
central (point) 253
rotation 239–40
similarity 122
singular 245
standard quadratic 124
translation 240–2
Transitivity 137–8, 315–22
Transversal 68
Triangle 8, 54
diagonal 55
of reference 29, 96, 227
self-polar 184
Triangular inscription, circumscription 199
Trilateral 54

Unboundedness 8, 266
Undefined elements, relations 12, 47–48, 62
Unified Field Theory 106

Veblen-Wedderburn system 317–8
Vector(s) 92, 139
spaces 92–93, 105, 106, 144–6, 190–8, 220, 223, 227, 238, 245–6, 273
Vish 37

Weierstrassian elliptic functions 204
Würfe 107, 163

Yaglom's examples 25–26